ARCTIC
OFFSHORE ENGINEERING

ARCTIC
OFFSHORE ENGINEERING

Andrew Palmer
National University of Singapore, Singapore

Ken Croasdale
K R Croasdale & Associates Ltd, Canada

World Scientific

NEW JERSEY • LONDON • SINGAPORE • BEIJING • SHANGHAI • HONG KONG • TAIPEI • CHENNAI

Published by

World Scientific Publishing Co. Pte. Ltd.
5 Toh Tuck Link, Singapore 596224
USA office: 27 Warren Street, Suite 401-402, Hackensack, NJ 07601
UK office: 57 Shelton Street, Covent Garden, London WC2H 9HE

British Library Cataloguing-in-Publication Data
A catalogue record for this book is available from the British Library.

ARTIC OFFSHORE ENGINEERING

ISBN 978-981-4368-77-3

Printed in Singapore by B & Jo Enterprise Pte Ltd

Foreword

The Arctic is important to all of us. It offers to humanity enormous resources of oil, gas, gas hydrates, gold, iron ore, diamonds, timber, fish and hydroelectric power. At the same time, it is a fragile human and biological environment. It has indigenous peoples with their own cultures and history, all too easily disrupted by outsiders. It has a wealth of animals and plants, each special in its own way but liable to damage.

Andrew Palmer and Ken Croasdale have written this book about Arctic Offshore Engineering, and examine how we can exploit the resources in and under the sea for human purposes, and can do so safely, economically and with minimal risk to the environment. Singapore may at first seem a surprising place to be writing such a book, but in fact we have a significant and growing interest in the Arctic, from several directions, among them shipping and petroleum production. At Keppel we are already active in more than one of those fields, and have a long term commitment to the area.

I welcome the book enthusiastically.

Choo Chiau Beng
CEO, Keppel Corporation
Singapore
June 2012

Preface

AA Milne wrote in the preface to a detective story that the only good reason for writing anything was that you wanted to, and that he would be prouder of a telephone directory written with love than of a tragedy in blank verse, written as a chore. We agree completely.

The Arctic is a region of endless fascination, beauty and excitement. Many writers more eloquent than we are have explored its biology, geography, history and environment. At the same time, it is the home of many people who want to pursue civilised and comfortable lives with the amenities of the 21st century. The Arctic seas and the lands bordering them are important to the wider human community as a source of raw materials, above all petroleum, but as well as for fish, timber and mineral ores, and important too for tourism and exploration. Many important reserves of petroleum have been found already, some of those reserves are in production, and the unexplored geology of this vast region has many promising structures: some estimates have it that as much as a third of the petroleum still to be discovered will be found in the Arctic. With that as one priority, we must not forget that the Arctic is uniquely vulnerable to damage: if we make a mess, the consequences will be with us for decades.

Oil and gas reserves close to shore can be produced by horizontal drilling, but beyond a few kilometres the petroleum industry will need platforms to drill from and produce to, and pipelines and other systems to bring the petroleum ashore and transport it to markets. Those structures have to operate safely in an extremely demanding environment. The ocean is cold and rough, and for much of the year it will be covered with ice, often in very large pieces. Ice pushes against structures with great force, and drags along the seabed, strongly enough to cut huge gouges. Sometimes the seabed will be partly frozen, and anything we do may alter the thermal regime and thaw or freeze the seabed soil, greatly modifying its physical properties. Almost all the problems of offshore construction in lower latitudes are still present, and there are storm waves and high winds, tidal currents, shifting seabeds and the added challenge of long periods of winter darkness. Most importantly, anything we are going to build has to be constructible, at an acceptable cost, within a reasonable timeframe, safely and without environmental damage.

Serious thinking about offshore structures in the Arctic seas began more than forty years ago, initially in Cook Inlet in southern Alaska, then in the Beaufort Sea and the Canadian Arctic Islands, and later in many other areas. One of the critical subjects was the level of ice forces on fixed structures, as opposed to ships, which are rather different and can move to avoid the worst ice features. That raised questions unfamiliar to civil engineers accustomed to steel, concrete, rock and soil, because ice is a totally different material of a remarkably unusual kind.

In this book we explore the many issues that an engineer designing for the Arctic offshore will encounter and will have to respond to. Without apology, and because engineering is about human needs and desires, we begin with the human context, and consider the people who will be affected by offshore construction, and their diverse priorities. Next we examine the physical and biological environment. We then move on to ice as a material and how it responds to stress, a deeply controversial and difficult subject where the development of knowledge is still at an early stage. That leads us to the many different kinds of offshore structures, and to the factors that influence a choice between them. From there we move on to pipelines and transportation by tanker, and then to safety, environmental factors and the ultimate decommissioning and removal of structures we no longer need. Finally, we consider the implications of human factors and the people who will build and operate systems in the Arctic offshore.

We are not the first to have written about these subjects. Four books have been a particular inspiration. Professor Bernard Michel of Université Laval, Quebec, devoted his career to ice research and was in the forefront of developing university courses in ice mechanics, and postgraduate research. His book *Ice mechanics* published in 1978 was a very valuable and comprehensive early contribution. Tim Sanderson's *Ice mechanics: risks to offshore structures* came out in 1988. Elegantly written, full of insight, not afraid to question the received opinion of the time, it remains a powerful source that repays rereading, even though the subject has moved on. Sadly, and to all our loss, Tim's health made it impossible for him to continue at the leading edge of ice mechanics. Peter Wadhams' *Ice in the ocean* carries with it his unrivalled knowledge of the Arctic oceans and his delight in their beauty and strangeness. More recently, Willy Weeks' magisterial *On Sea Ice* was published in 2010, and reflects the deep and broadly-based experience of a lifetime.

We are grateful to many people we have worked with over the years, in many different places and on many different projects. In addition to the colleagues mentioned above, we would particularly like to thank Ainur Abuova, Razek Abdelnour, Gray Alexander, Norm Allyn, Ricardo Argiolas, Mike Ashby, Ravi Aurora, Eleanor Bailey, Anne Barker, Ken Been, Frank Bercha, Morten Bjerkås, Denis Blanchet, Steve Blasco, Bob Brown, Tom Brown, Jim Bruce, Steve Bruneau, Gus Cammaert, George Comfort, Max Coon, Greg Crocker, John Dempsey, Dave Dickins, Ken Downie, Rod Edwards, Tony Evans, Simon Falser, John Fitzpatrick, Ginny Frankenstein, Lindsey Franklin, Bob Frederking, Mark Fuglem, Lorne Gold, Dougal Goodman, Jed Hamilton, Willott Heerde, Kevin Hewitt, Chris Hill, John Hutchinson, Knut Høyland, Chris Heuer, Hans Jahns, Catherine Jahre-Nilsen, Waleed Jazrawi, Bengt Johansson, Michelle Johnston, Ian Jordaan, Ollie Kaustinen, Tuomo Kärnä, Arno Keinonen, John Kenny, Hans Kivisild, Ibrahim Konuk, Austin Kovacs, Rick Kry, Ivana Kubat, Glenn Lanan, Arny Lengkeek, George Li, Pavel Liferov, Sveinung Løset, Mauri Määttänen, Glenn Mainland, Bob Marcellus, Dan Masterson, Dmitri Matskevitch, Derek Mayne, Dave McGonigal, Richard McKenna, Michel Metge, Per-Olav Moslet, Derek Muggeridge, Karen Muggeridge, Charlie Neill, Don Nevel, Rune Nilsen, Peter Noble, Ravi Perera, Roger Pilkington, Anatoly Polomoshnov, Alan Ponter, Bob Pritchard, Matthew Quah, Terry Ralston, Jim Rice, Ron Ritch, Mohamed Sayed, Joachim Schwarz, Karl Shkhinek, Victor Smirnov, Dev Sodhi, Walt Spring, Alan Strandberg, Paul Stuckey, Gennady Surkov, Rocky Taylor, Graham Thomas, Garry Timco, Hendrik Tjiawi, Pavel Truskov, Paul Verlaan, Sergey Vernyayev, Stanislav Vershinin, David Walker, Cynthia Wang, Jeff Weaver, Alexandra Weihrauch, Mitch Winkler, Brian Wright, Pete Wiebe, Yap Kim Thow, Abzal Yergaliyev and Zheng Jiexin.

We thank the many companies we have worked with, and particularly RJ Brown and Associates, Keppel Offshore, Panarctic Oils, Polar Gas, BP, ExxonMobil, Dome Petroleum, Imperial Oil, Elvary Neftegaz, AgipKCO, Total, Statoil and Shell. It should not be thought that those companies or the individuals listed would agree with all our conclusions. Some of this material was first put forward in the Arctic Engineering course OT5207 at the National University of Singapore, and we are grateful to the students on that course for their comments.

Our determination is to make this book stimulating and enjoyable, on principle and not merely for the utilitarian reason that people engage more with subjects they find enjoyable. The mistakes and misunderstandings are our own, and we shall be grateful to anyone who points them out, or – better still! – argues with us.

Most of all, we thank our wives Jane Palmer and Anne Murphy Croasdale for their infinite patience.

Andrew Palmer
Singapore, 16 March 2012

Ken Croasdale
Calgary, 16 March 2012

Contents

Chapter One

The Human Context

1.1 Introduction

Engineering and the rest of technology are about the application of science to human needs. People in the Arctic have needs for technological development. Moreover, the Arctic has huge natural resources of petroleum, minerals, timber, animals and fish. If those resources are to be developed wisely, the human context is central.

Technological needs do not exist in the abstract. We need to think about who the people are and what they want. It goes without saying that some of their perceived wants may be in conflict. They may 'want' economic growth, adequate education, food, warmth and clothing, the opportunity to develop culturally and to travel, and at the same time they may 'want' an undamaged natural environment, peace and quiet, and the absence of highways, dams and power lines (and of tourists, hunters, engineers, scientists and government officials).

Equally, there are many groups of people involved. Some have lived in the Arctic for thousands of years, and we begin with them. In the past there have been explorers, who came to the Arctic from further south, looking for resources and adventure. Then came developers, who arrived to look for minerals and fur. Finally, there are many people outside who have an interest in the Arctic. They often have strong opinions and legitimate disquiet, but they have no intention whatsoever of living in the Arctic.

It will be clear that these different groups have divergent interests and priorities. Inevitably, they do not agree among themselves.

1.2 The Peoples Native to the Arctic

Mankind is thought to have originated in Africa some fifty thousand years ago. Humans moved north into Asia and Europe, and progressively adapted to northern Asia. In glaciations, sea level was a hundred metres lower than it is today, and before 35,000 years BP (before present) and between 22,000 and 7,000 years BP there was a broad land bridge between Asia and North America where the Bering Strait now is. The most widely held opinion is that human beings crossed the land bridge into Alaska, but there are conflicting theories that they came across the Pacific or from Europe, while First Nation tradition has it that they have been there 'for ever'. The subject is outside the scope of this book, and remains political and bitterly controversial: one archaeologist remarked that '...the archaeology of America is more like a battlefield than a research topic'. Archaeological remains from the Swan Point and Broken Mammoth sites near the Tanana river tell us that human beings reached Alaska by at least 14,300 years BP. They continued south, and by 12,500 BP they had reached Patagonia at the southern end of South America. There is no evidence that Antarctica was ever reached.

Northern Asia is inhabited by many diverse peoples, with very different cultures and languages. Some of those peoples are large and strong, among them the Komi, the Yakuts and the Nenets, the latter one of the peoples within the linguistic group once called Samoyed, a word now abandoned because of its implication of cannibalism. Some are much smaller, and exist at the margin of language and cultural survival. Armstrong [1] lists the population groups, describes where each group lives, and gives their numbers in 1970. Only a few groups practised any kind of agriculture, and the others lived by fishing and hunting, eating wild berries and plants, and herds of domesticated horses and reindeer.

Russians and Scandinavians came up from the south, and encountered the non-Russian northern peoples at a very early stage. Often they fought. Often too they brought with them missionaries, whose goal was to convert the native peoples from the ancient shamanistic religion [2]. Much of the interaction was exploitative, the fighting was one-sided because the incomers had firearms and the native peoples did not, and alcohol and disease played a part.

After the Soviet revolution, people from the rest of Russia interacted with the native peoples from various interlinked motives. Some acted from missionary enthusiasm, and wanted to bring everyone the perceived benefits of a socialist ideology. They saw that northern peoples were poor, and that they lived a life that could be harsh and cruel, within a limited and static society that they felt to be dominated by superstition. Soviet power would bring increased happiness, they thought, no doubt sincerely. Others were looking for mineral resources. Others still were concerned that northern peoples that had not integrated into Soviet society might serve as a base for counter-revolution or for interference from outside.

Much important linguistic and sociological research was done. Many of the languages had no written form, and they were given alphabets. The different peoples were encouraged to cherish their heritage, but always provided that it did not turn into nationalism or into a rejection of Russian leadership. The cultural heritage was secured by the preservation of handicrafts, sometimes of extraordinary distinction: at the Second International Conference on Permafrost in Yakutsk in 1972, the delegates were shown a stunningly beautiful coat made from the skins of duck heads. At the same conference, the delegates had to listen to a somewhat creepy poem about how the Russians and Soviet power had brought to the Yakuts "the golden key to the future". A little earlier Mowat [3] wrote an upbeat account of two extensive journeys through Siberia in the late 1960s, in the depths of the Soviet period, though critics have suggested that he was inclined to believe whatever he was told. His account of a visit to Magadan enthuses about development but neglects to mention that it was centre of the Gulag chain of prison camps. At that time the region was essentially closed to foreigners unless they had been invited for specific and narrowly defined Soviet purposes.

It is not entirely clear what has happened in Arctic Russia since the collapse of the Soviet system in 1990. Many of the reports are far from encouraging. The region has become less significant to the new Russian state, and appears no longer to be thought so important militarily or strategically. The three biggest Arctic cities, Murmansk, Vorkuta and Norilsk are reported together to have lost a third of their inhabitants since 1989. Thubron [4] visited Dudinka and Potapovo on the Yenisei, and describes desperate poverty, social breakdown, and near-universal alcoholism. Bychkova-Jordan and Jordan-Bychkov [5] write eloquently about the former's home village, Djarkan (62º21'N 116º41' E) near Mirnyi in Sakha (Yakutia). They describe a collective decline, in which society aged as the younger people migrated away, the birthrate dropped (from 21 in the village in 1993 to seven in 1995 and six in 1999), the airline and bus services stopped, the airstrip was abandoned, and the collective and cooperative farms closed down. Livestock production had declined, agricultural mechanisation had been largely lost, and there had been a reversion to subsistence farming. 'The new millenium dawned bleak and ominous'. Vitebsky's finely-written account of many years of anthropological fieldwork with the Eveny in the Verkhoyansk region [6] describes their reindeer-based culture in affectionate and fascinating detail. He points out the changes that were brought by the collapse of the Soviet system, and describes various issues of societal breakdown and alcoholism, among them a horrifying statistic, that among the 180,000 people in thirty ethnic groups across the Russian North, one-third of all deaths are through accident, murder or suicide. He goes on to say that the high death and alcohol rates cannot be blamed entirely on Soviet rule, and that they 'are not very different in native Arctic communities in Scandinavia, Greenland and North America'. Wheeler [7] went to Anadyr in Chukotka and describes that city hesitatingly picking itself up in the aftermath of the collapse.

The region has to some extent opened up to travellers from outside, and much interesting and positive information is to be found in guidebooks (see, for example [8]), though they too point out the problems of alcoholism, crime and population decline. Inevitably that information is somewhat superficial, and it rapidly becomes outdated.

Northern America again has many different native groups. The Inuit (formerly 'Eskimo') live along the Arctic seaboard, and have a hunting culture based on fish, polar bears, and sea mammals, supplement by summer berries and roots. Southerners find their culture romantic, attractive and inspiring, and there is a huge literature. In the introduction to her anthology of writing about the Arctic, Kolbert [9] remarks that

> "--- I sometimes felt as if everyone who had ever visited the Arctic had left behind an account of his or her (usually his) experience. In one of his many Klondike tales, 'An Odyssey of the North', Jack London compares the Arctic whiteness to 'a mighty sheet of foolscap' ..."

Interaction with Canada, the US and Greenland has brought many changes. With it came opportunities for education and improved health, and the prospect of paid work on defence projects, mines and pipelines, as well as for the State. It has also brought alcohol, drugs, disease, and social disruption created by huge discrepancies in wealth and education. The culture has changed markedly. A hunter whose father used to go out with a harpoon, a dogsled and a kayak, now takes a rifle, a snowmobile and a motor boat. It is hard for any outsider to blame him, but he needs to find the money for the imported equipment and for cartridges and gasoline. Some communities have deliberately isolated themselves, in a partially successful attempt to maintain their traditional way of life. Most communities have not done so, and there are many social problems. Ertel [10], for example, contrasts modest prosperity on the west coast of Greenland with social breakdown in Tasiilaq on the east coast, where there had been an epidemic with 15 suicides or attempts within a few days. Suicide is an immense problem in Greenland as a whole, with a suicide rate of 100 per 100,000, compared with much lower numbers elsewhere, for instance 14 in Denmark. Ehrlich [11] eloquently describes a dogsled journey from Uummannaq to Niaqornat, and notes some of the impacts of 'progress'.

Some of the changes have been particularly artificial. The islands of the Canadian Arctic Archipelago have almost no indigenous population, because the climate is even more severe than that of the mainland coast further south. The federal government became concerned to replace the missing population, because it worried about sovereignty and strategic presence, particularly if the North-West Passage should come into use. It encouraged Inuit to move to new communities such as Resolute on Cornwallis Island, by providing housing and subsidy.

There is continuing and unstoppable progress towards a greater control of local affairs. Greenland (Kalaallisut) has had a degree of autonomy from Denmark since 1979, but Denmark has continued to support the island with a $550M /year subsidy. It has protected its fisheries against outsiders from the European Union. In 2008, more than three-quarters of the population voted for more autonomy, and many hope that the country will be completely independent in 2021, the 300th anniversary of the arrival of the first missionary, Hans Egede. Independence can hopefully be sustained by revenues from oil under Davis Strait to the west, which the US Geological Survey has estimated at 31 billion barrels. A huge amount of work, investment and time are required between such an estimate and the day when revenue begins to flow into the national economy. In Canada, Nunavut is a federal territory that separated from the remainder of the Northwest Territories in 1999. It has a population of 30,000 spread over 1.9 million km of land area^2. Almost all its budget comes from the federal government.

It is possible that some areas might later seek to become genuinely independent nations, but this seems unlikely in the case of Canada, and still more unlikely in Russia, unless the whole country should break apart.

Many questions to do with the native peoples of the Arctic are intensely political, sometimes in unexpected ways. An engineer needs to approach these questions with immense sensitivity and care. This is not merely a matter of good manners: the success or failure of a project may rest on it.

On the day this is written, Michaëlle Jean, the Governor-General of Canada, is in the news and being criticised by animal welfare groups because she ate raw seal heart when visiting an Inuit community. She pointed out that it would have been an insult not to eat the heart:

> "The heart is a delicacy. It is the best you can offer to your guest. It is the best that is offered to the elders. So, do you say no to that? You engage, and at the same time you are learning about a way of life, a civilisation, a tradition...
> It would have been an insult, and it's not in my nature to stay at a distance and not participate."

1.3 Explorers

The native peoples were in the Arctic first, but in recent centuries outsiders came. These people were hardly ever discoverers in any true sense, but they were explorers moving into unknown territory, with almost nothing to guide them. Navigators sailed up from the north Atlantic and the White Sea. On land Russians moved steadily outwards from the medieval heartland close to present-day Kiev, and soon reached the Arctic. The history is described by Armstrong [2].

This is not a history of Arctic exploration, and the subject can only be briefly touched upon, by a brief mention of a few of the explorers. There is a very extensive literature, and biographies of some of the key figures can be found in the Encyclopaedia of the Arctic [12].

The Cossack brigand leader Yermak (?–1585) advanced into Siberia with a small army on behalf of the Stroganov family, who had been given a patent to colonise the regions along the Kama river, beyond the Urals, and along the Ob' and Irtysh rivers. The Tartar Khan Kuchum defeated Yermak's army at Qashliq near Tobol'sk in 1585. In Russia Yermak has been regarded as a folk hero of conquest, and many legends sprang up after his death, but Thubron [4] sees him as 'a brutal mercenary, and his Cossacks an army of freebooters, itinerant labourers and criminals'.

Willem Barents (1550?–1597) was a Dutch navigator who sailed to Novaya Zeml'ya in 1594-95, and recognised it from the description of the earlier explorer Brunel. He thought that he had discovered a sea route to China and Japan. His ship was trapped in the Kara Sea and was forced to overwinter. The same thing happened in a second expedition in 1596-97. Later he sailed to Svalbard.

Martin Frobisher (1535–1594) made three voyages from England to look for a Northwest Passage to the Orient. He landed on Baffin Island in 1576, made contact with the Inuit, and brought back hundreds of tonnes of what was supposed to be gold ore but turned out to be worthless (thus unwittingly setting a precedent for enthusiastic but fruitless exploration). Many other explorers followed.

Bering (1681–1741) was a Dane from Horsens, and served in the Russian navy. He was the first European to discover Alaska and the Aleutians. Peter the Great had ambitious plans for the region and the Northeast Passage, and the Russian government supported Bering's Second Kamchatka Expedition. He went by land to Okhotsk on the Pacifc, sailed to Kamchatka and then northwards, sighting the Alaska mainland, landing on Kayak Island, and returning by way of the Aleutians. He died of scurvy, along with many of his men, on one of the islands of the Komandorskiye group, now named after him.

Franklin (1786–1847) was a British naval officer. After an exciting early career that included the battles of Copenhagen, Trafalgar and New Orleans, his first Arctic expedition went to Svalbard, looking for a direct passage to the Bering Strait across the Open Polar Sea that was believed to exist. That belief was illusory, though it persisted far into the nineteenth century. Franklin returned to the Arctic to chart the Canadian coast south of Victoria Island and to the Mackenzie River. He was governor of Tasmania for six years, and then in 1845 went again to chart the Northwest Passage. His expedition was in two ships that incorporated many innovations, among them steam engines, retractable rudders that could be lifted into wells that protected them for the ice, and a distillation system for fresh water. He took provisions for three years, but when he had not been heard from the Admiralty dispatched expeditions to look for him, vigorously encouraged and in part financed by Lady Franklin. Those expeditions played a large part in mapping the region. Many of the geographical names in the Canadian Arctic are those of commanders who searched for Franklin and found some relics but no survivors. Rae (1813-1893) found traces of Franklin's lost expedition, and made himself deeply unpopular by reporting the evidence of cannibalism, but Lady Franklin refused to accept that British seamen could sink to that, and Charles Dickens wrote pamphlets attacking him

Nansen (1861–1930) was educated as a zoologist. He made the first crossing of Greenland in 1887. He then decided to investigate a theory that there was a steady current across the Arctic Ocean that would carry him to the North Pole. He built an ice-strengthened ship, *Fram* (to be visited in a museum on Bygdoy close to Oslo), and deliberately let *Fram* be frozen into the ice north of Siberia. In 1895 he and a companion left the ship for the pole, and reached 86°14'N, but then had to turn back. They reached Franz Josef Land after five months on the ice, overwintered there, and by incredibly good luck found another ship and returned to Norway, arriving at almost the same time as *Fram.* His expedition had shown that there is no continent in the middle of the Arctic Ocean, as had been conjectured, and soundings demonstrated that the Ocean is very deep. He became a national hero, was much involved in the movement that secured Norwegian independence from Sweden, and after the First World War was engaged in humanitarian aid to Russian and Armenian refugees.

Amundsen (1872–1928) was arguably the greatest explorer of all. A memorial stone outside his birthplace near Fredrikstad in Norway records his achievements: 'Northwest Passage 1903–1907, South Pole 1910–1912, Northeast Passage 1918–1920, North Pole 1926, a hero's death in the Northern Ice Sea, 1928'. He was a member of the Belgian Antarctic expedition in 1897–99, and was the first to sail through the Northwest Passage, in a small seal hunting vessel, *Gjøa*, reaching the Bering Strait in 1906 after three winters in the ice. During that voyage the Inuit taught him survival skills that were to serve him well. He sailed in Nansen's ship *Fram,* initially intending to go to the North Pole, but when he learned that Peary and Cook had already got there, he changed his mind and sailed to Antarctica. His team of five men and 52 dogs left the coast in October 1911, and reached the South Pole on 14 December 1911.

Shackleton (1874–1922) was a British Merchant Navy officer who went to the Antarctic with the *Discovery* Expedition 1901–1904, and returned with the *Nimrod* expedition. In January 1909 he and three companions reached 88°23' S, only 155 km from the South Pole, but then had to turn back (and famously remarked to his wife: 'I thought you would rather have a live donkey than a dead lion'). He went south again to cross the Antarctic continent, but his ship was crushed in the ice. He and his crew reached Elephant Island, and he and a few others made a legendary small-boat voyage to South Georgia and returned to rescue the remainder of the crew.

The British explorer Scott (1868–1912) led the *Discovery* Expedition to the Antarctic, and reached 82°17'S. He set off from the Antarctic coast in November 1911, with a party of five, and arrived at the South Pole to find that Amundsen had been there first, 34 days earlier. The party marched north and died of hunger and exhaustion, the last three just 18 km short of a food and fuel dump. Meanwhile Amundsen's party skied back to the coast, and lost nobody. Scott's expedition exerts a hold over the British imagination, and used to be promoted as an exemplar of how people should behave: when one of the writers was at school, his class was only taken once to the cinema, to see *Scott of the Antarctic*, and the same thing happened to the other writer. It was rather ludicrously argued that Amundsen was "unsporting" because his objective was to be first at the Pole, rather than to conduct scientific exploration, and because he did not tell Scott until he was well on his way. The traditional sentimental interpretation was more recently and controversially questioned by Huntford [13], who pointed out that Scott had only become interested in the Polar regions when he was trying to reignite his faltering career as an officer in the Royal Navy. Rather than being a model of stoic heroism, his expedition was a model of amateurishness, incompetent management, inadequate planning, and rejection of the advice of experienced Polar travellers such as Nansen. The controversy continues (Fiennes [14], Barczewski [15], Crane [16]) and recently there has been a small shift back to a more positive assessment of Scott.

The age of exploration was almost over, but some adventures remained. In 1926 the Italian airship engineer Nobile (1885–1978) flew the airship *Norge* from Svalbard to the North Pole and on to Alaska, with a crew that include Amundsen. In 1928 Amundsen disappeared in the Barents Sea, in a flying boat searching for lost members of another Nobile expedition.

It still remains possible to try to imitate the explorers of the heroic age, but only by deliberately turning one's back on modern technology such as aircraft and global positioning systems (and relying on people with those devices for rescue when things go wrong).

The ease with which technology has now overcome most of the difficulties faced by the early explorers was brought home to one of the writers when in 1986 he visited the South Pole for the afternoon. This was courtesy of the US National Science Foundation which had sponsored a workshop in Antarctica on the future of the Antarctic Treaty which was coming up for renegotiation [17]. The workshop was held in very comfortable huts on the plateau above the Beardmore glacier (up which Scott's party had struggled in 1911/12). We even had a cocktail hour after the sessions each day before dinner. One day during the week of the workshop, two Hercules aircraft on skis came from McMurdo and took the participants to the South Pole for a few hours and later dropped us back at the Beardmore.

Of course, sometimes technology is pushed to its limits. Scott base at the South Pole is a year-round base, but in the winter no resupply takes place because in the bitter cold, the operation of flying and landing aircraft there is on the borderline of current equipment. Only in the case of severe medical emergency have flights been made by Twin Otters (which were brought all the way from Canada) to conduct emergency evacuations.

1.4 Developers

Once the exploration phase was largely complete, people came to the Arctic for economic reasons. There are many fascinating stories.

Fur from wild animals is a high-value low-volume low-weight product for which people are prepared to pay large sums. It had of course already been exploited by native peoples, and was immediately attractive because little technology is required to make it ready to transport. It was an essential component of indigenous clothing and footwear, and of the clothing of outsiders coming to the Arctic, almost without competition until the development of synthetics and of fur farms.

The northern forest (taiga) covers many millions of square kilometres in Arctic Russia, Canada and Alaska. The timber supply is effectively unlimited, but its economic value depends on transportation to users in the south. Only if it can be shipped out economically can timber from Siberia compete with more favourably sited areas such as Finland and Sweden. Timber shipments to western Europe from the White Sea began centuries ago, and later they came from further east, particularly from Igarka on the Ob' estuary.

In Arctic Russia, alluvial gold was found on the Lena in the nineteenth century, and later developed in other river valleys, among them the Aldan and the Vitim. Soviet Russia had no source of industrial diamonds, and was forced to buy them from South Africa. Diamonds are formed from carbon under high pressure and temperature in ultra-basic conditions, and are found in kimberlite 'pipes'. A legendary exploration effort led to the finding of diamonds in 1955 at Mirnyi on the Vil'yuy, and later at Udachnaya and Aykhal. The strategic and commercial importance of diamonds prompted substantial investment, because recovery of diamonds from kimberlite requires large amounts of electricity and water. Mirnyi had a population of 26000 in 1973 and 40000 in 2007. The original pipe is reported now to be exhausted.

In Alaska, in 1905 the explorer Leffingwell saw natural oil seeps in the cliffs on the Arctic shore close to the Canning River. Nothing was done for many years, but in the late 1960s Atlantic Richfield (ARCO), Exxon and British Petroleum (BP) went back to the area, and in 1968 found the Prudhoe Bay oilfield. It soon realised that the field is very large indeed, one of the largest in the world and by far the largest in the USA, with oil originally in place estimated at 2.5×10^{10} barrels (4×10^{9} m^3). In addition there is a equally large gas reserve (8×10^{11} m^3 recoverable, 26 Tcf) and smaller oil fields not far away, among them Milne Point, Northstar, Liberty and Badami. That discovery was so large that development was at once seen to be economically attractive, and it led to the construction of the Alaska pipeline from Prudhoe Bay to the ice-free port of Valdez on the south coast. Some of the argument surrounding that once-controversial project is described in another chapter.

The Prudhoe Bay development is not based on a permanent resident population with wives, children, and the amenities of a city. It is not clear whether or not that option was ever seriously considered. Instead people commute, on a two weeks on/two weeks off basis, and while they are at Prudhoe Bay they work, eat, read, surf the net, watch videos and exercise, and then fly home for their weeks off. That pattern is nearly universal for industrial developments in Arctic North America. Marginal exceptions are the sub-Arctic petroleum developments around Norman Wells on the Mackenzie River, and the tar sands region of northern Alberta. Oil at Norman Wells was discovered in 1920. It was strategically important in World War II, when gasoline was needed for vehicles on the Alaska Highway and for aircraft carrying Lend-Lease to Russia. The tar sands are estimated to contain 1.7×10^{12} barrels (2.7×10^{11} m^3) of oil.

In contrast, Arctic Russia has large industrial cities such as Mirnyi, Noril'sk and Vorkuta, and many smaller settlements, among them ports strung out along the shore of the Arctic Ocean, and inland where economic development needs a substantial labour force. A high proportion of the people who live in those settlements originally migrated from the south. In the Soviet period some were attracted by opportunity and a relative lack of constraint, and by financial compensation for harsh conditions, but many others went involuntarily. They are there all year round, have their families with them, and expect to enjoy the facilities of a modern city, though their compensation may include unusually generous vacations and subsidised airfares. Mirnyi is the largest centre for diamonds. Noril'sk is a mining centre for nickel, palladium, gold, platinum, copper and cobalt (and is notorious for pollution). Vorkuta has extensive coal reserves, and was a centre of the Gulag. Many of those cities have declined in population and importance since the political change in Russia.

1.5 Outsiders

Many people express strong opinions about the Arctic, and particularly about its technological development, but have not the slightest intention of devoting their lives to it. It would be wrong to argue that their interest is not legitimate. The Arctic areas are parts of nation-states, and a state and all its people have a real interest in what is going on in its territory, and in how best to administer it. There are conflicting demands from many directions, among them developers, the military, people with environmental concerns, and people with broader humanitarian interests. Moreover, we have all become accustomed to wanting to engage with what is happening in the rest of the world, whether dynamite fishing in Indonesia or cholera in Zimbabwe or child hunger in Darfur. We do not want to be told that what happens elsewhere is no concern of ours.

One aspect is psychological. The Arctic holds a special place in the hearts of many people who will never go there, except perhaps as wealthy short-term tourists, and the level of their concern is often proportional to their physical distance. Some wish to see the Arctic as a clean wilderness, free from the untidy presence of billions of people, and free too from the many messy compromises that all those people's desires and needs impose. There are traces of this notion in Lopez' wonderful book on the Arctic [18]. He remarks disdainfully that petroleum workers at Prudhoe Bay have pornographic magazines. Those magazines are available on every airport bookstall: they are part of life in 2012, however much we may dislike them, and it would only be remarkable if they were absent from the Arctic. An Arctic resident would be entitled to be outraged if an outsider told him that he ought not to have a magazine because it conflicted with a notion of a clean and perfect environment.

Resource extraction issues have always been controversial, in part because careless mining, petroleum and lumber developments are environmentally damaging (and often socially damaging). A separate article [18] discusses the circumstances in which petroleum development might be acceptable, and the central question of whom it should be acceptable to. The Alaska oil pipeline was the subject of a huge amount of controversy [20]. One of the writers worked in the early 1970s in a US university on one of the engineering questions raised by the design of the pipeline, and was roundly abused by some of his academic colleagues (and their wives), on the grounds that the pipeline would "ruin Alaska". He replied that he had thought about it carefully and did not agree that Alaska would be ruined, and that in any case his work was intended to study a possible source of damage and make sure that it was eliminated. Looking back, the argument that Alaska has been ruined is not sustainable, though there have been mistakes and there have been environmental impacts. Any environmental impact is unfortunate, but tourism and the military have created a much more extensive impact in Alaska than oil has. At the time of writing, most of the people of Alaska seem to want further development. One of the options for taking North Slope gas to markets in the south is to build a marine pipeline parallel to the Arctic Ocean shore and eastward into Canada, to link up with a pipeline south from the Mackenzie Delta. A newspaper in Fairbanks castigated that option as the worst for Alaska, an option that must be fought tooth and nail, not on environmental grounds but because it would minimise economic opportunities for Alaskans.

A further difficulty with the extraction of irreplaceable mineral resources is what to do when the resource has been depleted. It is difficult to clean up industrial sites in the Arctic, very expensive to do so, and the clean-up itself has environmental impacts. There is little local market for scrap materials, and the temptation is to carry out minimal clean-up and then walk away, possibly after closing off the site so that nobody can go by and observe what has been left. The many examples include abandoned gold dredges around Fairbanks in Alaska, the notorious beach of ammunition shells at Shemya Island in the Aleutians, the sites of nuclear weapons tests on Amchitka and Novaya Zeml'ya, and no doubt others. In one Arctic project the writer was concerned with, the responsible company cleaned up twenty years later, and got rid of one large piece of construction equipment, a highly specialised plough, by bulldozing a hole in the ground, pushing the plough into it, and backfilling the hole – an environmentally modestly benign solution that will be an intriguing puzzle to an archaeologist in a thousand years' time.

The answer is to establish mechanisms to make sure that the authorisation of an industrial development provides for secure funding for ultimate clean-up, and that that money is protected against the possible failure of the original enterprise.

It is not only industrial sites that might ultimately have to be cleared. A strict application of a sustainable civilisation philosophy would require that when humankind no longer needs Inuvik or Vorkuta (or for that matter Moscow or Chicago) they too ought to be removed.

Conclusion

Human beings have conflicting desires, and they cannot all be satisfied. That is as true in the Arctic as it is elsewhere. Communities in the Arctic are even less stable than elsewhere: some will boom, and some will disappear.

References

1 Armstrong, T., Rogers, G. and Rowley, G. (1978).*The circumpolar north.* (Methuen, London, UK)

2 Armstrong, T. (1958). *The Russians in the Arctic.* (Methuen, London, UK).

3 Mowat, F. (1971).*The Siberians.* (Little Brown , New York, USA).

4 Thubron, C. (1999).*In Siberia.* (Chatto & Windus, London , UK).

5 Bychkova-Jordan, B. and Jordan-Bychkov, T.G. (2001). *Siberian village: land and life in the Sakha Republic.* (University of Minnesota Press, Minneapolis, USA).

6 Vitebsky, P. (2005). *The reindeer people: living with animals and spirits in Siberia.* (Houghton Mifflin, Boston, USA).

7 Wheeler, S. (2009) *The magnetic north: notes from the Arctic Circle.* (Jonathan Cape, London, UK; Farrar, Straus & Giroux, New York, USA)

8 O'Carroll, E. and Elliott, M. (1999).*Greenland and the Arctic.* (Lonely Planet, Footscray, Australia).

9 Kolbert, G. (2008) *The Arctic: an anthology*, (Granta Books, London, UK).

10 Ertel, M. (2008). Aufbruch im Eis. Der Spiegel, (46) pp. 150-151

11 Ehrlich, G. (2008) Alberti's Ride. In *The Arctic: an anthology* (ed. E. Kolbert), (Granta Books, London, UK).

12 *Encyclopaedia of the Arctic* (2005). (ed. Nuttall, M.) (Routledge, Abingdon, UK)

13 Huntford, R. (1979) *Scott and Amundsen.* (Hodder & Stoughton, London, UK)

14 Fiennes, R. (2003) *Captain Scott.* (Hodder and Stoughton, London, UK).

15 Barczewski, S. (2007).*Antarctic destinies: Scott, Shackleton and the changing face of heroism.* (Hembledon Continuum, London, UK).

16 Crane, D. (2005) *Scott of the Antarctic: a life of courage and tragedy in the extreme south.* (Collins, London, UK).

17 *Antarctic Treaty System: An Assessment.* Proc of a Workshop held at Beardmore South Field Camp, Antarctica, Jan 7 -13, 1985. (1986).(Polar Research Board. National Academy Press, Washington, USA),

18 Lopez, B. (1986) *Arctic dreams; imagination and desire in a northern landscape.* (Scribner, New York, USA).

19 Palmer, A.C. (2008) Under what conditions can oil and gas development in the Arctic be acceptable, and to whom? Polar Record, 44, pp. 1-5.

20 Coates, P.A. (1993) *The Trans-Alaska pipeline controversy: technology, conservation and the frontier.* (University of Alaska Press, Fairbanks, USA).

Chapter Two

The Physical and Biological Environment

2.1 Climate

The Arctic covers a vast area, and has many diverse environments. Almost every physical environment found in other parts of the world can also be identified in the Arctic, where there are high mountains and featureless plains, deserts and rolling hills, cliffs, great rivers and huge deltas. It would make no more sense to think of a single undifferentiated 'Arctic' environment' than it would to think of a single 'tropical' environment.

One factor is climate, and climate too is diverse. Table 2.1 lists a series of Arctic places, and the temperature and wind at each place on December 8 2008 (morning in the eastern hemisphere, evening in the western hemisphere). Obviously another day and another year would be different, and the table is no more than a snapshot that does not replace the detailed and careful statistical work carried out by climatologists. Websites [1,2] give almost immediate information. Recall that the distances are immense: at 60°N, 10 degrees of longitude correspond to 556 km on the ground going east or west, so that many of these places are thousands of km apart.

Several striking features emerge. The northernmost places are not the coldest. Oymyakon is below the Arctic Circle, far to the south of many places along the Arctic Ocean coast, but it has the lowest temperature in the list, and is reputedly the coldest continuously-inhabited place in the world, a distinction it took over from Verkhoyansk when its weather station was moved to the airport, which lies in a frost hollow. The diamond-mining centre of Mirnyi has almost as low a temperature.

Pevek in Chukhotka is further north but much warmer. Similarly in Alaska, Fairbanks too is below the Arctic Circle, but its temperature is lower than Barrow on the Arctic Ocean. Places like Mirnyi and Fairbanks have continental climates with warm summers and extremely cold winters, whereas near the ocean the climate is moderated by the ocean, which has a large thermal inertia and can relatively exchange heat with the air.

It is also noticeable that at around the same latitude, temperatures fall as you go east in Russia, and fall as you go west in North America. In Russia, this reflects the effect of the North Atlantic Drift that brings warmer water north-east across the Atlantic, and skirts the north coast of Europe to bring heat to the Barents Sea. Temperatures fall much lower to the east of Novaya Zeml'ya. In North America, it is the moderating effects of the oceans that lead to warmer temperatures to the east and west and colder temperatures in the interior of the continent.

On this particular day, several places have very high winds. Danger and discomfort to human beings can be assessed by a semi-empirical wind chill index, which combines the effect of wind and low temperatures and can be linked to heat transfer models that calculate the likelihood of frostbite. Heat transfer between the air and a human body is plainly complex, and involves issues like whether one is primarily concerned with an exposed face or with cooling of the whole body. There are also dynamic effects: cooling is most rapid at the beginning of exposure, because the skin blood vessels have not had time to contract.

Table 2.1 lists the index for the temperatures and wind speeds quoted. There is a strong influence of wind speed, which accords with our day-to-day experience. Resolute is chillier than Alert, because Resolute has a high wind but Alert does not. Another day would be different. A strict application of the formula is misleading when the wind speed is very low, because the formula has an anomalous sensitivity at low wind speed. If there is no wind at all and the V terms in (2.1.1) are taken as zero, the wind chill at Oymyakon comes out at -17°C, but if the wind speed is taken as 5 km/h it becomes -56°C, which is far more reasonable.

Table 2.1 Weather on December 8 2008
*anomalous: see text

location	latitude	longitude	Temper-ature (°C)	wind direction	Wind-speed (km/hour)	windchill index (°C)
Murmansk, Russia	68°57' N	33°18' E	-5	N	4	-7
Archangel'sk, Russia	63°34' N	40°37' E	-1	SW	2	-1
Amderma, Russia	69°47' N	61°51' E	-6	SW	22	-13
Vorkuta, Russia	67°23' N	63°58' E	-19	SW	4	-22
Dudinka, Russia	67°24' N	86°11' E	-27	S	11	-36
Irkutsk, Russia	52°17' N	104°18' E	-17	NW	25	-28
Mirnyi, Russia	62°32' N	113°58' E	-45		7	-55
Oymyakon, Russia	63°28' N	142°40' E	-48		0	-17*
Magadan, Russia	59°34' N	150°46' E	-15	NW	4	-18*
Pevek, Russia	69°42' N	170°19' E	-15		43	-28
Wrangel' Island, Russia	71°14' N	179°25' E	-8	SE	11	-13
Nome, AK, USA	64°30' N	165°24' W	-13	NE	11	-19
Barrow, AK, USA	71°18' N	156°18' W	-24	SE	24	-37
Fairbanks, AK, USA	64°50' N	147°43' W	-26	NE	6	-32
Anchorage, AK, USA	71°18' N	156°18' W	0		0	13
Cambridge Bay, NU, Can.	69°07' N	105°03' W	-29	N	28	-44
Resolute, NU, Canada	74°41' N	90°50' W	-32	NE	46	-51
Qaanaaq, Greenland	77°29' N	69°20' W	-16	E	10	-22
Iqaluit, NU, Canada	63°45' N	68°31' W	-17	SE	17	-26
Alert, NU, Canada	82°28' N	62°30' W	-34	W	7	-42
Goose Bay, NFL, Canada	53°18' N	60°25' W	-5	NE	17	-11
Nuuk, Greenland	64°10' N	60°25' W	-5	S	37	-14

The US National Weather Service 2001 version of the wind chill index is

$$T_{wc} = 13.12 + 0.6215T_a - 11.37V_a^{0.16} + 0.3965T_aV^{0.16} \quad (2.1.1)$$

where

T_{wc} is wind chill index (°C)

T_a is air temperature (°C)

V is wind speed (km/hour), measured at the standard reference height of 10 m.

Some places have fog, which usually occurs when intense radiative cooling at the ground surface chills the air close to the ground, and water condenses to form droplets. There is a temperature inversion, and the temperature increases with increasing height. Particulates in the air act as nucleation sites and encourage fog to form, as we know from notoriously foggy industrial cities such as Lanzhou (and London before burning coal was made illegal). Continuous fog disrupts transportation and construction, and may be psychologically discouraging.

Precipitation usually falls as snow, but the depth of snow is much less than one might expect. Along the Arctic Ocean the accumulated depth is less than 250 mm, and over the Arctic as a whole it rarely reaches 750 mm. Often the snow falls in one place and is picked up by the wind and dropped somewhere else These depths are small by comparison with some inland areas such as the Finger Lakes region of NY, the European Alps, and the mountains of western Canada. There is more snow further to the south, particularly on mountains and on exposed coasts in the sub-Arctic: Valdez on the Gulf of Alaska has an annual snowfall of 20 m. If we estimate the snow density as 100 kg/m^3, this corresponds to 2000 mm of rain, typical of a fairly wet climate.

It would be a mistake to think that these places are invariably cold. Those with continental climates away from the coast can have summer temperatures that approach those in European Russia and the central USA.

2.2 Permafrost and Land Ice

Cold air cools the ground surface. Heat flows from the earth to the air, and the ground freezes, forming permanently frozen ground. Permafrost, or permanently frozen ground, is ground (soil, sediment, or rock) that remains at or below 0°C for at least two years. It occurs both on land and beneath offshore Arctic continental shelves, and underlies about 22 per cent of the Earth's land surface, some in unexpected locations such as the Cairngorm summits in Scotland.

The depth to which the ground freezes is determined by a balance between insolation from the sun, cooling at the surface and geothermal heat flux from the interior of the earth, approximately 0.075 W/m^2, though a little larger at tectonic plate boundaries. In areas where the climate is very cold, permafrost reaches 700 m down, as it does in Yakutsk in Siberia, where in 1837 Fedor Shergin dug a shaft 116 m down and found that he was still in frozen ground.

In less severe climates, the permafrost base extends less far down. Further south still, permafrost occurs in discontinuous patches. Discontinuous permafrost is much influenced by the interactions between the geotechnics, the vegetation and the surface drainage. Inadvertent or deliberate removal of vegetation removes a natural insulating layer, and tends to extend the permafrost boundary, but that is a complicated issue also influenced by changes in surface drainage. One engineering implication is that driving a wheeled or tracked vehicle across frozen ground can damage the vegetation enough to alert the thermal balance and lead to the formation of unsightly water-filled ruts that persist for years.

In most places the summer temperature is above freezing, and a surface layer thaws. The depth of this active layer is quite limited, and usually it extends no more than a metre or so. Johnston [3] gives the thickness of the active layer as 0.5 m at Resolute, NU (74°N), 1 to 2 m at Norman Wells, NWT (65°N) and 2 to 3 m at Hay River, NWT (61°N). If we adopt a much simplified idealisation and treat the heat transfer mechanism as one-dimensional pure conduction (therefore neglecting the influence of freezing, thawing, thermal diffusion, water migration and convection), the time variation of temperature θ is governed by

$$\frac{\partial \theta}{\partial t} = \frac{k}{\rho c} \frac{\partial^2 \theta}{\partial z^2} \tag{2.2.1}$$

where

k	is the thermal conductivity
ρ	is the soil density
c	is the heat capacity per unit mass (specific heat) , and
t	is time
z	is depth.

If we go on to idealise the surface temperature variation as an sinusoidal oscillation with peak-to peak amplitude θ_{max}-θ_{min} about a mean θ_m with period T and take the density and conductivity as uniform with depth z the temperature variation at depth z is the relevant solution of (2.2.1) and is

$$\theta(z,t) = \theta_m + (1/2)(\theta_{\max} - \theta_{\min})\exp(-z\sqrt{\frac{\pi\rho c}{kT}})\sin\left(\frac{2\pi t}{T} - z\sqrt{\frac{\pi\rho c}{kT}}\right) \tag{2.2.2}$$

and so increasing the depth z by a factor β multiplies the amplitude of the temperature variation by $\exp(-\beta\sqrt{\pi\rho c/kT})$. Because of the second part of the sine term, the temperature maxima and minima within the soil lag behind the maxima and minima at the surface.

In a typical saturated soil, k is 2.5 W/m K. Johnston [3] gives empirical formulas for ρc ('volumetric heat capacity' in his terminology) for frozen and unfrozen mineral soils and peat. Taking ρc as 1.8×10^6 J/m^3 K, the thermal diffusivity $k/\rho c$ is 1.39×10^{-6} m^2/s (0.12 m^2/day). If the annual surface temperature range is 60°C (from -30 to +30, typical of an extreme continental climate), then from (2.2.2) the annual range at a depth of 2 m is 35°C and the range at 4 m is 20°C. Diurnal temperature variations have a period of 1 day instead of 365 days, and their influence is limited to only a metre or so. These oversimplified calculations can obviously be refined, but the conclusion is straightforward. Freeze/thaw phase transitions have an effect qualitatively similar to increasing the heat capacity, and therefore further reduce the temperature variations below the surface.

Water has a huge influence, as in geotechnics it does almost always and almost everywhere. The ground that freezes is usually saturated. If the particles are large, like gravel or broken rock, the pore water freezes close to 0°C and the ice fills almost the whole pore space, except for thin surface layers. If the particles are much smaller, as they are in silt or clay, the pore spaces are tiny by comparison. Much of the water remains unfrozen, even if the temperature falls well below 0°C, so that the pores are not full of ice and it remains possible for water to migrate through the soil. Nersesova [4] showed that in fine-grained soils a substantial fraction of the pore water remains unfrozen at -10°C. Temperature gradients themselves induce the migration of pore water.

Often the water does not remain evenly distributed within the soil. When the soil freezes, the ice segregates on many different scales. The thinnest forms of segregated ice are lenses less than 1 mm thick, sometimes called 'sirloin' by analogy with the marbling on high-quality beef. The thickest lenses may be 1 m thick, and may be oriented horizontally as lenses or vertically as wedges.

An even more dramatic formation is the pingo (Inuvialuit; Russian/Yakut bulganyakh), a striking ice-filled hill that may contain a lens 50 m thick. The best-known location of pingoes in North America is the Tuktoyaktuk area of Canada, which has over a thousand. More than one mechanism of formation is possible. Closed-system (hydrostatic-system) pingos form when a mass of unfrozen soil lies beneath a lake, with permafrost beneath it. The lake then dries up and the soil begins to freezing, trapping the remnants of the unfrozen soil and driving it upwards. Open-system pingos (hydraulic-system) form when permafrost overlies unfrozen soil, and a crack in the permafrost allows water to flow upward, raise the surface layer of soil, and form the hill. Some pingos remain stable for hundreds of years, but others can be recent. At the Yakutsk permafrost conference referred to in chapter 1, a very old man accompanied the delegates on a field trip. It was unclear what his role was, but when we arrived at a pingo it emerged: he was to say that when he was a boy the pingo was not there. It was proposed that Canada deliberately initiate formation of a pingo to celebrate the centenary of Confederation, but nothing seems to have come of this idea. Pingos eventually degrade and collapse. Relic pingos from the last glaciation are found in the Netherlands and the UK.

Frozen ground slowly deforms under stress, partly because of slow migration of water, partly because of melting and refreezing, and partly because of creep in solid ice. A slope in ice-rich soil can slowly creep downhill, a process called solifluction. The deforming layer may crumple and fold. The slow movements deform structures such as pipelines and highways across the slope, and may lead to damage.

Some permafrost contains hydrates. Hydrates are solid, rather like wet snow, and form at low temperatures by chemical combination of water and hydrocarbon gases such as methane and ethane [5]. Figure 2.1 is a phase diagram that shows the conditions under which methane and methane/ethane hydrates can be stable. Hydrates are stable above and to the left of the two different phase boundaries shown. The source of the hydrocarbon gases is low-temperature decomposition of organic matter, and perhaps also high-temperature decomposition, the principal mechanism by which petroleum is formed from organic matter cooked by heat and pressure. (Hunt [6]). If the hydrate temperature rises, as a result of surface disturbance, construction, or global warming, the hydrates dissociate back into their constituents. Methane is a potent greenhouse gas, and it is thought that methane released by hydrate dissociation may create a positive feedback that reinforces global warming. Hydrates are a major problem in petroleum pipelines and petrochemical plants, and much effort is given to controlling them. Ground hydrates are a valuable source of energy for the future.

Permafrost can also extend under the sea [3, 7], often as relic ground permafrost originally formed when the sea level was lower during the last glaciation. Going out to sea, the upper permafrost boundary drops away and the lower boundary rises. Seawater freezes at -1.8°C, and so if the sea ice does not freeze to the bottom the ground beneath the bottom does not freeze further. The water in undersea soils is generally saline, and the presence of salt further increases the amount of unfrozen water and makes saline permafrost mechanically weaker [8].

There are many good books on permafrost [3, 9, 10], particularly in Russia. Much painstaking research has mapped the extent of permafrost, and a useful starting point is the International Permafrost Association Circum-Arctic map [7], whose circumpolar permafrost and ground ice data contribute to a unified international data set that depicts the distribution and properties of permafrost and ground ice in the Northern Hemisphere (20°N to 90°N). The original paper map includes information on the relative abundance of ice wedges, massive ice bodies and pingos, ranges of permafrost temperature and thickness. The map is

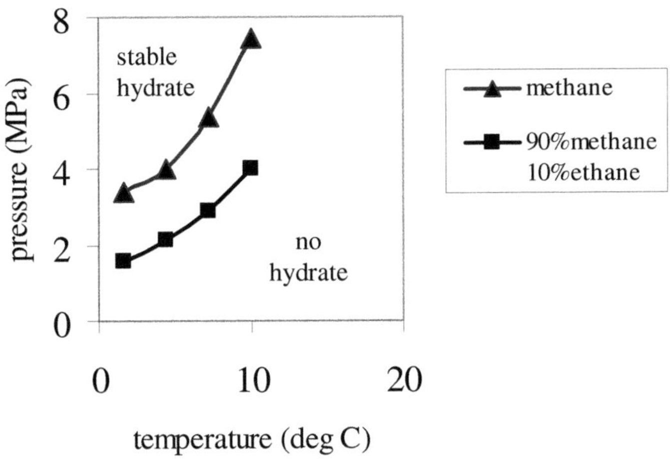

Figure 2.1 Conditions under which gas hydrates can be stable

now available from the US National Snow and Ice Data Center The re-gridded digitised and extended data set shows discontinuous, sporadic, or isolated permafrost boundaries. Permafrost extent is estimated in percent area (90-100%, 50-90%, 10-50%, <10%, and no permafrost). Relative abundance of ground ice in the upper 20 m is estimated in percent volume (>20%, 10-20%, <10%, and 0%). The data set also contains the location of subsea and relict permafrost. The data are gridded at 12.5 km, 25 km, and 0.5 degree resolution. The shapefiles were derived from the original 1:10,000,000 paper map.

However, that map is to such a small scale that it can only provide a general overview, though it may be extremely useful in gauging the severity of the permafrost issue, and in planning a proper survey. An investigation for a specific project requires much closer detail, particularly because permafrost varies rapidly over short distances, both horizontally and vertically [11]. Careful site investigation is important, and this is an issue whose importance has often been underestimated.

Permafrost can be a major problem in Arctic construction. Some of the engineering problems are examined in chapter 6.

A related problem is the formation of surface icings of solid ice. Something obstructs the free flow of surface water, the water freezes, more water arrives and freezes, and a large irregular block of ice builds up. Icings of this kind can block highways and railroads.

2.3 Sea Ice

2.3.1 *Introduction*

The sea freezes if it is cold enough.

Sea ice is hugely important it many contexts. Native peoples use the sea ice to travel, and to reach hunting grounds for seals, polar bears and fish. The first navigators from the south soon encountered sea ice, and found it the major obstacle to further progress to the north and east. It continues to be the principal obstacle to the development of shipping to the northern coasts, though much has been done to learn how to operate ships safely and economically, first through the development of the North-East Passage (North Sea Route; Russian: SevMorPut) and more recently through the development of more powerful icebreakers. Sea ice is also a major factor in the design of systems to exploit offshore petroleum, and this issue is considered in greater detail in later chapters.

Sea ice has far wider implications. It plays a large part in the dynamics of the Arctic Ocean, and its presence alters the transfer of heat and momentum between the wind and the sea. That in turn influences the world as a whole, because there are large exchanges of heat between the Arctic and the other oceans, by water flows through the Bering Strait and the Fram Strait. The water carries nutrients, and those nutrients determine the biological productivity of both the deep ocean and the shallow seas that border the continents.

Ice reflects most of the solar energy that falls on it, whereas open water absorbs almost all of it. The majority of climate scientists accept the reality of global warming, though there are some who argue that there is no longterm effect, or that change is due to natural processes and is not anthropogenic. Global warming will lead to a longer period of open water, and therefore over the year less heat will be reflected and more will be absorbed. That creates a positive feedback mechanism that will further enhance global warming, and has implications for the rest of the world.

Global warming will reduce the ice thickness and the length of the ice season from today's levels, but the methods set out in this book will remain applicable. Projections indicate that the area of open water will much increase, and that by 2050 the Arctic Ocean may be completely ice-free in summer. In turn that will increase the practicability of Arctic shipping, and some enthusiasts believe that the North-East Passage will become a major shipping route between Europe and East Asia, though there are many practical snags, and political and environmental issues. The notion is currently attracting great interest, though it is almost certainly premature and perhaps fanciful.

There are several books on sea ice. The best books on the genesis and morphology of sea ice are by Wadhams [12] and Weeks [13], both beautiful and inspiring labours of love. The best book on sea ice mechanics is by Sanderson [14], and another useful book was written at about the same time by Cammaert and Muggeridge [15], but they are to some extent out-of-date and do not reflect the last 30 years of research. Two further books have been written recently by a Norwegian/Russian group [16,17]. The International Standards Organisation (ISO) has for many years been developing a standard and it appeared in 2010 [18]. It represents a consensus view of an immature and incomplete subject. Det Norske Veritas (DNV) has begun to work towards a recommended practice. All that research remains the subject of lively dispute and vigorously conflicting opinions. The reader will find good articles on these ongoing controversies in the proceedings of the major series of conferences on this subject (POAC, Port and Ocean Engineering under Arctic Conditions, held biannually, [11,20,21,22] most recently in Montreal in 2011 [22]), and in the Cold Regions Science and Technology Journal, the American Society of Civil Engineers Cold Regions Journal, and the proceedings of the annual conferences on Offshore Mechanics and Arctic Engineering, (most recently in Rio de Janeiro in 2012) and of the International Society for Offshore and Polar Engineering (in Rodos in 2012).

This sub-chapter describes sea ice formation processes in general. For specific engineering projects more detail is required than is given here for the region of interest. Information is generally needed on; freeze up and break up dates; ice thickness; types of ice and their morphologies;

ice movement patterns and rates; and so on. A good overview of ice conditions by region is contained in Annexe B of ISO 19906 [18]. This covers 21 northern regions from Baffin Bay to the Sea of Azov. References are supplied together with relevant sources of more detail such as the Canadian Ice Service (http://ice-glaces.ec.gc.ca).

For detailed engineering design even more detail is often required on ice conditions and features than can be obtained from the literature. This is why there has also been, and will continue to be, focused field work as is referred to later in this book.

2.3.2 *Oceanographic Context*

Ice forms on the surface of the ocean. It is therefore influenced by the topography of the seabed, by the marine currents that form part of the worldwide pattern of flow in the oceans, by river flows from the land, and by the atmosphere.

The Arctic Ocean consists of two deep basins, the Canadian and Eurasian Basins, each some 4000 m deep, separated by the Lomonosov Ridge, a relatively steep-sided high linear feature over which depths are much smaller, some 1000 m. The Ridge is not a classical mid-ocean ridge at the meeting of two tectonic plates, but is thought to be a section of continental crust that split away from Siberia when the Eurasian Basin was formed. The continental margins are much shallower, particularly on the Siberian side, where water no more than 50 m deep extends hundreds of km from shore. On the Greenland and Canadian sides deep water is reached much more rapidly.

The Arctic Ocean is almost enclosed by land, with one major exception, the Fram Strait between Svalbard and Greenland. Most of the exchange of water between the Arctic Ocean and the other oceans is through the Fram Strait, which is more than 1000 m deep. The other connections to the oceans are through the Bering Strait between Alaska and Russia, the Barents Sea between Svalbard and Scandinavia, and Nares Strait between Greenland and Ellesmere Island, but all of them are narrow and very much shallower.

Wadhams [12] identifies three layers in vertical temperature and salinity profiles osf the Arctic Ocean. The uppermost layer, the polar surface water, is up to 200 m thick, close to the freezing point, and has a low salinity between 30 and 32‰ (parts per thousand), to be compared with the higher salinity of most seas, typically 34‰. Below that surface layer comes a significantly warmer and more saline layer called Atlantic water. Much of it has flowed in through Fråm Strait, as an extension of first the Gulf Stream (which rounds Florida and then turns north-east), then the North Atlantic Current, and finally the West Spitzbergen current running northward on the west side of Svalbard. It sinks below the polar surface water because it is denser. At a depth of 200 m or so, the Atlantic water is at between 1 and 3°C, but with increasing depth its temperature decreases, while its salinity remains about the same. Below 900 m, the temperature is below 0°C and continues to decrease: this layer is called the Arctic Ocean deep water, and it has slightly different properties on either side of the Lomonosov Ridge.

The surface flow in the Arctic Ocean is dominated by two large systems. The first is the Beaufort Gyre, a clockwise rotating flow, some 2000 km across and centred north of Alaska around 160°W. The second is the Transpolar Drift, which starts around 140°E in the Laptev Sea north of eastern Siberia and flows westward, to the north of Franz Josef Land and Svalbard and down the east coast of Greenland, where it is called the East Greenland Current. That south-going current carries with it large quantities of ice, and the net flow of water balances the water that came into the Arctic Basins as a deep flow of Atlantic water through Fram Strait. This is of course a simplified picture: just as in other parts of the oceans, the currents have branches and eddies whose positions change with time.

Salinity is influenced by the large flows of the rivers that flow northward. On the Siberian side, the huge Lena, Yenesei and the Ob' rivers each flow more than 5×10^8 km^3/year into the ocean. Together with other north-flowing rivers, they drain half the land surface of Asia. The inflow of fresh water contributes to the low salinity along the coast of Siberia, in places as low as 27‰. A similar salinity is reached on the North American side, where the flow of the Mackenzie is 2.6×10^8 km^3/year. In some sub-Arctic seas, the salinity is very low indeed. The Baltic Sea is often said to be more like a lake than a sea, and in the northern part of the Gulf of Bothnia the salinity is below 10‰.

Similarly the northern part of the Caspian Sea has low salinity owing to the discharges of the Ural and Volga.

The atmospheric conditions change with the seasons, but the long-term averages show a barometric high centred around 80°N 180°W. In the northern hemisphere, winds circulate clockwise around a high, roughly parallel to the isobars, so that the pressure gradient is balanced by the Coriolis acceleration induced by the interaction between flow and the Earth's rotation. This is consistent with Nansen's rule (alternatively called Zubov's Law), which says that the ice moves parallel to the isobars, just as the wind direction is almost parallel to the isobars, in the northern hemisphere anti-clockwise around an area of low pressure. Ice moves clockwise around the Beaufort Sea, and the current moves ice from the seas north of Russia over the North Pole and downwards Fram Strait. Nansen discovered one of the earliest pieces of evidence for this, when he found off south Greenland wreckage from the ship *Jeanette*, which had been crushed in the ice off Wrangel Island to the north of eastern Siberia.

2.3.3 *The Structure of Ice*

Ice is an unusual material in many ways. Water is one of the few substances that are less dense in the solid state than in the liquid state. Water is H_2O, and so a water molecule has one oxygen atom and two hydrogen atoms: the relative positions of the three atoms are not fixed.

Ice can have at least nine structures, and each of them is more ordered than the loose and changing structure of liquid water, but all but one of them are stable only at higher pressures and lower temperatures than occur in nature. Several texts on ice examine its microstructure in detail, and relate it to crystallography and materials sciences. An engineer does not normally need that level of detail, and this section is intentionally restricted to a minimal description that sets out in qualitative terms the microstructural properties that influence engineering decisions. This is not to minimise the scientific significance of microstructure: for a deeper appreciation the reader is referred to Schulson and Duval [22], Wadhams [12] and Paterson [24].

The form of ice that occurs in nature at the surface of the Earth is called ice Ih: the other forms are found only at much lower temperatures and higher pressures. Each oxygen atom O is at the centre of a tetrahedron with four more O atoms at the apices of the tetrahedron. The distance between the O atoms is 0.276 nm. In Wadhams' words [12], the hexagonal structure looks rather like a honeycomb composed of parallel layers of slightly crumpled hexagons. The structure is reminiscent of close-packed hexagonal structure of some metals, but is not exactly the same. Figure 2.2(a) is what we see if we look at right angles to the layers, as if we were looking along the cells in the honeycomb: crystallography calls this the **c**-direction or the **c**-axis. The positions of the O atoms are indicated by circles. Figure 2.2(b) is what we see if we look at right angles to the c-direction, so that we see the slightly crumpled hexagons edge-on from the side. The layers are often called basal planes, though they are not exactly planes. At right angles to the c-direction, there are three **a**-directions at 120° to each other.

(a)

(b)

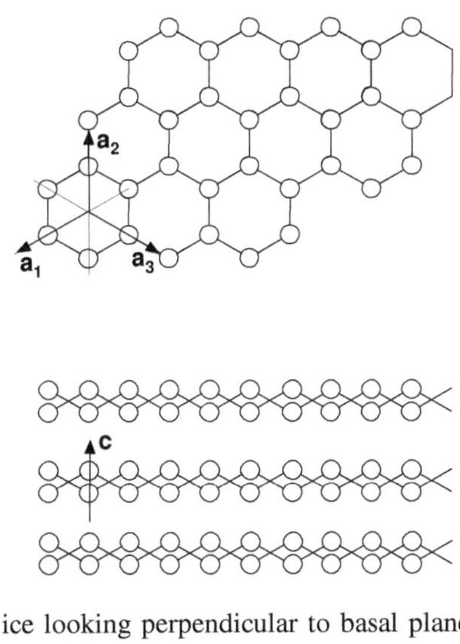

Figure 2.2 Structure of ice looking perpendicular to basal planes (b) looking parallel to basal planes Circles represent oxygen atoms

If you want to understand the structure, by far the best way is to make a model of it, using grapes (or pingpong balls or balls of modelling clay) to represent the O atoms, and toothpicks (or chopsticks) to represent the bonds between them. Start by making a tetrahedron with one O atom at the centre and four more O atoms at the vertices. Then build on by adding more O atoms to make the hexagonal layers, always respecting the condition that every O atom is at the centre of a tetrahedron. Having made one hexagonal layer, add the next one.

Water is H_2O, and so there are two hydrogen atoms H for each O atom. Each O has four closest O neighbours, but each bond has an O at either end, and so there is one H along each bond, nearer one end than the other. There are strong bonds between each O and the nearest H, and weaker bonds between the H atoms. It is because ice has such an open structure that ice is less dense than water, which is unusual, because in most substances the solid phase is denser than the liquid.

The structure influences the growth and mechanical behaviour of a single crystal. It is easier for a crystal to grow by adding atoms to a existing basal plane than by starting a new place, and so crystals prefer grow in the **a** directions rather than in the **c** direction. That is why snowflakes and ice crystals on the surface of the sea have hexagonal symmetry (though the details of the shapes vary enormously. If the ice is forced to deform, deformation parallel to the basal planes disrupts fewer than half as many bonds as deformation across the basal planes does, as you can see by looking at the model. An ice crystal therefore shears relatively easily in directions at right angles to the **c**-direction.

Like grains in metals, single crystals of ice vary greatly in size. Often they are small, a fraction of a mm across. In a group of neighbouring crystals, the axes may be randomly oriented, so that a group of neighbouring crystals has randomly oriented **c**-axis directions. Sometimes on the other hand, the individual crystals all have the same or nearly the same c-axis direction. How this comes to happen depends on how the ice was formed, and that is described in the next section.

When crystals grow in flowing water, it often happens that the c-axes are aligned with the flow direction.

2.3.4 *Ice Formation*

Fresh water and salt water freeze in slightly different ways.

If the water is fresh, salinity 0, its density is a maximum when the temperature is +4°C, and water that is either colder or warmer than +4°C has a smaller density than the maximum. If the water has a low salinity, the temperature at which the density is a maximum decreases towards 0°C, and there is a narrowing region in which the density increases with decreasing temperature. If the salinity reaches 24.7‰, that region disappears, and beyond that salinity the density always increases as the temperature falls. Water with a salinity greater than 24.7‰ is called seawater, and water with a salinity less than 24.7‰ is called brackish. Brackish water is unusual, and occurs only where the ocean is diluted by large volumes of river water. Locations where the sea is brackish are in the Kara Sea off the mouth of the Ob', in the Laptev sea off the mouths of the Yenisei and the Lena, and in the northern Baltic (Wadhams [12]).

Imagine first fresh water in a lake, uniformly at a temperature of say +6°C, and that the air is below freezing, at say -10°C. The surface layer of the water cools by conduction of heat from the water to the air (and to a small extent also by radiation). As long as its temperature is above +4°C, the cooled surface layer is denser than the warmer water below. This situation is unstable. The cooler water sinks and is replaced by warmer water. This process continues until all the water is at +4°C. Further cooling creates a less dense surface layer, because now a reduction in temperature leads to a reduction in density. A less dense colder surface layer below +4° overlying a denser body of water at +4°C is stable, convective interchange of water between the layers comes to a stop. The surface layer continues to cool and ice can form on the surface, even though the water below is still at +4°C.

In the sea with a salinity greater than 24.7‰, on the other hand, there is no density minimum. A cooler surface layer would always be denser than a deeper warmer layer, and so convection continues until the whole water column has cooled down, although at some depth there is often a density increase (called a pycnocline), and convection may cease at that level. For those reasons, if a sea and a neighbouring lake have the same climatic conditions and air temperature, the lake freezes first and the sea freezes later. An additional factor is that the presence of salt lowers the freezing point, so that typical sea water with a salinity of 35‰ freezes at -1.8°C.

Ice crystals cannot form until they have been nucleated. If the water is clean and has no particles that can initiate nucleation, the temperature can fall well below 0°C before nucleation occurs. This is called supercooling. It can be demonstrated by taking a sealed bottle of clean water free of particles, such as commercially-prepared tonic water or soda water, and placing it in the freezing compartment of a domestic refrigerator. After some time, the temperature of the water in the bottle is below 0°C, as can be checked by looking at ice cubes in the same freezing compartment, but the water in the bottle is still not frozen. If the bottle is opened and left for a minute or so, however, dust particles in the air act as nuclei, and much of the water can be seen to freeze rapidly. In nature, on the other hand, there are many nuclei, among them dust particles, bacteria, and bits of plant debris, supercooling is unusual, and ice nucleates only a little below 0°.

Sea ice has a complex structure, because of its history, temperature gradient, and the influence of salt. Figure 2.3 is a simplified schematic cross-section of typical first-year ice, and reflects the processes that have formed it.

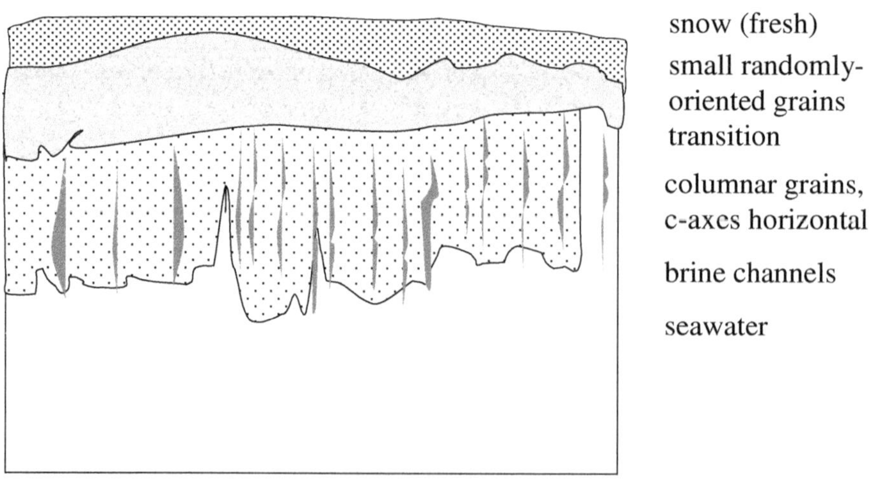

Figure 2.3 Structure of sea ice

Ice formation begins with the formation of small separate crystals, which give the water surface a greasy appearance. Each crystal is a small disc 1 or 2 mm across, with its **c**-axis vertical and growing horizontally in the **a** directions. Symmetrical growth becomes unstable, and long horizontal arms appear and extend. The arms are fragile and break off, and together the fragments form a suspension in the water: that is called frazil or grease ice. As freezing continues, and if the water is calm, the fragments freeze together to form a continuous sheet, called nilas, transparent as long as it is still thin, but becoming darker and less transparent as it thickens. In rough water, on the other hand, the waves prevent the formation of nilas, but eventually the small crystals freeze together into clumps, and then the clumps freeze together into horizontal floating discs. Wave action causes water containing more ice crystals to slop over the sides of the discs, and some of the ice crystals freeze in place and give the discs a raised edge: this is called pancake ice. The pancake discs can grow until they are several metres in diameter and perhaps half a metre thick [12], but the precise sequence depends on the seastate and temperature history. The pancakes freeze together in larger and larger groups, and finally coalesce and form a continuous sheet of consolidated pancake ice. There is no longer any open water, and further growth of ice is by conduction of heat through the consolidated layer.

A complication is the presence of snow, which often but by no means always forms a surface layer. Its grains are randomly oriented, and it does not include salt. It is a poor thermal conductor, with a thermal conductivity 0.086 W/m K for loose new snow and 0.34 W/m K for dense compacted snow [3], both very much lower than 2.2 W/m K for solid ice, and so the snow layer tends to insulate the ice from air temperature fluctuations. This effect may be modified if the snow partially melts and then refreezes, or if rain falls on it.

Below the snow comes the ice that formed first. The ice grains are small and randomly oriented. Below that the grains are columnar, elongated vertically, and much larger than they are closer to the surface. The crystallographic *c*-axes are horizontal and aligned with the under-ice water current that flowed when the ice formed. Between the columnar grains are vertical brine channels and pockets that formed when the grains grew horizontally and progressively excluded salt. The brine in the pockets is denser than sea water, and so if an open channel connects the pocket to the sea beneath, the density difference leads the brine to drain out. (It is because of this that multi-year ice is less salty than first-year ice, and that is why you are told to choose multi-year ice to thaw if you find yourself on the sea ice without drinking water). The brine channels play an important role in ice mechanics. The channels are also home to marine organisms, among them microalgae, diatoms, bacterial that prey on the diatoms, and bacteriophage that preys on the bacteria; this point is examined in section 2.4.

In some seas such as the northern Baltic and the Caspian, no ice lasts through the summer and so there is only first-year ice. In some parts of the Arctic the ice survives from one winter to the next. It is important to be able to distinguish multi-year ice from first-year ice, because its mechanical properties are different: Johnson and Timco [25] have written an instructive guide to visual identification of different kinds of ice.

The ice is rarely stationary. Wind drag on the upper surface creates compression in the ice, and it can begin to fracture and break up. Fractures form when one area of ice is pushed against another, and the accumulating fragments form a pressure ridge. Some of the fragments are pushed down into the sea to form a 'keel', and some are lifted to form a 'sail'. The deepest keels reach down 40 m, and the largest sails are 10 m high, but many keels and sails are very much smaller than this, often no more than a metre in height or depth. The fragments progressively freeze together to form a ridge. The ridge may be compact and nearly solid, but it may also have a loose open structure, in which the fragments nearer the bottom are only weakly attached to the ice above, or perhaps held in place only by their buoyancy. Some ridges only survive for one winter ('first-year ridges'), but others last longer ('multi-year ridges') and they consolidate and become stronger. A ridge structure changes with time: it tends to freeze more firmly together, the ice becomes less saline (because brine drains downward and is excluded as fragments freeze together), summer warmth may partially melt the sail (so that the keel moves upward when the sail weight is reduced), and loosely-attached fragments may break away from the keel base.

Ice breaks easily, as we shall see in chapter 3, and so there are ice fragments on many different scales. It is important to have standardised descriptive words, because one person's 'big ice floe' may be another person's 'small ice floe'.A well-established agreed terminology is set out in the World Meteorological Organisation (WMO) standard Sea Ice Nomenclature [26], and the Arctic and Antarctic Research Institute in St. Petersburg has prepared a consistent Russian glossary [27,28].

Icebergs are not sea ice but fresh-water ice, composed of glacier and ice-sheet ice that originally fell as snow. When the ice reaches the sea, large pieces break off and drift away. As the ice floats south, driven by wind and current, it progressively melts, and it may roll over. Icebergs are not found in all Arctic seas, but there are large sources on the coast of Greenland, on Axel Heiberg Island, on Svalbard and on Novaya Zeml'ya. Icebergs have been the subject of much research, originally stimulated by the *Titanic* disaster in 1912, which led to the establishment of the International Ice Patrol.

Ice islands are related to icebergs, and like them are made of fresh-water ice. They are broken-away fragments of ice shelves, such as the shelves around Svalbard and Axel Heiberg Island. They are not found in all Arctic seas. Large ice islands occasionally drift through the channels between the Canadian Arctic Islands, and an island grounded in 45 m of water was found 3 km away from the Panarctic Drake F76 project described in chapter 7.

2.4 Gathering Data about Sea Ice

2.4.1 *Identifying Needs*

The various methods described in this book for the quantification of ice interaction with platforms and the sea floor obviously require the input of various ice data for the region of interest. Fortunately, during past decades there has been a considerable amount of activity solely dedicated to gathering ice data in key Arctic regions. These activities and the resulting knowledge have been well documented in conference and journal papers, and a search engine can be used to identify them. In addition, good summaries of what is known about key ice parameters are given in ISO 19906 Annexe B [18]. The quality and authenticity of the data vary, and the user needs to understand the sources of the data and methods of data collection.

In new areas of activity and when getting into detailed analysis of ice interaction and operations in ice, existing knowledge and data may be insufficient and additional data must be gathered either by field expeditions or by remote sensing or both. The planning process and methods for these activities need to be discussed, because they can be very costly and potentially hazardous. The safety aspects of on-ice data gathering are reviewed in Chapter 9.

2.4.2 *Planning*

In planning, the first steps are to look at the calculation methods and determine the critical inputs, then to rank the inputs in order of importance and the state of knowledge. For example, for limit stress ice loads we know that ice thickness is very important. We may know a lot about level ice thickness in the region, but it will be the thickness of extreme ice features such as pressure ridges that will control the design, not level ice. For ridges, sample calculations will usually show that the thickness of the refrozen layer is a dominant input. What data do we have on this quantity? The keel depths and shapes of the keels are also important, but also the strength of the keel rubble which is a very difficult parameter to measure and so there are usually insufficient data. In such a situation, one should ask whether the data on ridge keel strengths from other regions can be used, and do we understand enough about the physics to make this call? Could large-scale controlled tests on ice rubble substitute for in-situ tests in a new region?

The above discussion is simply an example, but demonstrates the thought process that should be undertaken in first deciding on important parameters to consider in a new field programme. Operations can also be considered and could require a different set of important parameters. It makes sense to gather them at the same time if it is feasible to do so.

Once a list of important parameters has been compiled and filtered with a discussion of the state of knowledge, the parameters that require additional data will be clearer and if possible should also be ranked in importance. The next step in planning is to assess methods of gathering the data.

2.4.3 *Methods for Ice Thickness*

In the old days, the only method to obtain ice thickness was to drill through the ice with a human-powered auger and use a measuring stick or tape inserted through the hole to get the thickness. This is an excellent method and is still used for spot checks which are carried out for example before landing a helicopter as well as to calibrate other methods. Fortunately, methods have improved during the past few decades. A natural progression has been to the power auger which speeds up the process and can allow thicker ice features to be drilled. Power augers can be unwieldy, so that in thick ice two or three people can be needed, and they can be hazardous. Care is needed to avoid loose clothing and gloves from being caught in the rotating parts. Stringent safety procedures and training are needed.

Remote-sensing methods such as ground penetrating radar (GPR) and other wave propagation methods have now been perfected [29,30]. They can be used from the ice surface or from helicopters flying just above the ice. Since 2002 for example, surveys using GPR from a helicopter have provided considerable statistical data on level and rafted ice thickness in the North Caspian Sea [31] which in turn have been used in ice load prediction methods. Prinsenberg et al. [32] have used similar devices to obtain extensive ice thickness data for Canada's ice-covered regions. These methods are excellent for level ice up to a certain thickness, depending on the device. The GPR used in the Caspian was good to about 1.8m. GPR is not proven for thick multi-year ice, first-year ridges and ice rubble.

For ridges and rubble, there are two options, thermal drilling and upward looking sonar. Thermal drills melt their way through the ice, either with an electrically heated tip or with a jet of hot water. The hot water system is faster, and 2 m of ice can be penetrated in less than a minute. The hot water drill consists of a metal hollow wand about 1m long and about 3 cm in diameter attached to a long flexible hose (Figure 2.4). Hot water is generated in a small kerosene fuelled boiler and is pumped through the hose (Figure 2.5). The water supply is usually the water under the ice, so that its discharge is essentially environmentally neutral. The drill is simply a way of penetrating the ice, and the thickness can still be measured with a tape. More usually the hose is marked off, and once the operator senses that the drill has no further resistance from the ice calls out the distance marked on the hose. A more sophisticated system runs the hose over a pulley which is instrumented to count rotations and hence distance. In a thick ridge or when drilling though grounded rubble, once the drill has been removed, a water level detector can be run down the hole to give the water line from the surface. When drilling through ice rubble, the operator can call out when voids are detected and porosity can be determined. With a pulley system, the change in speed of the drill's penetration can be recorded and calibrated against solid ice, porous ice and voids.

Hot water drills have been used extensively off Sakhalin to study extreme first-year ridges [33] and in the Caspian to study stamukhi [34]. In the latter case, the emphasis has been on detecting pits in the sea floor under the grounded ice. In this application care is needed in designing a tip with a low velocity jet so that jetting into soft soils in the sea floor is

Figure 2.4 Thermal drill probe (Photo by KRCA)

minimal. An electric tip drill is less likely to penetrate soft soils but it is slower. The electric tip drill can also better detect small changes in porosity based on penetration rate.

The other approach to measuring the thickness of ridges and hummocks is the sea-floor mounted upward-looking sonar, which has been used very successfully for decades. Prominent in its development was the Ocean Sciences group in the Department of Fisheries and Oceans, Canada [35] and it was first deployed in the Beaufort Sea. Before this, upwards looking sonars had been used from nuclear submarines traversing under the ice to get ridge depths [36]. Today's bottom-founded devices also incorporate an acoustic Doppler system to measure ice speed and direction. During the past two decades or so, these devices have been deployed world-wide to give data on ridge depths, keel shapes and ice thickness. They can be deployed through the ice (using helicopters to access the ice) or during the open water season from ships. They are self recording and the data is stored until recovery several months later, recovery is in open water. Real time devices can be deployed in front of platforms using a cable to bring the data back to the

Figure 2.5 Thermal drill in use on small stamukha (Photo by KRCA)

platform. In this case, there may be a risk of scouring keels severing the cable, so redundancy is recommended.

Upward looking devices can only give underwater thickness. To get total thickness, the correlations obtained between sails and keels obtained from on-ice drilling can be applied. Another approach is to run aerial surveys with a laser profiling device at set times, but correlation with the underwater record, especially in terms of ice movement direction, is difficult.

The laser profiling device is useful in its own right as an alternative to on-ice surveys of stamukhi and grounded ice rubble in front of platforms. In recent years this method has been used extensively in the North Caspian [29] for stamukhi and in the Beaufort to look at grounded ice rubble [37].

2.4.4 *Ice Movement*

Drift buoys reporting back through satellites are good to map general ice drift [31]. The upward-looking devices with acoustic Doppler will give instantaneous ice speeds over a given point. Marine radars operating from a platform or from shore stations can also be tuned to give ice speed and direction. In landfast ice, small movements can be recorded using wire-reel devices referenced to the sea floor [38,39].

For floe size and ridge length, as might be needed in limit force estimates, aerial photography has been used extensively in the past and still gives excellent information. Satellite imagery is now superseding photos from aircraft, but high resolution images are expensive and usually have to be ordered ahead of time. Radarsat images have the advantage of penetrating clouds, and experts in interpretation can detect ridges, large stamukhi and even icebergs.

In addition to the above important parameters, on-ice activities to obtain ice physical, chemical and mechanical properties can be extensive and deserve some discussion.

2.4.5 *Ice Strength and Related Parameters*

Ice strength is problematic. For the reasons discussed in Chapters 3 and 4, ice strength on a large scale can be dominated by fracture processes and natural flaws and this creates the "size effect". Therefore, small-scale strength tests as input into ice load methods are of limited value. So why collect them? One answer is that large scale strengths are very difficult to collect unless an instrumented structure is available; methods of instrumenting platforms follow in this Chapter. This is particularly true of crushing, so historically small scale strengths were the only available input, and ice load methods for structures such as bridge piers were related to small scale crushing strengths in equations such as Korzahvin's. Small-scale strengths became embedded in the codes, especially in the Russian codes, and an operator was required to measure them.

Looking back at the old literature it can be seen that extensive references are made to crushing strengths obtained from ice cores. The original approach was to take a cylindrical core from the ice cover, cut it into test lengths and test them right away with a hand operated test machine (sometimes powered to give a relatively constant strain rate). If the tests are done quickly enough, the original temperature and salinity gradient through cover might be retained and the strength as a function of these parameters also obtained. But clearly if the testing took some time, then the ice core would trend to the air temperature and the brine would start to drain from the core. Later approaches, simply took the core, measured the temperature gradient then sealed it and shipped it back to a laboratory for testing, packing it in "dry ice" (solid CO_2) to minimize loss of brine. This approach had the advantage that the test samples could be better prepared and a more sophisticated test machine could be used. Unfortunately apparent strengths increased because, as is now well known, the measured strengths are quite sensitive to machine stiffness and how well the end of the specimen is machined flat [40]. Laboratory measurements on carefully prepared samples in sophisticated test machines were generally higher than obtained in the field and insertion into the same equations led to high predicted loads.

During this period of looking at the strength of small ice samples with intense scrutiny, it was usefully established that compressive strength varied with ice temperature, ice salinity (porosity) and strain rate. These trends are useful to know, especially the effects of strain rate because as discussed in Chapter 4, ice can load a structure at low strain rates and in creep its properties may not be that different between small and large volumes. As a result of this work, the strength of small ice samples in compression as a function of porosity (temperature and salinity) is well-established and is referred to in ISO 19906 [18]. It is therefore quite possible to predict small-scale compressive strengths in an ice cover by simply measuring ice temperature and salinity profiles.

Based on the above discussion, the authors see no need to continue to collect small-scale ice compressive ice strengths in an area of interest. If the codes require reference to small-scale strengths, these are reliably predictable using ice temperature and salinity. Ice temperature can be measured by either taking a core and inserting a thermometer in small holes along its length (which has to be done quickly). The core can then be sliced and packaged in plastic bags to measure salinity after the ice has melted back on shore. Alternatively for temperature, a thermistor can be taped on the end of a plastic tube and inserted progressively through the ice sheet as a dry hole is augured. A more sophisticated approach is to freeze a thermistor string into the ice sheet and send the results to shore by satellite. Such a method can also give ice growth with time because the bottom of the ice sheet will always be at the water temperature. This method can be usefully combined with an ice movement buoy.

Using temperature and salinity to predict small scale strength is a reasonable approach except that the method is not very reliable at warm temperatures (close to melting). The method is not reliable if there is porosity in the ice for other reasons (e.g. trapped air and unfrozen regions in a pressure ridge consolidated layer). Another approach that gives an index of strength is the borehole jack; see also [31]. A hole is created through the ice sheet (progressively, to avoid immediate water influx) and a cylindrical device just fitting the hole is lowered into it and the two halves are jacked apart using a hydraulic pump. It is a confined test and the measured confined strength can be related to the unconfined strength using a factor of about 3 [41]. If the borehole jack measurement will be used as a strength index, this conversion is not necessary and is discouraged. The borehole jack can also be useful to define the effective thickness of the consolidated layer of a first-year pressure ridge based on the definition that the thickness of interest is the thickness that has some significant strength and does not include the ice slush or keel blocks below. Other ways of obtaining the keel strength are reviewed in Chapter 5.

Finally, in terms of compressive strength, it should be emphasized again, that the current methods in ISO 19906 [18] for both global and local ice pressures on platforms do not require the input of ice compressive strength. The ice pressure equation (equation 4.4.2) does have the "strength coefficient" C_r in it which is given different values for different regions. ISO suggests 2.8 for the Arctic, 2.4 for the sub-Arctic and 1.8 for temperate regions including the Baltic. It is presumed that the effects of temperature are implicitly included in these coefficients. ISO defines the Arctic has having, on average, more that 4,000 freezing degree days; the sub-Arctic as having more than 2,000 degree days and the temperate regions as having about 1,000 degree days. The authors of this book have some reservations about linking the value of C_r (and hence the ice loads) to freezing degree days. Freezing degree days certainly affect the ice thickness but this is already in equation (4.4.2) as a separate term. We agree that the higher the freezing degree day value, the more likely that the ice will be colder. But if it is temperature that controls the value of C_r, why not use this parameter directly?

ISO also allows two other methods to derive the value for C_r. Both essentially use small-scale strengths to ratio from the Arctic value or the Baltic value (where the ice pressures on instrumented structures were measured). For example, the small-scale strength ratios can be obtained using the borehole jack and compared to Arctic values for the borehole jack [42]. Using small scale strengths based on temperature and salinity is also allowed and the average of a strength profile is recommended. Obviously, if these methods are used then on- ice measurements of ice strength (borehole jack), ice temperature and salinity will be required.

However, the authors have some doubts about this approach because it pre-supposes that large scale ice failure processes are similar to those on a small scale and that macro cracks and flaws play no role. Such cracks and flaws are not captured in small-scale tests or by predicting ice strength based on temperature and salinity. Moreover, during the measurement programmes on the Molikpaq there seemed to be no temperature effect detected as the ice warmed up in the Spring, and neither has there been any distinction between high salinity first-year ice and multi-year ice. On the other hand, a paper by Schwarz and Jochmann [43] suggests that temperature effects may be causing some of the trends in the ice pressures measured in the Baltic which are currently attributed to a thickness effect. Kärnä and Masterson[44] suggest that there are insufficient data to include temperature effects and argue essentially for lumping all the crushing data into either a "Beaufort bin" or a "Baltic bin".

In summary, the usefulness of small scale compressive strength measurements is not proven and the whole topic of how large scale ice pressures vary with ice temperature and salinity, if at all, requires additional investigation.

The ice strength parameter used in load predictions on sloping structures is the bending strength. It can be argued that bending or flexural strength is not a fundamental property and that we should simply use tensile strength because a beam fails when the tensile strength in the extreme fibre is reached. However, tensile strengths are not particularly easy to measure because of the effects of stress concentrations, and are usually conducted on small samples in a laboratory. On the other hand, the testing of full thickness cantilever beams is relatively easy for a well equipped field expedition, and gives a direct measurement of the bending moment to fail the beam. The flexural strength can be deduced from simple beam theory. The beam can be cut using either chain saws or ditching machines. The end of the beam is then either pulled up or pushed down using a loading frame.

Flexural strengths determined by these means are usually in the range of 0.3 to 0.5 MPa. ISO 19906 gives an expression for flexural strength based on brine volume that is based on analysis of a large number of full thickness beam tests.,

$$\sigma_f = 1.76 \exp(-5.88\sqrt{\nu_b}) \quad (2.4.1)$$

where υ_b is the brine volume fraction, calculated from the temperature and salinity by

$$v_b = S\left(\frac{49.18}{|T|} + 0.53\right) \tag{2.4.2}$$

where

v	is the brine volume fraction
S	is the salinity
T	is the temperature in °C.

ISO 19906 does not propose any size effect for flexural strength, but it can be argued that the effect has not been recognised because nobody has looked for it. In the experience of one of the authors, small beams tested from the fragments of large beams on both first –year and multi-year ice in the Beaufort Sea, always gave higher strengths [45]. This suggests that strengths from small beam tests will be too conservative.

However, observations of real-world interactions suggest that simultaneous failure in bending across the full width of a wide structure is unlikely, and that this should be accounted for, as is suggested for very thick ice on an Arctic sloping platform in Chapter 5. Furthermore, the review of loads on sloping structures in Chapter 5 shows that the contribution of the breaking term on wide structures with ride up is usually limited to about 20% of the total, so the need to obtain additional field data on flexural strength is perhaps a low priority. Justification based on the ranking process suggested earlier in planning field expeditions should be rigorously applied. An exception is the full thickness flexural strength of thick multi-year ridges because in this scenario, calculations suggest that the load is dominated by the breaking terms.

The topic of the strength of ice rubble is discussed in Chapter 5 including a description of test methods which have been used on first-year ridges. Very heavy equipment is needed to conduct such tests and the preparation procedures require methods to cut through thick refrozen layers of ridges. New approaches to reduce the complexity of such tests would be welcome, as would more test data.

2.5 Biology

Almost nowhere in the Arctic is a sterile desert empty of life.

The brine channels in sea ice described in section 2.3 are cold, dark, saltier than the sea, and partially secluded from exchange of oxygen and nutrients. One would hardly expect to find life in such a hostile and extreme environment. On the contrary, however, Thomas [46], Junge [47], Deming [48] and others have shown that the brine channels are home to a rich community of living organisms. They include copepods, diatoms, bacteria, bacteriophage and virusses. The most conspicuous are pennate diatoms (unicellular photosynthetic microalgae), and there can be so many of them that their photosynthetic pigments colour the undersurface brown. Junge's study focused on Arctic wintertime sea ice, the coldest marine habitat on Earth, and collected samples were collected during the coldest period of the Arctic winter, in March from two sites accessed from Barrow, AK, one on the coastal fast ice of the Chukchi Sea and the other in nearby Elson Lagoon. Microscopic observations of intact ice sections revealed numerous liquid brine inclusions that were inhabited by bacteria, even at -20ºC. On the scale of a bacterium, there turns out to be lots of habitable brine-filled pore space within the ice matrix, with both isolated and fully connected brine tubes and veins.

Extremophiles of this kind are of great interest to biologists, particularly to those concerned with other extreme environments such as deserts and thermal vents, and with the possibility of extraterrestrial life in places like Jupiter's moons Europa and Ganymede. The contribution they make to biological productivity is remarkably high; in the Antarctic sea ice, primary production is estimated to be 63 to 70 million tonnes of carbon a year, about 5 per cent of total primary production in the Southern Ocean, and especially significant because it is out of phase with production in the water column. Their engineering significance is that they are part of the food web, leading through protozoans and metazoans to krill, and thence to squid and other fish, birds, seals, whales, sea mammals and people, and so they have a strong environmental impact. It has also been suggested [48] that their presence may influence the mechanical properties of sea ice, perhaps through proteins that they secrete into brine channels.

Organisms exist on many scales and in almost equally severe environments. During construction of the Panarctic Drake 76 gas well in 55 m of water under 2 m of sea ice off the coast of Melville Island, a fish more than 1 m long swam vigorously past the wellhead television camera, unfortunately too fast for identification or photography. Fish are an important component of the food web. It is possible that as the more easily accessible seas and oceans become depleted of fish, as is already happening to a frightening extent, the fishing industry will turn more of its attention to the Arctic seas, despite the obvious difficulties.

Sea mammals and polar bears play a role in the lives of people living along Arctic coasts, are environmentally important, and are a conspicuous and attractive focus of concern of the environmental movement. As in other parts of the world, there is concern about disturbance from snowmobiles and aircraft, from underwater sound created by seismic survey and drilling operations, and from unauthorised 'sport' hunting by incomers. Hunting for food is central to the lives of coastal peoples.

Land mammals range widely over the whole area [49]. They range from deer mice (Peromyscus *maniculatus*, found in the Yukon and along the Labrador coast as far north as Ungava Bay), through lemmings (Greenland Collared Lemming, *Discrontonyx greenlandicus*, found in the Canadian Arctic Islands) to large animals such as Barren Ground Caribou (*Rangifer arcticus,*along the coast from Baffin Island to Alaska), moose (*Alces alces*), muskox (*Ovibos moschatus*) and Arctic Fox (*Alopex lagopus*). Together they form a complex food web: moose eat woody plants and saplings, whereas Arctic fox is a scavenger, eating anything it can find, from the leavings of polar bears to lemmings, hares, birds and eggs.

Fox is important as a source of valuable fur. Caribou can be domesticated, and reindeer herds play a central part in the lives of many native peoples in Siberia, Canada and Alaska. Attempts have been made to introduce them to Greenland. There has been much concern about the effect on free migration of caribou of man-made linear features such as roads and above-ground pipelines, and care has been taken to raise pipelines high enough for caribou to walk beneath them. Some people think that concern not to take account of the animals' intelligence. Wild horses are significant in the extremely severe climate of Sakha.

The Arctic has an extraordinarily rich bird life [50]. Many birds are only present in the Arctic in the summer, and migrate huge distances for the winter. The Arctic Tern (*Sterna paradisaea*) is found everywhere along the circumpolar coast of America and Asia, and feeds on shrimp and smaller marine organisms. Each northern autumn it migrates to the Antarctic Ocean. It returns the following spring, a round trip of some 35000 km. The Snowy Owl (*Nycteaa scandiaca*) does not migrate so far: it lives on lemmings, breeds in the Arctic Islands and along the Alaska coast, and winters somewhat to the south.

Insects must not be forgotten [51]. In many places in the Arctic, there are astonishing numbers of mosquitoes, midges and black flies. They can be maddeningly annoying, and are not just a minor nuisance but a serious threat to efficient working.

Plant life is equally important, and is found everywhere. On the ground surface in the Arctic Islands, it looks at first as if there is nothing but snow, but if you kick the snow aside you find gleaming vermillion-red lichen. Going south, there is larger vegetation, at first grass and low shrubs and then small trees. Vegetation is important to the thermal balance at the ground surface and thence to ground temperatures, and many examples show that careless disturbance to vegetation can have important engineering effects. Arctic vegetation is slow to recover from damage, and the effects may last for many years. Driving one vehicle across the tundra crushes the vegetation, the altered thermal balance creates thawing, the wheel track turns into a water-filled rut, and the rut creates a channel that alters the pattern of surface drainage (and may lead to wider erosion).

Engineers engaged with the Arctic need to be sensitive to the biological context, and to seek advice.

References

1 wunderground.com/global/RS.

2 US National Oceamic and Atmospheric Administration, National Weather Service, www.weather.gov.

3 Johnston, G.H. (1981).*Permafrost: engineering design and construction.* (Wiley,

New York, USA)

4 Nersesova, Z.A. (1951). Melting of ice in soils at negative temperatures. *Doklady, USSR Academy of Scienes*, 4 (3)

5 Sloan, E.D. and Koh, C. (2008) *Clathrate hydrates of natural gases.* (CRC Press, Boca Raton, USA) .

6 Hunt, J.M. Petroleum geochemistry and geology. W.H. Freeman and Company (1996).

7 Brown, J., Ferrians, O.J. Jr., Heginbottom, J.A. and Melnikov, E.S. (1998, revised 2001) *Circum-Arctic map of permafrost and ground ice conditions.* Boulder, CO, USA: National Snow and Ice Data Center/World Data Center for Glaciology. Digital media

8 Walker, D.B.L., Hayley, D.W. and Palmer A.C. (1983) The influence of subsea permafrost on offshore pipeline design, *Proceedings, Fourth International Conference on Permafrost,* Fairbanks, National Academy Press, 1338-1343 (1983).

9 Tsytovich, N.A. (1973) *Mekhanika merzlykh gruntov*, Vyshkaya Shkola Press, Moscow (English translation: *The mechanics of frozen ground* (1975). Scripta Book Company, Washington, DC.

10 Mel'nikov, E.S. and Grechnishchev, S.E. (2002) *Vechnaya merzlota i osvoyenie neftegasonosnykh raionov* (Permafrost and development of oil and gas bearing regions) GEOS, Moscow.

11 Palmer, A.C. (1972) Settlement of a pipeline on thawing permafrost, *Transportation Engineering Journal, Proceedings of the American Society of Civil Engineers*, TE3, pp. 477-491.

12 Wadhams, P. (2000) *Ice in the ocean.* Gordon & Breach, London (2000).

13 Weeks, W.F. (2010) *On sea ice.* University of Alaska Press, Fairbanks, AK.

14 Sanderson, T.J.O. (1988) *Ice mechanics: risks to offshore structures.* Graham & Trotman, London.

15 Cammaert, A.B. and Muggeridge, D.B. (1988) *Ice interaction with offshore structures.* Van Nostrand Reinhold, New York.

16 Løset, S., Shkhinek, K.N., Gudmestad, O.T. and Høyland, K.V. (2006) *Actions from ice on Arctic coastal and offshore structures.* Lan', St. Petersburg.

17 Gudmestad, O.T., Løset, S., Alhimenko, A.I., Shkhinek, K.N., Tørum, A. and Jensen, A. (2007) *Engineering aspects related to Arctic offshore developments.* Lan', St. Petersburg..

18 International Standards Organisation. (2010) *Petroleum and natural gas industries – Arctic Offshore Structures.* ISO/CD 19906.

19 *POAC 2005: Proceedings, 18th International Conference on Port and Ocean Engineering under Arctic conditions*, Potsdam, NY, USA (2005).

20 *POAC 2007: Proceedings, 19th International Conference on Port and Ocean Engineering under Arctic conditions*, Dalian, China (2007).

21 POAC 2009: Proceedings, 20th International Conference on Port and Ocean Engineering under Arctic conditions, Luleå, Sweden (2009).

22 *POAC 2011: Proceedings, 21st International Conference on Port and Ocean Engineering under Arctic conditions*, Montréal, Canada (2011).
23 Schulson, E.M. and Duval, P. (2009) *Creep and fracture of ice*, Cambridge University Press, Cambridge.
24 Paterson, W.S.B. (1969) *The physics of glaciers*. Pergamon Press, Oxford, UK.
25 Johnston, M.E. and Timco, G.W. (2008) *Understanding and identifying old ice in summer*. Canadian Hydraulics Centre..
26 ice-glaces.ec.gc.ca/App/WsvPageDsp.cfm?ID=11743&Lang=eng 26
27 www.aari.nw.ru/gdsidb/glossary/glos_21_2.htm
28 Krutskii, B.A., Loschilv, V.S. and Shirokov, K.P. (1984) *International symbols on sea ice charts and sea ice nomenclature*, Gizmeteoizdat, Leningrad
29 Comfort, G., Metge, M., Liddiard, A., Vincent, P. (2002) Ice environmental data collection for the North Caspian Sea. *Proceedings, IAHR Ice Symposium 2002*, Dunedin, NZ.
30 Comfort, G. (1998) *Investigation of techniques for continuously profiling ice features*. PERD/CHC Report 5-91, 1998-03 National Research Council, Ottawa, Canada.
31 Nilsen, R., and Verlaan, P. (2011) The North Caspian Sea ice conditions and how key ice data is gathered. *POAC 2011: Proceedings, 21st International Conference on Port and Ocean Engineering under Arctic conditions,* Montréal, Canada.
32 Prinsenberg, S.J., Peterson, I. K., Holladay J. S., and Lalumiere, L. (2011) Labrador Shelf pack ice and iceberg survey, March 2009. *Canadian Technical Report of Hydrography and Ocean Sciences*, 269.
33 Kharitonov, V. V. (2008) Internal structure of ice ridges and stamukhas based on thermal drilling data. Cold Regions Science and Technology. 52 (3) pp. 302-325.
34 Crocker, G., Ritch, R. and Nilsen, R. (2011) Some observations of ice features in the North Caspian Sea. *Proceedings, 21st International Conference on Port and Ocean Engineering under Arctic conditions*, Montréal, Canada.
35 Melling, H. and Riedel, D.A.. (1993) Draft and movement of pack ice in the Beaufort Sea April 1990-March 1991. *Canadian Technical Report of Hydrography and Ocean Sciences*. 151.
36 Wadhams, P., and Horne, R.J. (1980) An analysis of ice profiles obtained by submarine sonar in the Beaufort Sea, *Journal of Glaciology*, 25 (93)..
37 Barker, A., and Flynn, M. (2011) A comparative analysis of rubble field data collection techniques. *Proceedings, 21st International Conference on Port and Ocean Engineering under Arctic conditions*, Montréal, Canada.
38 Croasdale, K. R. (1973) The movement of Arctic landfast ice and influence on offshore drilling. *Proceedings, Second International Conference on Port and Ocean Engineering under Arctic conditions*, Reykjavik, Iceland (1973).
39 Spedding, L. G. (1981) 1974/1975 Landfast ice motion observations for the Mackenzie Delta region of the Beaufort Sea. Proceedings, ASME ETCE Conference, Houston, TX.

40 Schwarz, J., Frederking, R., Gavrillo, V., Petrov, I. G. Mellor, M., Tryde, P., and Vaudrey, K. D. (1981) Standardized testing methods for measuring mechanical properties of ice. *Cold Regions Science and Technology* 4, pp. 245 – 253.

41 Masterson, D. (1992) Interpretation of *in-situ* borehole ice strength measurements tests, *IAHR Ice Symposium 1992*, Banff, Alberta..

42 Johnston, M., Timco G. W., Frederking, R. (2003) In situ borehole strength measurements on multi-year sea ice. *Proceedings,Thirteenth International Offshore and Polar Engineering Conference* Honolulu, Hawaii, USA (2003).

43 Schwarz, J. and Jochmann, P. (2009) Ice forces affected by temperature and thickness of the ice. *POAC 2009: Proceedings, 20th International Conference on Port and Ocean Engineering under Arctic conditions*, Luleå, Sweden.

44 Kärnä, T. and Masterson, D. Data for crushing formula. (2011) *Proceedings, 21st International Conference on Port and Ocean Engineering under Arctic conditions*, Montréal, Canada.

45 Gladwell, R. W. (1977) *Field studies of the strength and physical properties of a multi-year ice pressure ridge in the Southern Beaufort Sea.* APOA Project 91; Imperial Oil Report No. IPRT – 3ME -77.

46 Thomas, D.N. and Dieckmann, G.S. (2002) Antarctic sea ice – a habitat for extremophiles. *Science*, 295, pp. 641-664.

47 Junge, K. Eicken, H. and Deming, J.W. Bacterial activity at -2 to -20°C in Arctic wintertime sea ice. Applied and Environmental Microbiology, 70, 550-557 (2004).

48 Deming, J. (2007) *Proceedings, Arctic Frontiers Conference*, Tromsø, Norway

49 Burt, W.H. and Grossenheider, R.P. (1964) *A field guide to the mammals.* Houghton Mifflin, Boston, MA.

50 Bull, J. and Farrand, J. (1994) *National Audubon Society Field Guide to North American Birds.* Alfred A. Knopf, New York.

51 Milne, L. and Milne, M.(1980) *National Audubon Society Field Guide to insects and spiders.* Alfred A. Knopf, New York.

Chapter Three

Ice Mechanics

3.1 Introduction

Most technology in the Arctic involves ice in some form, sometimes as icebergs and large sheets of sea ice, sometimes as tiny ice lenses in permafrost, and everything in between. Ice may deform naturally, as it does when a glacier slowly creeps down a valley, or it may be deformed by some engineering object, as it does when ice crushes against an offshore platform or is squeezed between soil particles in permafrost. Ice force is usually a critical factor that governs much of the design of an offshore structure. The formation and morphology of ice are described in chapter 2. This chapter goes on to describe and quantify its mechanical behaviour. Ice mechanics is part of the broader subject of solid mechanics: it tells us how the ice will deform, and what forces and deformations are likely to occur. Deformation is often – but by no means always - the limiting factor that determines how much force ice can apply to a structure.

It is useful to know about the microstructure of ice, and that is briefly described in section 3.2. It forms the background to the subject. However, the overwhelming majority of ice engineering calculations and assessments do not start with the microstructure, but instead adopt a viewpoint that treats the ice as a continuum with mechanical properties described by constitutive equations. This is as true of ice as it is of other engineering calculations. If an engineer wants to design a steel girder, she does not begin with dislocations and grain boundaries, still less with atoms, but instead brings to bear the language and apparatus of continuum mechanics and structural mechanics, reflected in continuum descriptions like strain, yield stress and elastic modulus. In the same way, if we want to design a tunnel through soil, we use the geotechnical solid mechanics concepts of strength, pore pressure and critical state, rather than starting with individual soil particles.

How ice behaves depends markedly on how fast it deforms. A rough first classification between mechanisms is summarised in table 3.1. It lists a number of contexts in which ice deforms, and for each one identifies a typical length scale *L*, a typical velocity *U,* and the ratio *U/L* between them. The numerical values are very rough round numbers. There is often uncertainty about which length scale is the appropriate one, and the choice is to some extent arbitrary, and so the table identifies which dimension has been selected. *U/L* has the same dimension as strain rate, but does not necessarily need to be interpreted as a strain rate.

If ice deforms very slowly, it deforms in creep, it does not break up into fragments, and it remains a more or less continuous solid. That is the dominant mode of deformation in very large ice sheets, such as the Greenland ice cap, and in many valley glaciers. Slow thermal expansion of landfast ice can also load fixed structures such as platforms. Creep is described and quantified in section 3.2.

However, that behaviour is markedly atypical, and in most engineering contexts the ice is compelled to deform far more rapidly. It then deforms in a completely different way, by fracturing into broken pieces. The formation, growth and interaction of cracks are the dominant mechanisms, though creep and other microstructural processes are significant on a small scale, especially around the ends of cracks.

That ice is proverbially brittle is part of our day-to-day experience in cold climates, and has entered our language through phrases like "skating on thin ice" and "breaking the ice". We know that brittle materials crack easily, and we are careful to avoid situations where cracks might grow, for instance by throwing away a cracked wineglass. Fracture is the dominant mode of deformation when wind and current drive ice against fixed structures, when ships sail through ice, when we cut and drill ice, when icebergs break away from ice shelves, and when fast-moving glaciers surge through valleys. When we come to quantify the strength of ice against fracture, we find it to be tiny, even by comparison with other materials like glass that we are accustomed to think of as brittle. The simplest way to quantify resistance to fracture is a parameter called

Table 3.1 Ice deformation modes

	length velocity				mode of
	L (m)		U (m/s)	U/L (/s)	deformation
Antarctic ice cap	10^3	(depth)	3×10^{-8}	3×10^{-11}	creep
Alpine valley glacier	10^2	(depth)	3×10^{-7}	3×10^{-9}	creep
Surging glacier	10^2	(depth)	3×10^{-6}	3×10^{-8}	fracture/creep
Iceberg calving	10^2	(thickness)	10^{-3}	10^{-5}	fracture/creep
Laboratory compression test	10^{-1}	(length)	10^{-5}	10^{-4}	fracture/creep
Laboratory indentation test	10^{-3}	(indenter)	10^{-5}	10^{-2}	fracture
Arctic artificial island	10^2	(diameter)	0.1	10^{-3}	fracture
Rubble field	10^2	(breadth)	0.1	10^{-3}	fracture
Icebreaker	20	(beam)	2	10^{-1}	fracture
Monopod offshore structure	5	(diameter)	0.1	2×10^{-2}	fracture
Bridge pier	5	(diameter)	0.1	2×10^{-2}	fracture
Ice ditcher	10^{-2}	(cut)	1	10^3	fracture

fracture toughness or critical stress intensity, described in section 3.3. The fracture toughness of ice is about 0.1 $MPa/m^{3.2}$, ten times smaller than glass at 1 $MPa/m^{3.2}$ and a thousand times smaller than structural steel at 100 $MPa/m^{3.2}$ [1]. Lawn [2] gives an extended list of parameter for selected brittle materials, every one of them tougher than ice.

If ice is loaded and unloaded very quickly, as for instance when a brief impact lasting milliseconds sends small-amplitude stress waves through the ice, and if the stresses are small, then the response is elastic and reversible, so that when the stress returns to its original level the strain too returns to what it was before the impact. Elastic deformation is usually less significant than creep or fracture, because elastic strains remain small. Elastic deformations are part of the background to fracture, when energy flows towards a crack from a still-elastic hinterland, and they can be important in the response to loads that act briefly, such as an aircraft landing on ice or fast-moving ice running up against a conical offshore platform. Ice elasticity is described and quantified in section 3.4.

Another possible idealisation is to treat ice as a perfectly-plastic material. This is easy, and allows the analyst to bring to bear the whole apparatus of plasticity theory, including the limit theorems, slip-line field analysis and well-established and verified numerical methods implemented by finite-element software: see, for example. However, the perfectly-plastic idealisation is seriously misleading and is likely to give results that are incorrect. It is briefly discussed in section 3.5, but generally its use is to be discouraged, and it should never be applied unthinkingly as an easy option.

Continuous ice often breaks up when it comes into contact with a structure. A structure may be loaded indirectly, through an accumulation of fragments piled up around the structure, both above and below the waterline. A component of the broader subject of ice mechanics is the mechanics of ice rubble, a particulate material composed of slippery fragments of many different sizes. Not much is known about it, because of the extreme difficulty of making field measurements. It is discussed in sections 3.6 and 3.7.

3.2 Creep

At low stresses, ice deforms in power-law secondary creep. The deformation is at constant volume, and the shear strain rate is proportional to a power n of the shear stress.

Creep needs to be described by a relationship between stress and strain rate. In the simplest version of power-law creep, imagine a unidirectional flow of ice with velocity u_1 in the 1-direction (referred to Cartesian reference axes 1 2 and 3), and suppose that the only non-zero stress component is σ_{12} (equal to σ_{21}) This might represents an idealisation of flow in a broad valley glacier or in the laboratory in a simple shear test (although many laboratory shear tests are in reality far from simple). The only non-zero components of the strain are ε_{12} and ε_{21}, and the strain rate is

$$\frac{\partial u_1}{\partial x_2} = 2\dot{\varepsilon}_{12} = C\sigma_{12}^{n} \qquad (3.2.1)$$

Now suppose that the flow is still unidirectional in the 1-direction, but that there are two non-zero stress components, σ_{12} and σ_{13}. This is like flow in a valley glacier of uniform cross-section, where the ice velocity varies horizontally as well as vertically. The generalisation of (3.2.1) to that case is

$$\frac{\partial u_1}{\partial x_2} = 2\dot{\varepsilon}_{12} = C\tau^{n-1}\sigma_{12} \quad (3.2.2)$$

$$\frac{\partial u_1}{\partial x_3} = 2\dot{\varepsilon}_{13} = C\tau^{n-1}\sigma_{13} \quad (3.2.3)$$

where

$$\tau = (\sigma_{12}^2 + \sigma_{13}^2)^{1/2} \quad (3.2.4)$$

The stress state is usually more complex than pure shear, and the stress then needs to be described in the standard description of continuum mechanics, by a second-order tensor σ_{ij} (see, for example, [3]). The general constitutive equation for power-law creep at constant volume is

$$\dot{\varepsilon}_{ij} = \frac{1}{2}C(\frac{1}{2}s_{kl}s_{kl})^{(n-1)/2}s_{ij} \quad (3.2.5)$$

where C is the same as in (3.2.2), s_{ij} is the deviatoric stress defined by

$$s_{ij} = \sigma_{ij} - (1/3)\sigma_{kk}\delta_{ij} \quad (3.2.6)$$

and δ_{ij} is the Kronecker delta, equal to 1 if i and j are the same and to 0 if i and j are different, and σ_{kk} is the first invariant of the stress tensor. This is not the only generalisation possible. (3.2.6) can be inverted to give the stress as a function of the strain rate

$$s_{ij} = \left(\frac{2}{C^2}\dot{\varepsilon}_{kl}\,\dot{\varepsilon}_{kl}\right)^{\frac{1}{2n}-\frac{1}{2}}\frac{C}{2}\dot{\varepsilon}_{ij} \quad (3.2.7)$$

Figure 3.1, originally compiled by Hallam [4] and frequently reproduced [5], plots the results of compression and tension tests on ice, as strain rate against stress on logarithmic scales, for polycrystalline ice with different grain sizes and at different temperatures. It shows that n is 3 as long as the stress is smaller than about 2 MPa: glaciologists call this Glen's law. Increasing the temperature increases the strain rate for a given stress.

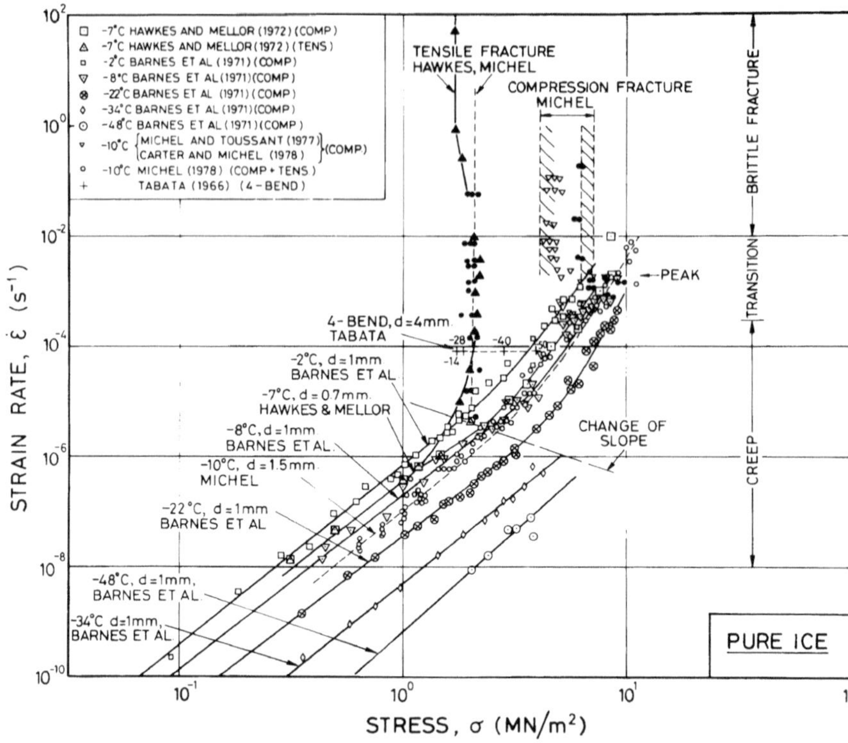

Figure 3.1 Uniaxial loading of pure polycrystalline ice: relation between stress and strain rate (Hallam [4]), originally published in Sanderson, T.J.O., Ice mechanics: risks to offshore structures, Graham & Trotman (1988), with kind permission from Springer Science+Business Media B.V.

Power-law creep is important in mechanical engineering, because it is a good idealisation of the behaviour of high-temperature steel and alloy components such as turbine blades and pressure vessels. Approximate solutions can be derived by linearly interpolating on $1/n$, between n equal to 1, which corresponds to linear elasticity, and n tending to infinity, which corresponds to perfect plasticity. Both those extremes have a wealth of theory and many solutions. There are several theorems that can be used to derive bounds on deformation rates. Nye [6] found power-law creep solutions for creeping flow in valley glaciers with circular and parabolic cross-sections uniform along the glacier length, and Palmer [7] showed that the bound theorems [8] can provide useful bounds for glaciers with any cross-section.

Creep can readily be analysed by numerical finite-element or finite-difference methods. It has the advantage that the deformation varies smoothly and without the cracks and strain singularities that are always present in fracture. That makes the finite elements well-behaved.

3.3 Fracture

3.3.1 *Introduction*

Fracture is much more important than creep, and more difficult and interesting. It lies at the heart of ice mechanics – though it ought to be added that not everyone would agree with this statement.

Fracture mechanics has been extensively developed since the early work of Griffiths [9], Obreimoff [10] and Irwin [11]. It is technologically important in many contexts. There are several good books, and those recommended by a group of specialists in this field are by Lawn [2], Broek [12], Anderson [13], Bažant [14], Hellan [15] and a classical paper by Rice [16].

Fracture is resisted by the energy need to create new surface in growing cracks, and by the plastic deformation that occurs around the ends of those cracks. The driving force and source of that energy is the external loading system, together with elastic energy released by the material around the crack.

3.3.2 *Linear Elastic Fracture Mechanics*

The simplest version of the theory is linear elastic fracture mechanics (LEFM), which treats all the material outside a crack as elastic, and assumes that any inelastic deformation is completely localised in a small region close to the end of a crack.

The following introduction is based on a neat and elegantly simplified analysis by Ashby and Jones [1]. Imagine a long flat plate (Figure 3.2(a)), with uniform breadth and uniform thickness t, held between fixed rigid grips in a testing machine. The plate is under tension, and the tension generates a uniform tensile stress σ: there are no other stress components. The strain energy per unit volume in the plate is $\sigma^2/2E$, where E is the elastic modulus. This can be checked by thinking of a tensile stress that increases from 0 to σ in a unit cube, so that the tensile strain increase linearly from 0 to σ/E (since the material is elastic), and the work done by the tensile force on one face of the unit cube is half the stress σ multiplied by the elongation σ/E.

Now suppose that a crack of length a forms at one edge of the plate, perpendicular to the edge and to the faces of the plate (Figure 3.2(b)). The formation of the crack releases some of the stress, and so allows some of the strain energy around the crack to be released. Imagine then that the crack formation releases *all* the energy in a semi-circle of radius a centred on the outer end of the crack: that semicircle is white in Figure 3.2 (c). The crack releases *none* of the energy outside the circle.

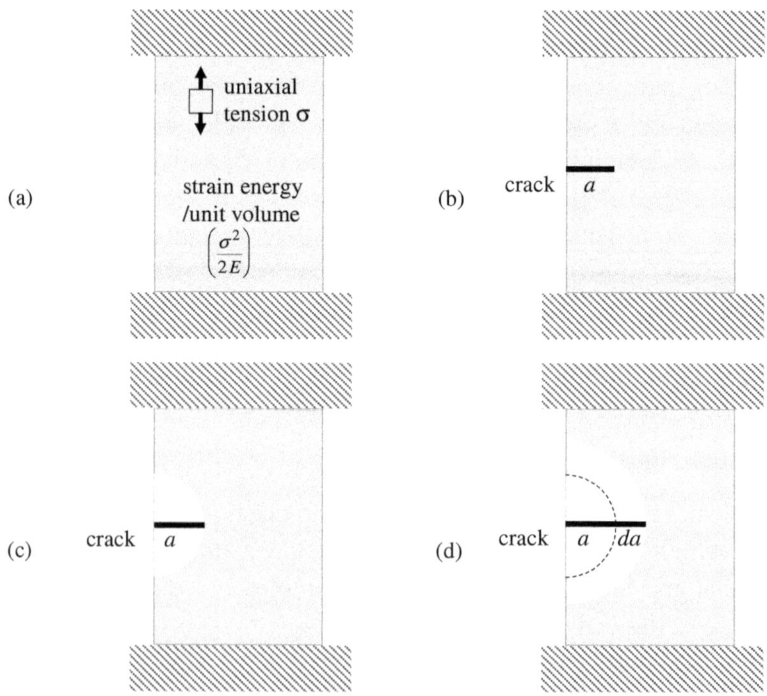

Figure 3.2 Growth of a crack in a plate under tension
(a) uncracked state (b) a side crack (c) energy release by formation of crack
(d) energy release by extension of crack from a to $a+da$

Next suppose the crack to grow a little further in the same direction, so that its length extends from a to $a+da$ (in Figure 3.2 (d)); da is small compared to a. That crack growth releases all the energy in a half-ring, white in the Figure, whose inner radius is a and outer radius is $a+da$, but releases none of the energy outside the outer radius. The energy released is the strain energy $\sigma^2/2E$ per unit volume multiplied by the volume of material in the half-ring, which is the area $\pi a da$ multiplied by the thickness t. It is therefore

$$\text{energy released} \qquad \left(\frac{\sigma^2}{2E}\right)(\pi a t da) \quad = \quad \frac{\pi\sigma^2 at}{2E}da \qquad (3.3.1)$$

The growth of the crack by the length da has created two new surfaces, with areas tda on the top side and tda on the bottom side. Some energy is required to form new surface: we can think of a loose analogy with surface tension in liquids, and recall that small droplets form as spheres because a sphere has the smallest energy for a given volume. Call the total energy to create the new surfaces G per unit area (counting both the top and bottom surfaces together). The energy requited to create the new surface created during the crack growth from a to $a+da$ is

$$\text{energy to create new surface} \qquad G(\text{new area}) \quad = \quad Gt\,da \qquad (3.3.2)$$

The crack will grow if the energy released is greater than the energy required to create the new surfaces, that is, if

$$\frac{\pi\sigma^2 at}{2E}da \quad > \quad Gt\,da \qquad (3.3.3)$$

which rearranges into

$$\sigma\sqrt{\pi a} \quad > \quad \sqrt{2}\sqrt{EG} \qquad (3.3.4)$$

$$\sigma \quad > \quad \sqrt{\frac{2}{\pi}} \frac{\sqrt{EG}}{\sqrt{a}} \qquad (3.3.4)$$

and that gives us the stress required to make the crack grow longer.

The longer the crack is, the smaller the stress needed to make it grow. That fits in with our day-to-day experience. Think of trying to tear open plastic packaging: it may be hard to start the first crack in the plastic, but once the crack has started it becomes relatively easy to make it tear some more, and the longer the crack gets the easier it is to tear it further. Many examples of plastic packaging include a notch that provides a stress concentration that allows the user to start and lengthen a crack more easily. Or think of a cracked wineglass: we are careful of a glass with a short crack, and very careful indeed of a glass with a long crack. Or think of splitting wood with an axe: it is easier to extend a long crack than to extend a short crack. The last example shows that it is not just a matter of surface energy, because if it were it would be just as easy to split off from a log a wide chunk (say 100 mm wide) as to split off a narrow sliver (say 2 mm wide), whereas we know from experience that the narrow sliver splits off far more easily.

In fact (3.3.4) is not quite right. The analysis assumes that all the energy in the half-ring of breadth da is released, but that none of the energy outside the half-ring is released. That cannot be exact, because it would imply that the outer surface of the half-ring would no longer fit the inner semi-circular surface of the material outside it: in continuum mechanics language, the deformations would not be compatible. That discrepancy can be corrected with elastic theory, and when we do so we get a slightly different inequality

$$\sigma\sqrt{\pi a} \quad > \quad \sqrt{EG} \qquad (3.3.5)$$

which has the same general form as (3.2.4), the same dependence of the critical stress on a, and the same material parameter $\sqrt{(EG)}$, but has a different multiplier on the right-hand side, 1 instead of $\sqrt{2}$ (which confirms that bold but carefully-chosen approximate solutions often give quite good answers).

The left-hand side of inequality (3.3.5) is called the 'stress intensity': it is totally different from 'stress' and 'stress concentration', and needs to be distinguished from them. The right-hand side is a material parameter $\sqrt{(EG)}$, the square root of the product of the elastic modulus E (which has the dimension of stress, which is force per unit area) and the surface energy G per unit area (energy, force times length, divided by area, and therefore has the dimension of force divided by area). $\sqrt{(EG)}$ is *not* a "strength" with the dimensions of a stress, but instead a quantity different in kind, with dimension force/(length)$^{3/2}$ or equivalently stress$\times$ (length)$^{1/2}$. Fracture mechanics calls $\sqrt{(EG)}$ the 'fracture toughness' or the 'critical stress intensity': this book uses the first name. Some books and papers instead call G fracture toughness: clearly one needs to know how the term is being used. Once we come to nonlinear fracture mechanics, though, we shall find that fracture toughness is not a complete description, and that we may need other parameters too.

Fracture mechanics denotes fracture toughness by K_{Ic}. The subscript c stands for critical. The subscript I refers to mode I, because fracture mechanics makes a distinction between different kinds of cracks. Figure 3.3 illustrates different crack modes. The crack in Figure 3.2 is mode I, driven by tension. Mode II is a sliding crack driven by shear parallel to the crack growth direction. Mode III is another sliding crack, but now propagating perpendicular to the plane in which the driving shear stress lies. Each mode has a different fracture toughness.

Fracture toughness is measured in kPa $\sqrt{m}$ or MPa $\sqrt{m}$. Ashby and Jones [1] have a list of mode I fracture toughnesses of different materials of many kinds, both man-made and biological. Ice has the lowest fracture toughness in their table, 0.2 MPa $\sqrt{m}$ (200 kPa $\sqrt{m}$), and high-strength steels for rotors the highest, 210 MPa $\sqrt{m}$. Comparing ice with other materials that we think of as brittle, they give the fracture toughness of glass as 0.7 to 0.8 MPa $\sqrt{m}$ and of polymethymethacrylate ('Lucite','Perspex') as 0.9 to 1.4 MPa $\sqrt{m}$.

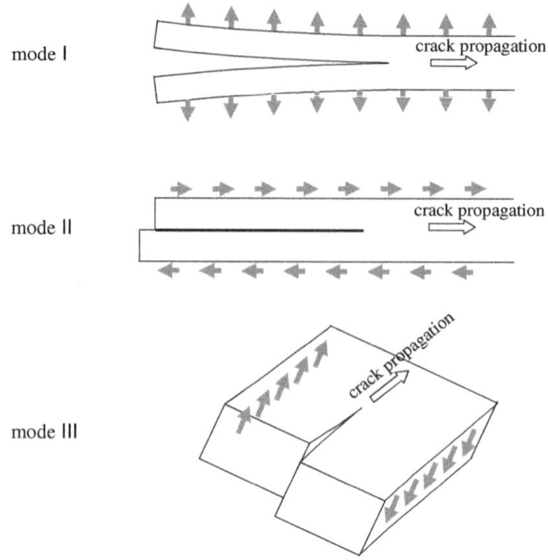

Figure 3.3 Modes I II and III

The analysis took the grips in Figure 3.2 as fixed. An alternative analysis takes the grips as loaded by a deadweight load, so that the grips can move and the force applied by the grips remains constant. The condition (3.2.6) comes out exactly the same.

This argument has enabled us to quantify an intuitive idea of brittleness, and to relate it to mechanics. In the distant past, engineers were reluctant to give up the idea of a stress as a measure of strength. They would assign strengths to different brittle materials, and would for example say that some particular cast iron has a strength of 350 MPa. That approach is unreliable, because - as the above analysis shows - it does not properly reflect the factors that govern the growth of cracks. In particular, fracture gives rise to a size effect (returned to in chapter 4). If strength is determined by a small-scale laboratory test and then applied to calculations about a big component such as a casting or a ship, the calculated strength turns out to be too large, and therefore unsafe. The same difficulty applies to an ice ridge or an iceberg, but there the calculated force is more likely to be too large rather than too small.

How can we measure fracture toughness? One option is to make a beam, to saw a crack into the tension side, and to load it in 4-point or 3-point bending, as shown in Figure 3.4. Many tests of this kind have been carried out in the laboratory, and some in the field. A better option for field tests is to use a self-equilibrated loading system, in which a triangular or square specimen is separated from the surrounding ice by sawing, a crack is sawn, and a flat jack forces the crack open. Mulmule and Dempsey [17] used this scheme to carry out tests for a wide range of scales, using square specimens ranging from 0.5 m side to 80 m side. They found a significant size effect, and that the apparent fracture toughness increased with specimen dimensions. The reasons for this are examined in section 3.3.3.

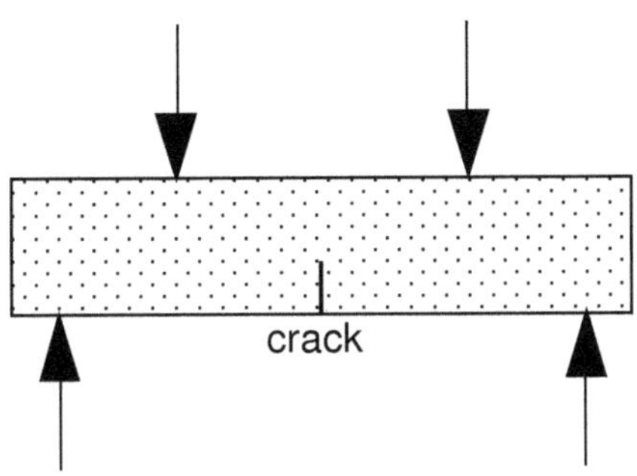

Figure 3.4 Four-point bending test to determine fracture toughness

3.3.3 *Nonlinear Fracture Mechanics*

Unfortunately things are not so simple as section 3.3.2 suggests. LEFM contains only one material property, but that property is not a complete mechanics characterisation of ice fracture. A typical fracture test program on an ice plate or beam proceeds as follows:

choose a geometry
choose a specimen size, characterised by a length ℓ and an ice thickness h
cut a traction-free crack of length a
conduct the test under load control

observe unstable fracture at a peak load P_{maz}
evaluate fracture toughness K_{Ic} from

$$K_{Ic} = \frac{P_{\max}}{h\sqrt{\ell}} f(a/\ell) \tag{3.3.6}$$

where the function $f(a/\ell)$ reflects the particular specimen geometry that is being used. Tests on ice at different scales then show that the apparent fracture toughness K_Q is dependent on the specimen size, and varies from about 90 kPa √m for laboratory-scale tests to about 230 kPa √m for specimens 30 m and larger [17,18]. That observation is not consistent with a fixed size-independent fracture toughness being the only significant material parameter. Similar behaviour is observed in concrete and rock.

The explanation is that the crack terminates in a fracture process zone (FPZ), within which the ice begins to separate and as it separates the stress transmitted across the FPZ progressively diminishes. Figure 3.5(a) illustrates the FPZ, which is not small by comparison with the specimen size or the initial crack length. The effect of the FPZ can be analysed by the Hillerborg fictitious crack model [19] sketched in Figure 3.3(b). Mechanical engineers call it the cohesive zone model, ice engineers call it the viscoelastic fictitious crack model [20], and it can be thought of as a generalisation of the models of Dugdale [21] and Barenblatt [22].

Outside the FPZ the stress-strain behaviour is that of the intact bulk material. The FPZ itself is narrow, and the tensile stress σ_{fpz} transmitted across it at a particular point is a function of the local separation w. At the end of the fictitious crack the tensile stress is the maximum stress σ_t that can be carried by the intact material. Beyond some critical separation w_c no tensile stress can be transmitted. The particular shape of the stress-separation curve (Figure 3.5(c)) is the governing material property: the energy G_c absorbed per unit advance of the traction-free crack is the area under the curve. Once the stress-separation behaviour has been determined, the fracture toughness can be predicted for any test geometry, size and crack length. The discussion above applies to mode I fracture, but it can obviously be generalised to modes II and III.

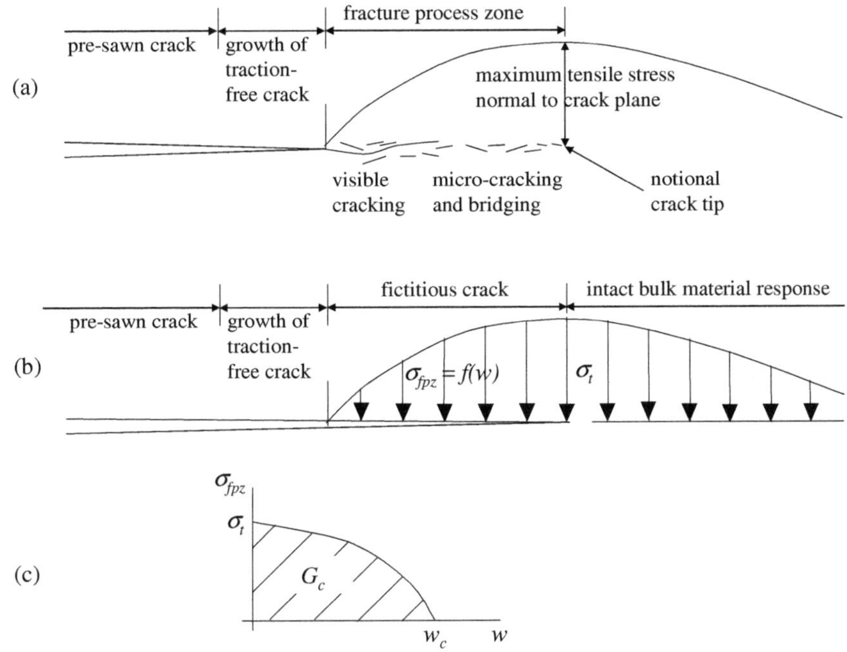

Figure 3.5 Fracture process zone

(a) description (b) stress across process zone (c) stress-separation curve

Uenishi and Rice [23] examine the nucleation of an unstable mode II slip in a material where the shear stress τ and the relative slip δ are related by a linear slip-weakening constitutive relationship

$$\tau = \tau_p - W\delta \tag{3.3.8}$$

where W is a constant slip-weakening rate; τ_p and δ correspond to σ_p and w above. They show, first by a dimensional argument and then by numerical calculation, that the critical slip size is proportional to μ/W, where μ is the shear modulus.

The stress-separation curve can be back-analysed from fracture measurements. Large-scale measurements on first-year sea ice at Resolute indicate that in that instance σ_t is 0.5 MPa, w_c 40 μm and G_c 15

J/m^2. It remains a research challenge to find ways of measuring stress-separation reliably by small-scale laboratory tests.

3.4 Elasticity

Ice responds elastically and linearly to low stresses applied for a short time. 'Elastic' means that when the stress is removed the deformation returns to its original state before the stress was applied. 'Linear' means that the strain is proportional to the stress. The linear elastic idealisation is a useful one, but only in a highly restricted range of circumstances. One situation when the response is linear is a stress wave induced by an explosion or an impact: the stress is too low for the ice to crack, and the time over which the stress is applied is too short for creep deformations to occur. Another such situation is an aircraft landing on ice well able to bear its weight and the impact loads: the ice deforms under the travelling load applied by the aircraft wheels or skis, and elastic waves radiate into the surrounding ice.

The general relationship between stress and strain for a linear elastic material in tensor notation is

$$\sigma_{ij} = C_{ijkl}\varepsilon_{kl} \qquad (3.4.1)$$

where

σ_{ij} is stress, a symmetric second order Cartesian tensor with six independent components

ε_{ij} is strain, a symmetric second order Cartesian tensor with six independent components

C_{ijkl} is a fourth order elasticity tensor with 36 components, only 21 of which are independent,

and applying the repeated subscript summation convention of continuum mechanics [3]: if a subscript is repeated, the expression with the repeated subscript is taken to mean the sum of all the terms that correspond to admissible values, so that σ_{kk} means the sum of σ_{11} σ_{22} and σ_{33}.

The background and derivation of equation (3.4.1) are described in standard texts on elasticity: see, for example, Prager [3], Barber [24] and many other books. That relation allows for the material to have different properties in different directions, in the jargon of solid mechanics the most general kind of anisotropy. We would need that idealisation to examine the local behaviour of the columnar ice described in section 3.2, which has elongated crystals with their *c*-axes aligned with the current. A simpler version describes a material that has the same properties in every horizontal direction but different properties in a vertical direction: solid mechanics calls such a material orthotropic.

For some purposes, ice can be idealised as an isotropic solid, one that has the same properties in all directions, and (3.4.1) then takes the much simpler form

$$s_{ij} = 2\mu\varepsilon_{ij} + \lambda\delta_{ij}\varepsilon_{kk} \tag{3.4.2}$$

where δ_{ij} is the Kronecker delta, which is 1 if i and j are the same and 0 if they are different. The elastic response of the ice is then completely described by two parameters, λ and μ, which can be related to the more commonly used Young's modulus E and Poisson's ratio ν by

$$\mu = \frac{E}{2(1+\nu)} \tag{3.4.3}$$

$$\lambda = \frac{E\nu}{(1+\nu)(1-2\nu)} \tag{3.4.4}$$

Schulson and Duval [25] discuss the stress-strain behaviour of ice in much greater detail, and consider anisotropic behaviour. They emphasise that elastic moduli cannot reliably be found by measuring the strain under and applied stress, unless the stress is low and the loading very brief indeed, because creep deformation contributes significantly. The apparent moduli measured in such a quasi-static test are significantly lower than the true elastic moduli measured in a dynamic test, by a factor of two or more. Values of E and ν for homogeneous polycrystals of isotropic ice Ih at -16°C are 9.33 GPa and 0.325 [25].

The elastic idealisation allows the whole vast apparatus of more than two hundred years of elasticity theory to be brought to bear, including a great number of standard solutions (see, for example, Barber [24]), general theorems such as the reciprocal theorem, methods such as complex variable solutions of two-dimensional problems, and powerful and well-validated numerical solutions by finite elements and finite differences.

None of that will help if the idealisation is not appropriate.

3.5 Plasticity

The simplest version of plasticity is the rigid/plastic idealisation, in which the material does not deform at all if the stress is low, but once it reaches a yield condition (defined in terms of stress) the material can continue to deform indefinitely without any further change of stress. Plasticity theory is described in many accessible books: see, for example, Prager and Hodge [26] and Calladine [27].

The simplest version is the von Mises idealisation, which says that the material yields if the second invariant of the deviatoric stress tensor reaches a critical value. An invariant is a property of the stress tensor that does not change if the reference axes change. The deviatoric stress is the stress with the mean principal effective stress subtracted. If stress is σ_{ij}, deviatoric stress s_{ij} is

$$s_{ij} = \sigma_{ij} - (1/3)\delta_{ij}\sigma_{kk} \qquad (3.5.1)$$

and the second invariant J_2 of the deviatoric stress is

$$J_2 = s_{ij}s_{ij} \qquad (3.5.2)$$

in both cases again applying the repeated subscript summation and using the Kronecker delta δ_{ij}, which is 1 if i and j are the same and 0 if they are different. If, for example, an ice specimen in a uniaxial compression test yields when the compressive stress is Y, if reference axis 1 is the sample axis, and the other two axes are perpendicular to it, then the stress components are

$$\begin{aligned} \sigma_{11} &= -Y \\ \sigma_{22} &= 0 \\ \sigma_{33} &= 0 \\ \sigma_{12} &= \sigma_{21} = \sigma_{23} = \sigma_{32} = \sigma_{31} = \sigma_{13} = 0 \end{aligned} \tag{3.5.3}$$

the deviatoric stress components are

$$\begin{aligned} s_{11} &= -(2/3)Y \\ s_{22} &= +(1/3)Y \\ s_{33} &= +(1/3)Y \end{aligned} \tag{3.5.4}$$

and

$$J_2 = \frac{4Y^2}{9} + \frac{Y^2}{9} + \frac{Y^2}{9} = \frac{2Y^2}{3} \tag{3.5.5}$$

This result can then be used to find the conditions under which yield will occur under some other kind of stress. Imagine, for example, a combination of an isotropic compression of magnitude p and a pure shear of magnitude τ in the 1-2 plane. The nine components of stress are then

$$\begin{aligned} \sigma_{11} &= \sigma_{22} = \sigma_{33} = -p \\ \sigma_{12} &= \sigma_{21} = \tau \\ \sigma_{23} &= \sigma_{32} = \sigma_{31} = \sigma_{13} = 0 \end{aligned} \tag{3.5.6}$$

and

$$J_2 = 2\tau^2 \tag{3.5.7}$$

and therefore yield will occur when τ is $Y/\sqrt{3}$, independently of the value of p.

That simple version of the rigid-plastic idealisation can be generalised. The yield condition can incorporate the first invariant J_1

$$J_1 = \sigma_{kk} \tag{3.5.8}$$

which is the sum of the principal stresses, and that allows a dependence on the mean principal stress, as in the Mohr-Coulomb condition widely applied in geotechnics. The yield condition can also include the third invariant J_3, which is the determinant of the matrix of stress components, and that allows a representation of the Tresca yield condition. Further generalisations account for strain-hardening: the yield stress can be made to depend on the amount of plastic strain that has occurred, and various hardening rules can be used to represent the response that occurs when the material is loaded in one combination of stress components, unloaded, and then reloaded with a different combination. An elastic component of deformation can be added in, by saying that the elastic strain depends on stress in the usual way. The restriction to isotropic materials can be dropped, so that the response to vertical compression can be different from that to horizontal compression: see, for example, Sand [28]. Much effort has been given to the development of generalisations, and some of them are located in available software.

Plasticity theory is seductive and tempting because it is so straightforward, but the plasticity idealisation is scarcely ever a good representation of the behaviour of ice. In particular, we know from observation that in ice there is a strong size effect, so that the force per unit area between ice and a structure decreases with increasing area. Plasticity does not allow for a size effect of this kind, though a size effect can be artificially forced into the theory by asserting that the 'strength' of the ice somehow depends on the size of the piece of ice that is involved. This point is further discussed in chapter 4.

Plasticity does of course remain a reasonable description of the behaviour of ductile metals such as steel, and in that context is much used in structural design.

3.6 Broken Ice

Many interactions between ice and structures involve broken ice rather than continuous masses and sheets. Collisions between ice sheets form pressure ridges composed of a rubble of fragments of different sizes. Continuing freezing forms freeze bonds between the fragments. In time the water in all the gaps between the fragments may freeze solid, but the structure retains some of the character of the original fragments.

Broken ice rubble is a particulate material like soil or broken rock. Understanding has progressed slowly, because of the general difficulty of dealing with particulate materials, further complicated by the fact that the particles are weak and can break easily, by the low level of friction between them, and by the difficulty of making full-scale measurements. Liferov [29,30] has recently reviewed the state of the subject.

Soil mechanics is another primitive and immature subject, and it has proved difficult to arrive at convincing constitutive models that relate the mechanical properties of a assemblage of particles to the microstructure, the strengths of individual particles, and the contacts between the particles. One of the difficulties is the evidence that within loaded particulate materials the forces are far from evenly distributed. Most of the force is transmitted through a relatively small number of 'chains' of loaded particles. The particles on either side of the chains carry relatively small forces, but their presence stabilises the chains.

One approach is to idealise ice rubble as a Mohr-Coulomb material, an idealisation often applied in soil mechanics. The shear stress that can be transmitted across an intersecting plane is then proportional to the normal stress at right angles to the plane, and the limiting ratio between them is the tangent of an angle of internal friction (not the same as the angle of friction between two particles in contact). The corresponding failure surface in a three-dimensional principal stress space is that sketched in Figure 3.6. If the compressive stress is large enough, the fragments can shift positions and move closer together, or the fragments can break. Those conditions limit the extent of the failure surface, and generate the outer cap shown in the Figure.

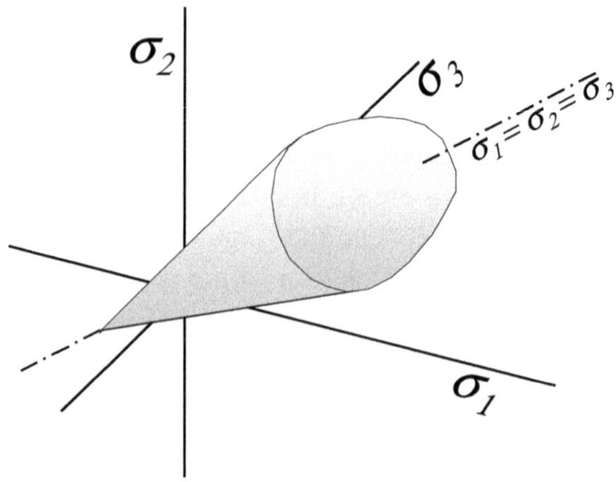

Figure 3.6 Yield surface for Mohr-Coulomb idealisation with cap

Figure 3.7 illustrates different schemes that could be used to determine the constitutive relationships for ice rubble.

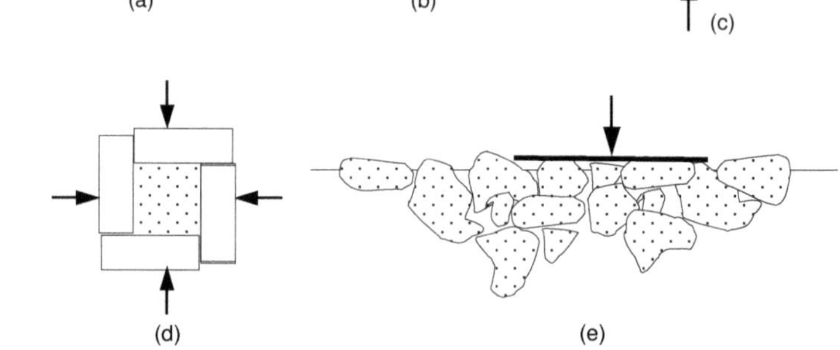

Figure 3.7 Alternative ways to measure the strength of ice rubble

One scheme that has been used to try to quantify the Mohr-Coulomb idealisation is the shear box, a device illustrated schematically in Figure 3.7(a) and first used by Coulomb more than two hundred years ago. A rectangular box is split horizontally into two halves. The box is filled with the material to be tested, and a flat piston is set on top of the material. The piston is loaded to provide a vertical compressive stress. The force required to move the top half of the box horizontally with respect to the bottom half is measured, and the mean shear stress is the force divided by the area. Prodanović [31] carried out shear box tests on artificial rubble prepared in a saline ice tank, and found that the limiting average shear stress was approximately proportional to the normal stress. If the limiting shear stress τ is related to the normal stress σ by

$$\tau = c + \sigma \tan \phi \tag{3.6.1}$$

Prodanović found that the cohesion c was 0.25 kPa for rubble prepared from 19 mm ice and 0.54 kPa for rubble prepared from 38 mm ice, and that the corresponding values of the angle of friction ϕ were 47° and 54°. He pointed out that the cohesion values are in marked contrast to Keinonen's 0.0113 kPa measured on a different kind of rubble ice, but that his ϕ was about the same. Liferov [30] suggested that very high values of ϕ sometimes arise from frictional idealisations that take c as zero.

Shafrova [32] describes shear box experiments on a coarse rubble of laboratory-prepared ice blocks. She found a weak dependence of shear stress on normal stress, corresponding to an internal friction angle from 3° to 8° that increased as the size of the contacts between the blocks increased.

The direct shear box is an inherently bad way of determining constitutive relationships. The box compels the sample to shear along a plane that lies in the direction of relative motion between the two halves of the box, and that is not necessarily either the plane on which the shear stress is a maximum or the plane on which the stress obliquity is a maximum. The stresses at right angles to the vertical stress are not known. The distribution of shear within the sample is very far from uniform, so that there are severe strain concentrations that propagate inwards from the ends of the box. Different regions of the sample reach the maximum shear stress at different stages of the deformation, so that the maximum value of the mean shear stress is lower than the maximum value reached locally: that is why shear box experiments in geotechnics find that the maximum mean stress depends on the length of the box. These grave difficulties could in part be avoided by using a ring shear apparatus, which can be thought of as an infinitely long shear box, but as far as is known that has not been done for ice.

Urroz and Ettema [33] made progress by developing a simple shear apparatus (Figure 3.7 (b)), like those that have been applied in soil mechanics.

Most serious work on the constitutive equations for particulate materials in geotechnics has been based on triaxial tests (Figure 3.7 (c)). A cylindrical sample is compressed axially, and subjected to a lateral pressure on the cylindrical surface. Different stress paths are possible: the lateral pressure can be held constant and the axial compression force increased ('triaxial compression'), or the axial force can be reduced ('triaxial tension'), or both the pressure and the axial force can be changed. The principal stress directions necessarily remain fixed. Gale et al. [34] and Wong et al. [35] report tests of this kind, on an artificial model broken ice made from freshwater ice cubes and with the pore space filled with air rather than water. They derived parameters for a Mohr-Coulomb constitutive model, though without including a cap. Their results show that the rate of dilation is much smaller than the normality condition of plasticity theory would suggest, so that the flow rule is not associated.

Another option is a 'true biaxial' (Figure 3.7(d)) or 'true triaxial' [36] apparatus in which a more uniform state of stress can be achieved, but even then there are deformation concentrations at the corners of the sample, and it is again impossible to rotate the principal axes of stress. Timco [37-39] and his co-workers devised a new test apparatus of this kind.

All the above refers to laboratory-scale tests, the majority on freshwater ice and necessarily composed of very small fragments prepared artificially. A real rubble field is a lot different. The ice is saline, the fragments are much larger, their sizes vary enormously, and the fragment geometry and relative positions themselves result from a fracture process that ultimately began with a more or less intact ice sheet. Ice formation in nature is not well understood, and so simulation in model and laboratory tests can be seriously misleading. For example, during ridge formation the ice blocks that are pushed downwards to form a keel will usually be colder than the water, and so that cold can freeze the some of the water between the blocks and create cohesion bonds. Buoyancy forces push the blocks together, and generate sintering and cohesion[40]. Croasdale et al. [41-43] show that the limiting strength of an ice rubble may be determined by that cohesion rather than by friction between the block. It seems clear that ice rubble should preferably not be modelled as a classical Mohr-Coulomb material in which cohesion and friction act together. In some cases, the modelling of ice rubble as a porous solid may be more appropriate, and this is an area that needs more research.

The interaction between broken ice and fixed structures is discussed in chapter 5 below, and interaction with floating vessels in chapter 6.

3.7 In-situ Rubble Tests

3.7.1 *Overview*

The difficulties recounted in section 3.6 suggest the possibility of field-scale tests, though the practical difficulties are substantial. There are only a limited number of in-situ tests worldwide, and they have generated about 40 – 50 data points altogether. The data exhibit significant variability of measured strengths. Another issue is how to apply in-situ test data from floating ridges to problems with different boundary conditions e.g. a grounded ice keel. We have little understanding of the effects of internal stress, thermal and aging effects on strength values obtained. Finally, measurements on real structures are usually handicapped by not being able to separate the load due to the keel from the load due to the refrozen layer. The placement of pressure panels has not been optimum for keel loads.

In situ measurements are considered important for keel strength, because it seems apparent that the strength of ice rubble is influenced by the processes which take place within a ridge, and they are not well understood. They range from the effects of temperature and salinity of the ice at formation, to aging processes with time, which in turn may be influenced by internal stress levels, as well as thermal and salinity gradients. Artificially created ice rubble can never hope to be representative of real ice rubble, and even real ice rubble properties might change if samples are recovered for testing. Thus, the best approach to the measurement and understanding of full-scale ice rubble is to obtain the properties *in situ* on actual ridges, preferably in the geographical area of interest.

In 1996, a study examined over 30 different methods in terms of their suitability for the *in situ* testing of ice rubble (Croasdale et al. [41], Bruneau et al [42]). The two most promising methods were carefully tested at model scale in an ice basin. Methods of analysis of the tests were developed in terms of rubble shear strength, and preliminary designs for their full-scale mobilization developed (Bruneau et al. [43]). The two methods were the "direct shear test" and the "punch shear test".

3.7.2 *The Direct Shear Test*

The direct shear test arrangement is shown in Figures 3.8 and Figure 3.9. The test is initiated by trenching through the refrozen layer of a ridge to isolate a rectangular slab. The load required to displace the slab horizontally is then measured and the corresponding displacement is recorded. In the shear test, the failure plane occurs at, or near, the bottom of the consolidated layer, in a way, which mimics the "shear plug" failure mode of a ridge keel (see chapter 5).

This test technique has been successfully used in four separate field projects, two in Canada [44-46] and two in Russia [47]. The test equipment consists of a 0.25 MN hydraulic ram with a 60 cm stroke, a compression load cell, and universal joints connecting each end of the ram to the two aluminium reaction panels. String potentiometers record the displacement of the ice slab. Power is supplied to the ram through a hydraulic power pack driven by a gasoline engine.

Site preparation requires that a number of cuts be made through the consolidated layer and some blocks removed to allow the apparatus to be lowered in place and to create room for the ice slab to be horizontally displaced. Ideally, a site is chosen on the shoulder of a ridge so that the upper surface is relatively smooth but is underlain by keel rubble. For tests conducted in the sail region of a ridge, the sail blocks are removed and the ice surface made as flat as possible to facilitate easy cutting. Cutting is performed with chain saws. In Canada, the main cuts around the slab, and the slot behind the slab, were made with a saw with a 1.51 m double bar, mounted on a sled. The double bar created a slot approximately 2.5 cm wide which eliminated friction and binding as the slab was being displaced. The sled provided a stable platform greatly reduced the effort required to operate the saw, and increased the safety of the operation. The tests in Russia used a much larger saw with a 3.2 m long blade [47].

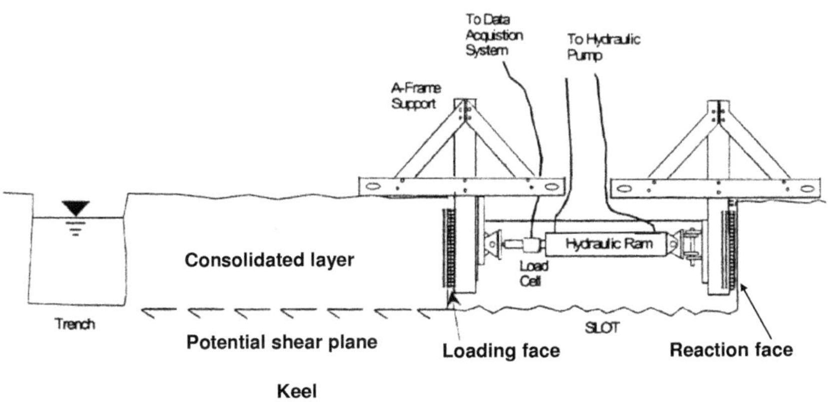

Figure 3.8 Direct shear test (cross-section)

The slab was tapered slightly from the front (where the load was applied) to back, to further ensure no friction and binding problems. The trench behind the slab was made by cutting the ice into smaller pieces and removing them using either, an "A" frame and block and tackle, or ice tongs. An I-shaped slot in front of the slab was required for the panels and ram, and was cut using the same methods.

Figure 3.9 Direct shear test (Confederation Bridge in background) (photo by KRCA)

The tests were recorded on video, and extensive measurements of block displacement and qualitative descriptions were recorded. In some tests, small diameter Styrofoam rods were placed vertically in drilled holes through the slab into the ice rubble. These gave an indication of the level of the shear plane that was usually directly at the base of the consolidated layer. Figure 3.9 is a photograph of the direct shear test apparatus.

Figure 3.10 is a typical load versus displacement plot for the direct shear test. It is believed that the initial peak load is indicative of a cohesive bond, which has to be overcome before the slab can be moved; the long flat portion is suggestive of a residual frictional strength.

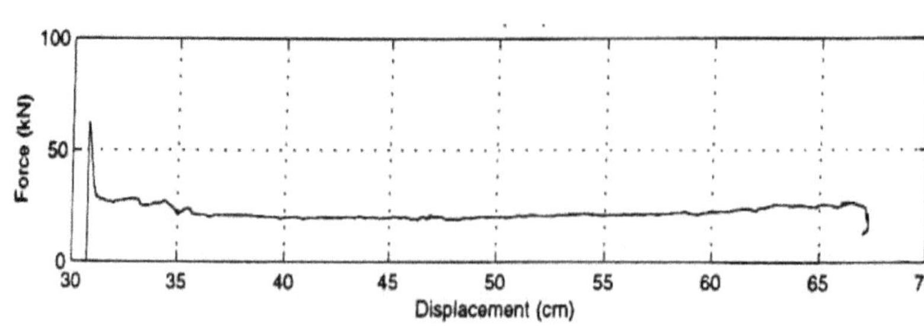

Figure 3.10 Direct shear test: typical load versus displacement

3.7.3 *The Punch Shear Test*

Figure 3.11 shows a punch shear test. A plug of the consolidated layer is cut through to the underlying rubble. A load is then applied to fail a plug of the keel downwards. This technique was first tried in the Baltic by Lepparanta and Hakala (1992) [48], with mixed success due to insufficient load capacity. In 1996, extensive testing of the technique was performed in conjunction with the model tests conducted as part of a study to investigate first-year ridge loads [40]. Over 100 tests were performed in model ice rubble. Initially, evaluation tests were performed to assess the effects of platen speed, platen (plug) diameter, ridge depth and state of the rubble (new or aged). Then, before each ridge test against a model structure, several punch shear tests were performed on the ridge. The purpose was to obtain *in situ* shear strength for the ridge to help in interpretation of the measured ridge load against the structure.

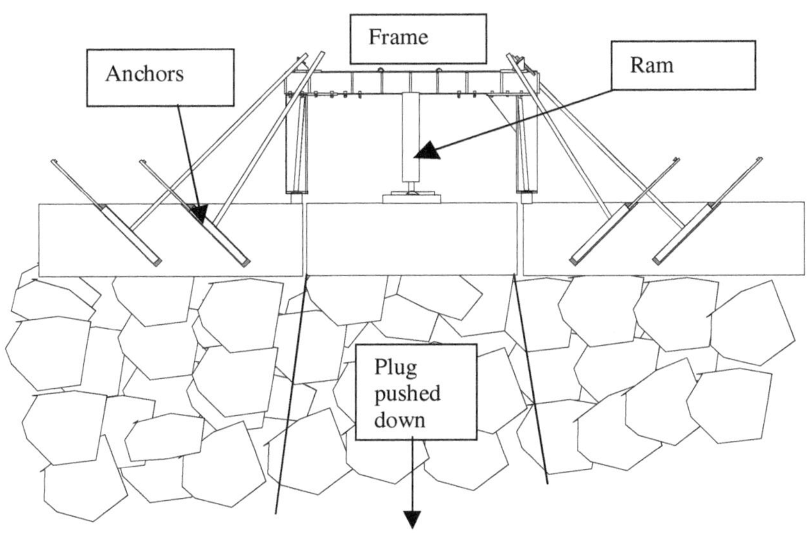

Figure 3.11 Configuration of the punch shear test

A scheme for interpretation of the punch shear test in terms of the friction angle and cohesion properties of the rubble was based on theories for a horizontal anchor plate in soil. The main advantage of the punch shear test is that the failure surfaces extend through the full depth of the ridge, as they do in the passive failure mode of a ridge keel, so that large-scale average values of keel shear strength are obtained.

Following this successful use in model tests, a full-scale test device was tested in Northumberland Strait in 1997 and 1998 [44,45]. In these tests, the apparatus consisted of a 30 tonne (0.3 MN) hydraulic ram mounted on an aluminum frame secured to the ice via four ice anchors. Following the use of the punch shear tests in Canada in 1997, larger equipment was developed for the subsequent year and for use in Russia. The equipment used in Russia was built to have a load capacity of 2 MN and designed to load a 3 m by 3 m block cut through the consolidated layer [47]. The equipment used in Canada I 1998 is shown as Figure 3.12.

Figure 3.12 Punch shear test apparatus as used in Canada in 1998 (photo by KRCA)

Typical load and displacement plots are shown in Figure 3.13. There is a distinct maximum strength reached after a very small displacement (as in the direct shear test), however in the punch tests, the drop in load with displacement in not as obvious as in the direct shear tests. This could be due to progressive failure or because of high residual friction. In many of the later tests, toggles and strings inserted into small holes drilled through the keel confirmed global plug failures with very little compression strain within the keel material.

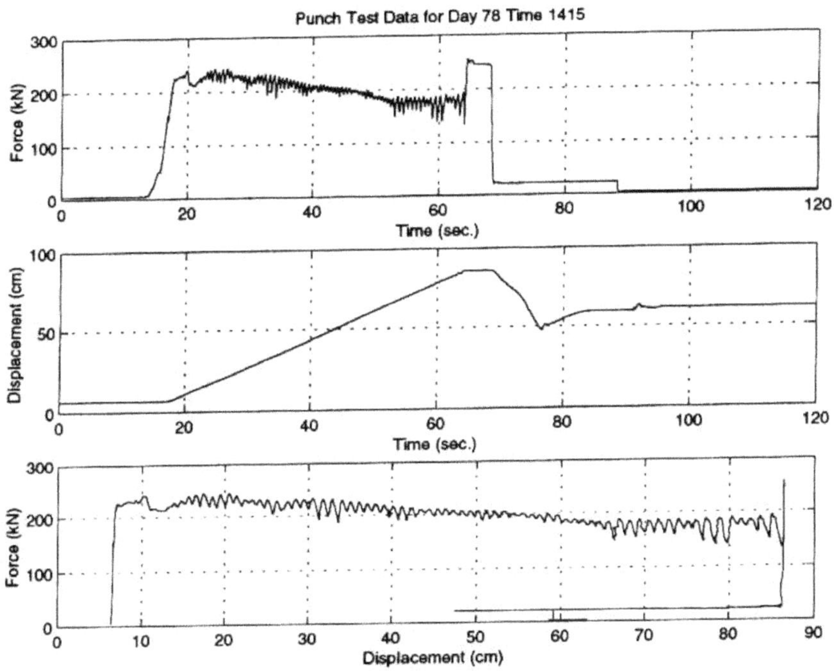

Figure 3.13 Typical load and displacement plots; punch shear test

3.7.4 *The Pull Up Test*

The objective of the pull up tests was to investigate the presence and nature of any bond that might exist between the consolidated layer and the underlying ice rubble. The idea arose when it was recognised that blocks would have to be cut and lifted from the refrozen layer in order to perform the direct shear tests. It was also realised that if there was no cohesive bond below the refrozen layer, then the load recorded in lifting the block would reflect the loss of buoyant support as it was raised, and should have a constant gradient.

On the other hand, any bond present would manifest itself as an increase in load superimposed on the buoyant gradient. Once the bond had been broken the load would revert to the buoyant gradient. It is suggested that the presence of a tensile bond indicates "cohesion" within the rubble.

The Canadian tests were carried out by cutting a rectangular slab through the consolidated layer, and inserting a toggle through a hole near the centre of the slab. A hydraulic ram and hinged beam apparatus was used to vertically lift the block of ice, thereby failing any bond between the slabs and underlying rubble. In later tests in Russia, the apparatus consisted of a gantry with a power winch connecting to a large ice screw in the ice block (Figure 3.14)

Figure 3.14 Apparatus used for pull up tests – Sakhalin Island 1997 and 1998 (the refrozen layer is cut through and ice lifted, so that the load cell measured tensile bond with ice rubble below) (photo by KRCA)

Detailed results from the Canadian tests conducted in 1997 and 1998 are now public and are reported in [44-46]. The tests conducted for Sakhalin 1 in 1997 and 1998 are still proprietary at the time of writing, though a general description is given in Smirnov and others [47]). Further tests were also conducted in the Baltic in 1999 under the LOLEIF project [49], but are generally proprietary.

Results from the Canadian tests are the main source of the data summarized below.

The strengths derived from the direct shear and punch tests are reported here on the basis of gross shear strength. The pull up test can only be calculated as a tensile "cohesive" strength. In both the 1997 and 1998 tests, the pull up tests show the presence of a distinct cohesive bond (Figure 3.15). Note that if there were no cohesive (tensile) bond between the keel blocks and the consolidated layer, the load trace would be similar to the second slope, to the right in the diagram, which plots the load required to lift the block through the water plane and represents the change in buoyancy.

Therefore, although the keel strengths may be interpreted as having both frictional and cohesive properties, the presence of this distinct cohesive bond suggests that we should probably put more weight on the cohesive strength for the initial shearing load. Later investigators [30] have similarly concluded that cohesion and frictional properties should not be additive, because the cohesion bonds have to be broken before a residual friction develops.

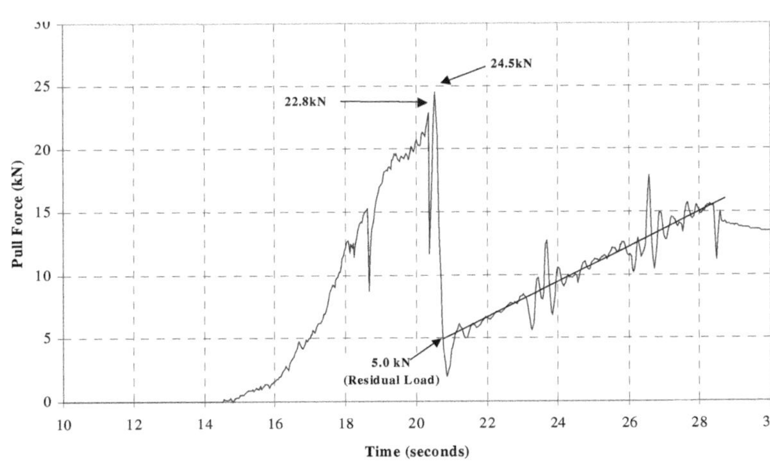

Figure 3.15 Typical pull up test load trace showing distinct tensile bond

3.7.5 *Summary of Results of in-situ Tests*

The results are summarised in tables and figures. The details are in the references cited.

Pull up strengths

For the pull up tests, the results obtained from all the Canadian tests are given in table 3.2. The so-called "tensile strengths" are the pull-up load divided by the area of the block being pulled. Significant scatter might be expected because the number of contact points between the underlying rubble and the block being pulled will not necessarily be the same (and not easily measured).

Table 3.2 Results from the Canadian pull up tests (1997 and 1998)

Year	"Tensile strength"	Air temp	Keel depth
	kPa	C	m
1997	9.50	-2.00	2.40
1997	13.50	-5.00	6.20
1997	17.90	-5.00	6.26
1997	11.20	-5.00	6.46
1997	17.10	-5.00	5.05
1997	12.20	-5.00	5.20
1998	6.10	-1.00	3.16
1998	9.00	-1.00	5.04
1998	8.90	-1.00	6.44
1998	3.60	-1.00	5.11
1998	2.70	-1.00	6.73
1998	6.80	-3.00	5.98
1998	6.30	-3.00	6.13
Average	9.60		
St Dev	4.70		
Ave + 3SD	23.70		

Correlation with keel depth was not observed, but there was a dependence on the air temperature when the tests were performed. 1998 was a warmer winter than 1997, and it can be seen that the strengths in 1998 were generally lower than in 1997. The dependence on the severity

Table 3.2 Results from punch and direct shear tests

Shear strength (kPa)	Test	Keel depth (m)	Air temp C
6.20	1997 PEI Punch	3.59	-4.00
7.64	1997 PEI Punch	3.29	-4.00
9.75	1997 PEI Punch	3.23	-6.00
6.08	1997 PEI Punch	3.28	-2.00
7.01	1997 PEI Punch	5.58	-7.00
9.33	1997 PEI Punch	4.98	-7.00
12.84	1997 PEI Punch	5.18	-10.00
9.48	1997 PEI Punch	5.43	-10.00
10.20	1997 PEI Direct Shear	3.56	1.00
9.00	1997 PEI Direct Shear	3.62	1.00
22.60	1997 PEI Direct Shear	3.75	1.60
11.60	1997 PEI Direct Shear	4.56	-4.00
12.70	1997 PEI Direct Shear	4.435	-4.00
14.80	1997 PEI Direct Shear	4.31	0.00
9.00	1997 PEI Direct Shear	4.06	-6.00
17.90	1997 PEI Direct Shear	4.3	-2.00
14.50	1997 PEI Direct Shear	7.06	-5.00
13.90	1997 PEI Direct Shear	5.9	-5.00
18.70	1997 PEI Direct Shear	5.6	-5.00
7.43	1999 Baltic Punch	5	-4
12.82	1999 Baltic Punch	6	-4
4.43	1998 PEI Punch	5.50	0.00
4.47	1998 PEI Punch	5.31	0.00
10.60	1998 PEI Direct Shear	4.39	-1.00
8.40	1998 PEI Direct Shear	6.30	-1.00
8.00	1998 PEI Direct Shear	4.40	0.00
13.20	1998 PEI Direct Shear	5.50	-2.00
9.40	1998 PEI Direct Shear	6.50	0.00
10.79	Average strength		
4.29	Standard deviation		
23.65	Extreme		

(PEI is short for Prince Edward Island – the test location close to Confederation Bridge)

of the winter is not surprising, because these "tensile" strengths are between ice blocks immediately beneath the consolidated layer. That region of the ridge is influenced by conduction through the blocks that are already embedded into the consolidated layer. A colder winter might be expected to lead to bigger areas of frozen contacts. Note that these "tensile strengths" do not have a direct influence on keel shear strengths being used for design (except for the beam failure mode as reviewed in 4.8.4). For most of the keel, the sintered or frozen contact points will be at a more constant temperature, controlled largely by the water temperature. Therefore we would not expect the shear strength of the bulk of the keel to be influenced significantly by air temperatures.

Punch and direct shear strengths

The data from these tests are shown in table 3.2; two data points made available from the Baltic tests are included, but in those cases the keel depths are estimated.

The direct shear tests measure the horizontal shear strength of the keel rubble just below the consolidated layer whereas the punch tests measure the average vertical shear strength through the thickness. An inspection of the table and of a plot of the data (Figure 3.16) indicates that the direct shear values are generally higher that the punch values. The average of the punch data is 8.86 kPa and the average of the direct shear is 14.08 kPa. For a purely cohesive material, this implies that the strength at the top of the keel is about 1.6 times the average. The plotted data also shows no obvious correlation with keel depth. Using the 1.6 factor we could also try plotting the direct shear data divided by 1.6 (i.e. multiplied by 0.63) together with the punch data. This is done in Figure 3.17; it shows a more distinct (but still slight) dependence on keel depth. Using the correlation shown, the average keel strength for a 23 m keel would be about 19 kPa.

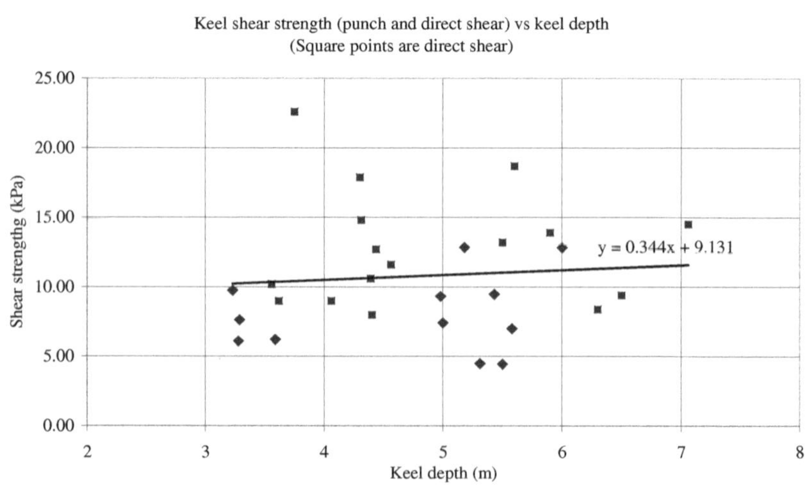

Figure 3.16 Plot of all available keel shear strength data (punch and direct shear together)

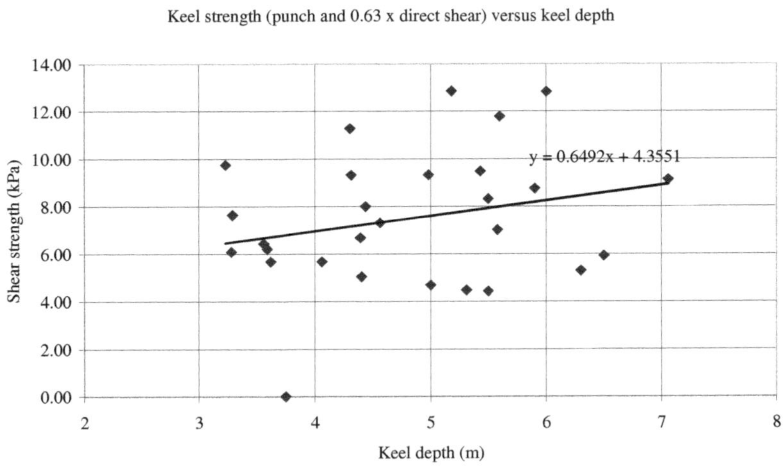

Figure 3.17 Plot of punch data and 0.63 × the direct shear data

In summary, all punch and direct shear strengths lumped together have an average strength of 10.79 kPa with a standard deviation of 4.29. An extreme based on the average plus 3 standard deviations would be 23.65. It will be noted that in Chapter 4 when sample keel loads are estimated, a typical shear strength value used is 24 kPa at the top of the keel with 12 kPa at the bottom. There are other ways of handling this data and it should be remembered that there does exist a data set from offshore Sakhalin which is still proprietary. In general, the values obtained off Sakhalin are slightly higher. Work performed by Croasdale, Comfort and Been [50] to investigate limits to ice gouge depths accounting for keel strengths used a typical maximum shear strength for keel rubble of 30 kPa with an extreme value of 55 kPa. (The latter being associated with features that had been grounded for some period of time).

3.7.6 *Translation of Rubble Shear Strength into a Bearing Pressure (or pseudo crushing strength)*

If ice rubble is subject to load over an area, the failure pressure over the loading area can be estimated in a similar way to a footing failure in soil foundation problems, or as will be discussed later, simply as an ice crushing failure. As a first approximation, it is assumed that the shear strength of the ice is constant along the failure plane and is independent of the internal stresses prevailing in the keel due to buoyancy or grounding forces (if the keel is gouging the sea floor). This is considered to be justified by the dominance of cohesive strength on the failure load as observed in the in-situ tests.

The force normal to the ice rubble to fail the ice (F_{ice}) is given by:

$$F_{ice} = qKA \qquad (3.7.1)$$

where:

q is the ice keel shear strength
K is a value based on the length of the failure plane
A is the contact area

and the normal pressure is given by

$$p_f = qK \qquad (3.7.2)$$

In soil mechanics, footing failures will occur for values of K in the range 5 – 9 depending on the shape and depth of penetration [51]. Those values are also compatible with plasticity theory. If we use the 30 kPa value given previously then p_f is 150 to 270 kPa

Combining a conservative, upper-bound, shear strength of 55 kPa with a K value of 9 gives a bearing pressure of about 500 kPa. This value has been used as an extreme to look at limits to gouge depths, pit depths and for direct ice loading on pipelines. This is a topic where additional data and understanding would be very useful. The single test described in the next subsection should be repeated with a range of platen sizes. However, the single result of 360 kPa bearing pressure does provide some support to the values and logic discussed above.

3.7.7 *Confined Compression Test (indentation test) on Ice Rubble*

A significant test performed in Canada in 1998 [45] was a vertical indentation test into a grounded pressure ridge. Figure 3.18 illustrates the principle of the test. The configuration is similar to a punch test except

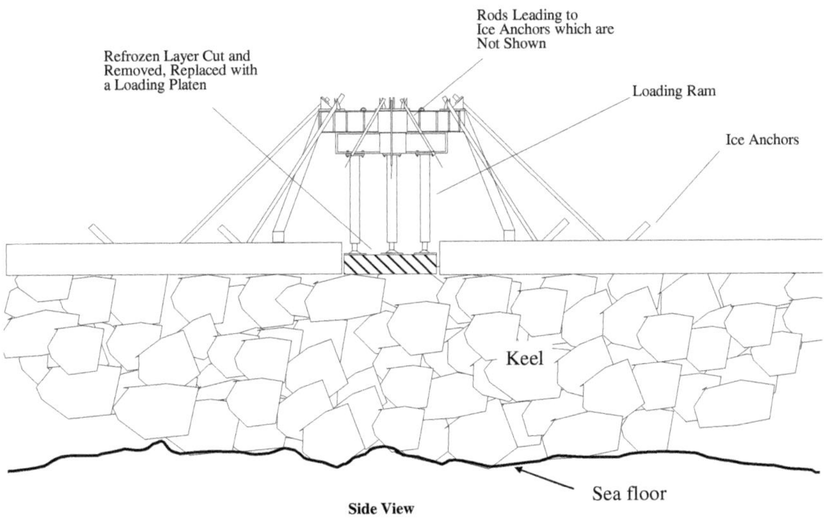

Figure 3.18 Indentation test on grounded rubble

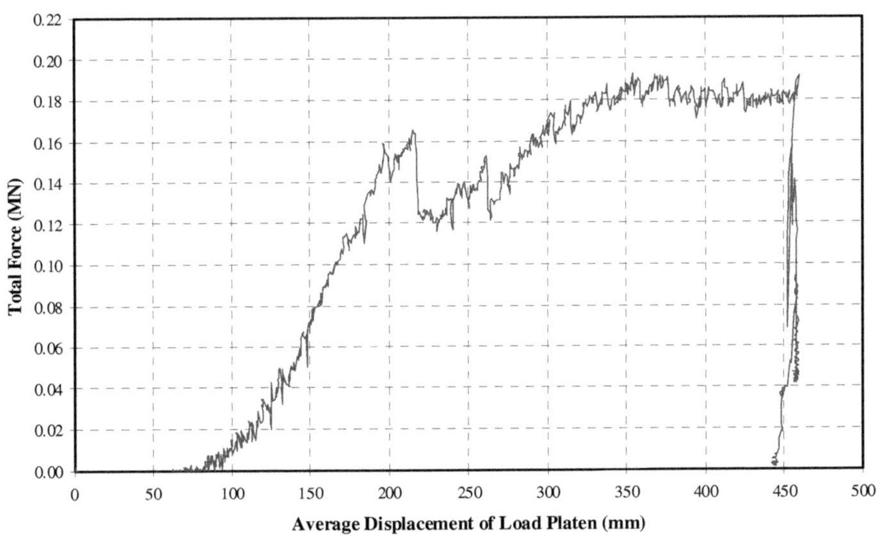

Figure 3.19 Indentation test – load trace

that a steel indenter is used and the ridge is grounded. The original goal of the test was to get an idea of the limiting local pressure that ice rubble could impose on a structure. It may also be thought of as a confined compression test on keel rubble. Such a compressive strength might be considered as a limiting frontal pressure that keel rubble could sustain during the gouging process.

Unfortunately, due to relative priorities established during the 1998 test program, only one of these tests was performed. Figure 3.19 is the load deflection trace. The maximum pressure sustained by the keel rubble was 360 kPa. The platen size was 0.5 m^2.

3.8 Model Ice

A gap remains between ice mechanics and the prediction of ice forces on full-size structures. Fracture is the central phenomenon that governs ice forces, but fracture mechanics is an incomplete subject. Almost all of it is concerned with the nucleation and growth of a single crack, whereas when Arctic ice pushes against a fixed structure there are tens of thousands of cracks, many of them quite short but a few much longer. In the present state of knowledge, fracture mechanics is not able to deal with that part of the problem.

Engineers still have to make design decisions about real structures, and cannot wait until the subject is fully understood. The strategies they have applied to that problem are discussed in chapter 5. One option is to make a physical model, but models that neglect to recognise the conditions necessary for similarity can be dangerously misleading [52]. Suppose that we want to model a full-scale structure 50 m wide in 2 m ice moving at 0.1 m/s, and that we decide to test a model 0.5 m wide in an ice tank with ice 0.02 m thick moving at 0.001 m/s, which is what we would get if we decide to scale down lengths by a factor of 100 and to scale down velocity in the same ratio. How can we be sure that the model test tells us *anything* useful about the prototype?

There has been much controversy about the scaling laws that ought to be applied, and about whether the ice in a model ought to be ‘real’ sea ice, or if it can be freshwater ice, or if the ice should have modified properties in order to obey scaling laws. A popular but far from universal strategy has been to decide that a model ought to satisfy both Froude scaling and Cauchy scaling. Froude scaling is almost universally adopted in ship model testing in towing tanks, and is intended to keep the ratio between forces related to inertia and forces related to gravity the same in the model as in the full-scale system. It is expressed by the requirement that Froude number Fr, $U/\sqrt{(g\ell)}$ be the same in the model as in the prototype, where U is velocity, g is gravitational acceleration, and ℓ is a characteristic length. Cauchy scaling is intended to keep the ratio between forces related to strength and forces related to gravity the same in the model as in the full-scale system. Cauchy number Ca $S/\rho g\ell$ must be the same, where in addition ρ is density and S is a strength with the dimensions of stress. If we have Cauchy similarity

$$\frac{S_{model}}{S_{prototype}} = \frac{\rho_{model}}{\rho_{prototype}} \frac{g_{model}}{g_{prototype}} \frac{\ell_{model}}{\ell_{prototype}} \tag{3.8.1}$$

Since it is not possible to change g, short of putting the model in a centrifuge, and not possible to change ρ significantly, the strength S has to be scaled down in the same ratio as the length ℓ, so a $1/10^{\text{th}}$ scale model requires ice weaker by a factor of 10. Much effort has been put into the development of model ice weakened by the addition of dopants such as urea, extra salt, ethylene glycol and detergents [53]. Lau [54] has reviewed the subject.

A different modelling strategy was proposed by Atkins [55,56], who argued that the ice forces depend not on a strength but on a LEFM fracture toughness K_{1c}, which has different dimensions and therefore implies a different similarity condition, represented by Atkins ice number At:

$$\text{At} = \frac{\rho U^2 \ell^{1/2}}{K_{1c}} = \frac{\rho U^2 \ell^2}{K_{1c} \ell^{3/2}} \tag{3.8.2}$$

which maintains the ratio between inertia forces that scale with $\rho U^2 \ell^2$ and fracture forces that scale with $K_{1c}\ell^{3/2}$. Then

$$\frac{U_{model}}{U_{prototype}} = \left(\frac{\rho_{prototype}}{\rho_{model}} \frac{K_{1cmodel}}{K_{1cprototype}} \right)^{1/2} \left(\frac{\ell_{prototype}}{\ell_{model}} \right)^{1/4} \tag{3.8.3}$$

and if the model ice has the same fracture toughness and density as the prototype ice, velocity scales with the ¼ power of length.

Many researchers have elected not to use artificially weakened ice in models, and instead use freshwater ice or saline ice. Palmer and Dempsey [57] support that position, and argue that neither Froude similarity nor Atkins similarity is necessary for modelling of ice movement against fixed structures, because the velocities are so low that inertia forces are negligible. They point out that LEFM is itself an incomplete representation of fracture, and conclude that it is important to have the correct ratio between length, shear modulus and a slip-weakening rate (Uenishi and Rice [23]), that

Froude scaling is an unnecessary distraction, that we do not need to experiment with model ice weakened by contaminants, and that sea ice is best modelled by real ice. Ships may be different, because the velocities are much higher. The debate continues.

Modelling is discussed in more detail in chapter 4.

References

1 Ashby, M.F. and Jones, D.R.H. (1980) *Engineering materials I*. Pergamon, Oxford.

2 Lawn, B. *Fracture of brittle solids*. (1993) Cambridge University Press, Cambridge.

3 Prager, W. (1961) *Introduction to mechanics of continua*. Ginn and Company, Boston.

4 Hallam, S.D. (1986) The role of fracture in limiting ice forces. *Proceedings, International Association of Hydraulics Research Conference 86*, 2, pp. 287-319.

5 Sanderson, T.J.O. (1988) *Ice mechanics: risks to offshore structures*. Graham & Trotman, London.

6 Nye, J.F. (1960) The flow of a glacier in a channel of rectangular, elliptic or parabolic cross-section. *Journal of Glaciology*, 5 (41) pp. 661-690.

7 Palmer, A.C. (1967) Creep-velocity bounds and glacier-flow problems. *Journal of Glaciology*, 6 (46) pp. 479-488.

8 Martin, J.B. (1964) A displacement bound technique for elastic continua subjected to a certain class of dynamic loading. *Journal of the Mechanics and Physics of Solids*, 12, pp.165-175.

9 Griffiths, A.A. (1920) The phenomena of rupture and flow in solids. *Philosophical Transactions of the Royal Society of London*, A221 pp.163-198.

10 Obreimoff, J.W. (1930) The splitting strength of mica. *Proceedings of the Royal Society of London*, A127 pp.290-297 (1930).

11 Irwin, G.R. (1958) Fracture. *Handbuch der Physik*, 6, pp.551-590, Springer-Verlag, Berlin.

12 Broek, D. (1982) *Elementary Engineering Fracture Mechanics*, 3rd revised edition, martinus Nijhoff, Boston.

13 Anderson, T.L. (2005) *Fracture Mechanics: fundamentals and applications*, 3rd edition., Taylor and Francis, London.

14 Bažant, Z. P. and Planas, J. (1998) *Fracture and Size Effect in Concrete and Other Quasibrittle Materials*. CRC Press, Boca Raton and London.

15 Hellan, K. (1984) *Introduction to Fracture Mechanics*. McGraw-Hill, New York (1984).

16 Rice, J.R. *Mathematical analysis in the mechanics of fracture*, downloadable at http://esag.harvard.edu/rice/018_Rice_MathAnalMechFract_68.pdf

17 Mulmule, S.V. and Dempsey, J.P. (1999) *International Journal of fracture*, 95, pp. 347-366.

18 Dempsey, J.P. (1991) The fracture toughness of ice. *Ice-Structure Interaction*, Springer-Verlag, Berlin, pp.109-145.

19 Hillerborg, A., Modeer, M. and Petersson, P.E. (1976) Analysis of crack information and crack growth in concrete by means of fracture mechanics and finite elements. *Cement and Concrete Research*, 6, pp. 773-781.

20 Mulmule, S.V. and Dempsey, J.P. (1997) Stress-separation curves for saline ice using fictitious crack model. *ASCE Journal of Engineering Mechanics*, 123 (8) pp.870-877.

21 Dugdale, D.S. (1960) Yielding in steel sheets containing slits. *Journal of the Mechanics and Physics of Solids*, 8, pp.100-104.

22 Barenblatt, G.I. (1962) The mathematical theory of equilibrium cracks in brittle fracture. *Advances in Applied Mechanics*, 7, pp.55-129.

23 Uenishi, K. and Rice, J.R. (2003) Universal nucleation length for slip-weakening rupture instability under nonuniform fault loading. *Journal of Geophysical Research*, 108 (B1), 2042, doi:10.1029/2001JB001681.

24 Barber, J.R. (1992) *Elasticity*. Kluwer, Dordrecht, Netherlands.

25 Schulson, E.M. and Duval, P. (2009) *Creep and fracture of ice*, Cambridge University Press, Cambridge.

26 Prager, W. and Hodge, P.J. (1951) *Theory of perfectly plastic solids*. Wiley, New York.

27 Calladine, C.R. (1969) *Engineering plasticity*. Pergamon Press, Oxford.

28 Sand, B. (2008) *Nonlinear finite element simulations of ice forces on offshore structures*. Doctoral thesis 2008:39, Luleå University of Technology, Sweden.

29 Liferov, P. (2005) *First-year ice ridge scour and some aspects of ice rubble behaviour.* Doctoral thesis, Norwegian University of Science and Technology, Trondheim, Norway.

30 Liferov, P. and Bonnemaire, B. (2005) Ice rubble behaviour and strength: Part I. Review of testing methods and interpretation of results. Part II. Modeling. *Cold Regions Science and Technology*, 41, pp.135-151 and 153-163.

31 Prodanović, A. (1979) Model tests of ice rubble strength. *Proceedings, Fifth International Conference on Port and Ocean Engineering under Arctic Conditions*, Trondheim, 1, pp.89-105.

32 Shafrova, S. (2007) Initial failure of the ice rubble in plane strain direct shear tests. *Proceedings, Nineteenth International Conference on Port and Ocean Enginering under Arctic Conditions*, Dalian, 1, pp.256-266.

33 Urroz, G.E. and Ettema, R. (1987) Simple shear box experiments with floating ice rubble. *Cold Regions Science and Technology*, 14, pp.185-199.

34 Gale, A.D., Wong, T.T., Sego, D.C., and Morgenstern, N.R. (1988) Stress-strain behaviour of cohesionless broken ice. *Proceedings, Ninth International Conference on Port and Ocean Engineering under Arctic Conditions*, Fairbanks, 3, pp.109-119.

35 Wong, T.T., Morgenstern, N.R. and Sego, D.C. (1990) A constitutive model for broken ice. *Cold Regions Science and Technology*, 17, pp.241-252.

36 Palmer, A.C. and Pearce, J.A. (1973) Plasticity without yield surfaces. *Plasticity and Soil Mechanics*, Proceedings of a Symposium on the Role of Plasticity in Soil Mechanics, Cambridge University Engineering Department, Cambridge, UK, pp.188-200.

37 Timco, G.W., Funke, E.R., Sayed, M. and Laurich, P.H. (1992) A laboratory apparatus to measure the strength of ice rubble. *Proceedings, Offshore Mechanics and Arctic Engineering Conference*, Calgary, 4, pp.369-375.

38 Sayed, M., Timco, G.W. and Lun, L. (1992) Testing ice rubble under proportional strains. *Proceedings, Offshore Mechanics and Arctic Engineering Conference*, Calgary, 4, pp.335-341.

39 Timco, G. and Cornett, A.M. (1999) Is ø a constant for broken ice rubble? *Proceedings, Tenth Workshop on River Ice*, Winnepeg.

40 Croasdale, K. R. (1999) A study of ice loads due to ridge keels. *Proceedings Fourth International. Conference on Development of Russian Offshore (RAO)*, St. Petersburg.

41 K.R. Croasdale & Associates Ltd. (1996) *In situ strength measurements of first year pressure ridges and rubble fields.* A study for the National Energy Board, Canada, supported by PERD.

42 Bruneau, S. E., McKenna, R. F., Croasdale, K. R. and Crocker G. C. (1996). In-situ direct shear of ice rubble in first year ridge keels. *Proceedings of 49th Canadian Geotechnical Conference*. St. John's.

43 Bruneau S. E., Crocker G. B., Croasdale K. R., McKenna R. F., Metge. M., Ritch, R. and Weaver, J. S. (1998). Development of techniques for measuring in situ ice

rubble shear strength. *Proc. 14th International Symposium on Ice*, Potsdam, N.Y.

44 K.R. Croasdale & Associates Ltd. (1997) *In-situ ridge strength measurements – 1997*. A study sponsored by NRC (PERD) and Exxon Production Research Co..

45 K.R.Croasdale & Associates Ltd. (1998) *In-situ ridge strength measurements - 1998*. A study sponsored by NRC (PERD) and Exxon Production Research Co. (1998).

46 Croasdale K.R., Bruneau, S, Christian, D. Crocker, G. English, J. Metge, M. and Ritch, R. (2001) In-situ measurements of the strength of first-year ridge keels. *Proceedings, Twelfth International Conference on Port and Ocean Engineering under Arctic Conditions*, Ottawa.

47 Smirnov,V.,Sheikin, I.B., Shushlebin, A., Kharitonov, V., Croasdale, K. R., Metge, M., Ritch, R., Polomoshnov, A., Surkov, G., Wang, A., Beketsky, S., and Weaver, J.S. (1999) Large scale strength measurements of ice ridges: Sakhalin, 1998. *Proceedings of 6th Intl. Conf. on Ships and Marine Structures in Cold Regions. ICETECH'2000*. St Petersburg Russia, 2000.

48 Lepparanta, M. and Hakala, R. (1992) The structure and strength of first year ice ridges in the Baltic Sea. *Cold Regions Science and Technology*, 20, pp.295-311.

49 Heinonen, J., and Mättänen, M. (2001).Full scale testing of ridge keel mechanical properties in LOLEIF Project, *Proceedings, Twelfth International Conference on Port and Ocean Engineering under Arctic Conditions*, Ottawa.

50 Croasdale, K. R., Comfort, G. and Been, K. (2005) Investigation of ice limits to ice gouging. *Proceedings, Eighteenth International Conference on Port and Ocean Engineering under Arctic Conditions*, Potsdam, NY, USA.

51 Chen, W. and McCarron, W.O. (1991) Bearing Capacity of shallow foundations. *Foundation Engineering Handbook*, 2nd edn., H.Y Fang (ed.), Chapter 4, Van Nostrand Reinhold Company, Inc, NY.

52 Palmer, A.C. (2008) *Dimensional analysis and intelligent experimentation*. World Scientific, Singapore.

53 Timco, G.W. (1986) EG/AD/S: a new type of model ice for refrigerated towing tanks. *Cold Regions Science and Technology*, 12, pp.175-195.

54 Lau, M, Wang, J. and Lee, C. (2007) Review of ice modeling methodology. *Proceedings, Nineteenth International Conference on Port and Ocean Engineering under Arctic Conditions*, Dalian, 1, pp.350-361.

55 Atkins, A.G. (1975) Icebreaking modeling. *Journal of Ship Research*, 18, pp. 40-43.

56 Atkins, A.G. and Caddell, R.M. (1974) The laws of similitude and crack propagation. *International Journal of Mechanical Sciences*, 16, 541-548.

57 Palmer, A.C. and Dempsey, J.P. (2009) Model tests in ice. *Proceedings, Twentieth International Conference on Port and Ocean Engineering under Arctic Conditions*, Luleå, Sweden, POAC09-40.

Chapter Four

Ice Forces on Structures in the Sea

4.1 Introduction

Several technologies need structures in the sea.

There are known to be huge reserves of petroleum under some of the Arctic seas. Figure 4.1 locates some of the areas where oil and gas have been found and are currently in production, together with areas where discoveries have been made but production systems are not yet in place. In other locations there has been seismic exploration but no drilling, and in others again there has as yet been no seismic but there are geological reasons for thinking them promising ('highly prospective' in the explorationists' jargon). It is often claimed that a large fraction of the petroleum resources yet to be discovered will be found in the Arctic, but the evidence for that is inevitably thin.

Some of the petroleum reserves may be unconventional: in particular, there are thought to be enormous reserves of solid natural gas hydrates [1-3] within the Arctic seabed, and they may be an important source of fuel for future generations, if the hydrates can be dissociated and the gas can be produced economically and safely. Hydrates under the Arctic seas are found at unusually shallow depths below the bottom, by comparison with seabed hydrates elsewhere, because the seabed temperature is only a little above the freezing point of seawater. That factor may make hydrates in the Arctic more attractive than elsewhere.

Petroleum close to shore can be reached by extended-reach horizontal drilling from the shore, but only over a limited distance and subject to geological difficulties and increased cost and well workover problems: the current world record for horizontal distance is 11 km. Figure 4.2 illustrates some of the options for production from fixed structures further from shore. The optimal production scheme depends on the water depth, as well as on other factors such as accessibility, the ice climate, and extreme waves and currents, just as it does in other petroleum producing areas such as the Gulf of Mexico and the North-West Shelf of Australia.

Figure 4.1 Oil and gas fields in the Arctic
(map courtesy of the University of Texas libraries, University of Texas at Austin)

Offshore structures in Arctic waters are needed for other purposes too. Ship navigation used to require lighthouses to mark shoals and to enable mariners to check their positions. Many lighthouses were built offshore in the northern Baltic, and some of them have been valuable as sites for systematic field measurement of ice forces, but the need for lighthouses has diminished with the advent of global positioning systems and other improved navigational systems.

Some long bridges cross ice-covered straits, and the bridge piers need to resist the ice. The best example is the Confederation Bridge between Prince Edward Island and mainland Canada, completed in 1997, and there are many others. Offshore sites for wind turbines are fashionable, because it becomes increasingly difficult to find land sites that are not objected to because of noise and visual intrusion. Gravesen describes research for wind turbines in the southern Baltic close to Denmark [4], where the highest recorded ice thickness is 0.48 m, and goes on to relate ice thickness to freezing degree-days and to examine the applicability of ISO code [5].

This chapter is arranged in the following way. Section 4.2 examines the many types of structures that have been proposed. Section 4.3 considers ice forces generally, and the different scenarios that limit the force that ice can apply to a structure. Section 4.4 looks at how to calculate the force on a structure with vertical sides, and section 4.5 does the same for structures with sloping sides, where the ice fracture mechanisms are different. Section 4.6 considers local ice pressure. For low freeboard platforms, ice encroachment is a risk and this topic is considered in section 4.7 ice encroachment. Section 4.8 is concerned with the controversial issue of physical model tests. Section 4.9 considers ice-induced vibrations. Many field measurements of forces have been made, and section 4.10 considers measurement methods and results. References are listed in section 4.11.

Design of an offshore structure engages many conflicting requirements. Ice action is far from being the only factor to be taken into account, and indeed it may not be the governing factor. The designer may be compelled to compromise on a design that is less than ideal from the viewpoint of ice, but that has to be preferred because of structural fitness, resistance to waves in open water, operability, safety, evacuation in emergency, storage and offloading.

4.2 Alternative Design Concepts

One option is a simple 'jacket' structure piled into the seabed, exactly like the thousands of jackets installed for offshore petroleum production and operation in almost every part of the world (Figure 4.2(a)). That option has the advantages of simplicity and the applicability of straightforward and well understood fabrication and installation technology. It has been applied in the enclosed and relatively shallow waters of the Bohai Gulf, where there is no multi-year ice and the maximum ice thickness is about 0.3 m. Some difficulties were encountered, and in the severe winter of 1969 the Bohai-2 platform was badly damaged by ice [6]. A concern with jacket and multi legs is that of potential ice jamming between the legs. This is addressed in chapter 5. An alternative is a monopod tower on a piled or gravity base. (Figure 4.2(b). Close to the waterline the tower is as narrow as structurally practicable, so as to minimise ice forces. The platform topsides are perched on top of the tower, and can be much wider. A wider base provides strength against overturning by ice forces. This option was chosen for several structures in Cook Inlet in southern Alaska, starting in 1962. The inlet is a macrotidal area with unusually high 3 m/s currents and a 6 m tidal range [7], but wave heights are limited and the maximum ice thickness is 0.6 m. The scheme was also applied to several lighthouses along the coast of the Baltic, but several were severely damaged [8].

An altogether different alternative is to build an artificial island from gravel and sand, and then to drill from the island as if one were drilling onshore (Figure 4.2(c)). Several islands off the north coast of Alaska have this form. The outer slopes cannot be made very steep, and the quantities of sand and gravel required increase very rapidly with increasing water depths. The slopes are subject to erosion from ice, breaking waves and currents, and have to be protected by rock rip-rap. It may be difficult to locate and recover the gravel and sand by dredging, and their mechanical properties may be disappointing.

In somewhat deeper water, a caisson-retained island has a caisson 'box' that sits directly on the natural seabed, or on a raised berm of gravel (Figure 4.2(d)). The sides of the caisson are steel and near-vertical, the vertical profile can be shaped to reduce ice forces, and the surface can be given a tough low-friction coating. The caisson can be filled with sand. The caisson can be deballasted, refloated, and towed to a different location. The much-studied Molikpaq caisson was originally deployed as an exploration structure in the Beaufort Sea, and later was moved to the east coast of Sakhalin, where it now serves as a production structure.

Yet another alternative is a concrete or steel platform like one of the gravity platforms applied in the North Sea (Figure 4.2 (e)). It rests on the seabed, perhaps in its natural state or perhaps after some preparatory dredging to uncover more competent lays of soil. Its lateral resistance can be enhanced by piles, or by a skirt, or by placing rock around it. The platform can be circular or octagonal or rectangular in plan. The vertical profile can be vertical or sloped. A sloped outer face complicates the structural design, but there is much evidence that sloped structures encounter significantly reduced ice forces compared to vertical-sided structures: this subject is examined in section 4.5. One other possibility is a steel and concrete composite structure, constructed of a sandwich of two relatively thin layers of steel separated by reinforced concrete. This kind of structure exploits the best properties of both steel and concrete, and tests have shown that a composite steel/concrete shell is very strong [9].

All these are surface-piercing structures fixed in position on the seabed. A surface-piercing structure is unavoidable if the structure is to support a wind turbine or a bridge. If it is part of a petroleum production system, and if the system is not too far from shore, a purely subsea option becomes feasible. Wells are drilled from a floating vessel or from stable sea ice at some convenient season, subsea wellheads and manifolds are installed, and the wells are tied back to the shore by seabed pipelines. This option was chosen for the Panarctic Drake F76 demonstration project off the coast of Melville Island, as long ago as 1978. [10, 11]. Its objective was to demonstrate that it was practicable to produce gas from the Hecla and Drake gasfields on either side of the Sabine Peninsula, in water up to 400 m deep, and that that could be accomplished without the need of expensive fixed offshore structures. The well was drilled from the sea ice, which had been artificially thickened to construct a floating ice platform 6 m thick. The pipeline part of the project is discussed in detail in chapter 7.

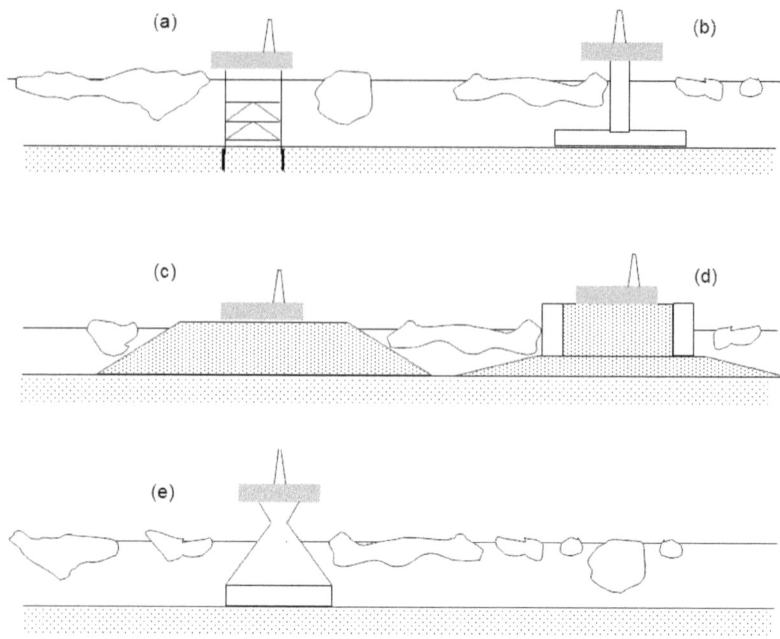

Figure 4.2 Alternative concepts for bottom-founded structures

Another and more ambitious alternative is to have an entirely subsea system, and to both drill and produce without any surface-piercing structure [12]. That option is within the reach of existing technology, but the drilling component would require substantial investment, and there are safety and environmental issues. It may become possible to have an entirely automatic system that does not require the presence of a human driller, but that technology lies some way into the future.

A structure does not have to be rigidly fixed to the seabed. Instead it can float. In conventional offshore petroleum production schemes, almost all structures in shallow water are fixed. The fixed platforms in the deepest water are the Troll platform in the North Sea (303 m water depth) and the Bullwinkle platform in the Gulf of Mexico (412 m). In water deeper than 400 m, and sometimes also in lesser depths, production uses a floating system.

Figure 4.3 illustrates some floating alternatives.

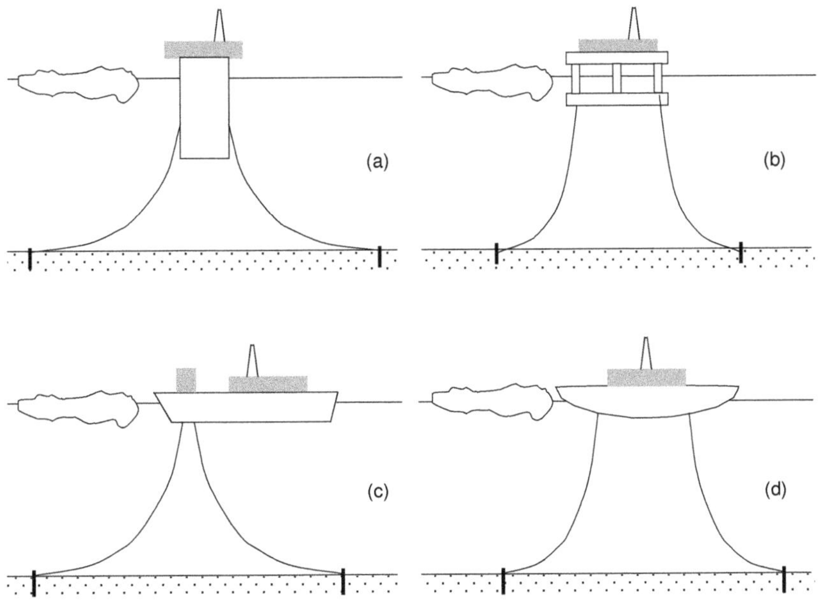

Figure 4.3 Alternative concepts for floating systems

A SPAR (Figure 4.3 (a)) is a cylinder floating vertically, with the topsides above it. The hull and the topsides are built separately. The hull is carried to location on a barge or a specialised ship, launched into the water, and up-ended by ballasting. The topsides are brought to location, and placed on the hull either by float-over or by a heavy-lift crane barge. The SPAR is held in position by mooring cables attached to anchors. In the diagram the anchors are shown as short anchor piles, but that is not the only possibility. Vertical risers (not shown) connect the hull to the wells below, and catenary risers or flexible risers connect the hull to the seabed export pipelines.

An alternative (b) is a semi-submersible configuration similar to that applied to many floating production systems around the world. A semi-submersible has most of its buoyancy in large pontoon some way below the water surface, and the topsides are supported above the surface on relatively slender columns. The advantage of this scheme is that it minimises motion in response to waves. The semi-submersible is held in place by cables to large anchors or anchor piles.

Another possibility is a ship-shaped floater (c) Finally, (d) is an unusually shaped floating vessel, circular in plan and with conical sides: the drilling vessel *Kulluk* is of this kind.

All the schemes (a) through (c) have been applied many times to offshore petroleum production in other parts of the world. There is much argument about which is to be preferred in the Arctic. The argument is sometimes distorted by fashion and prejudice, but there will certainly not be a single 'right' solution. Many factors are involved, among them water depths, ice climate, wave climate, the extent of the fields to be produced, the expected drilling sequence, the size and weight of the equipment that will be needed for onboard processing of the wellstream, safety and emergency evacuation, and the number of people that have to be present during operations.

Ice forces on floating systems are considered in more detail in chapter 6.

4.3 Ice Forces

Moving ice pushes against structures, and can damage them or destroy them. A rough back-of-the-envelope calculation confirms that the forces can be large. If we measure the compressive strength of sea ice in a laboratory test, we find the strength to be about 5 MPa (725 lb/in^2), defining the strength as the compressive force at which the ice crushes divided by the cross-section of the test specimen. If uniform ice 2 m thick pushes against a structure 90 m wide, the Molikpaq caisson-retained island platform illustrated in Figure 4.4, the projected contact area is 180 m^2, the contact breadth multiplied by the ice thickness. That suggests that the force that the ice can exert on the structure is the compressive strength multiplied by the contact area, which is 900 MN (90,000 tonnes).

This simple calculation can be criticised in many ways, but clearly the ice force has to be taken seriously: it might well be enough to move the whole structure sideways, or to destroy the walls, or to damage the structure locally. Ice forces are therefore a major concern to the designer of a marine structure in the Arctic, and will often be the principal factor that governs the structural design. Occasionally other factors may govern: if there are severe storms in the open-water season, and if large waves can reach the structure, the strength design may be governed by hydrodynamics rather than ice mechanics. Under extreme storm conditions, the hydrodynamic forces can be comparable with ice forces, and they may be calculated by the standard methods of offshore engineering: see, for example, Sarpkaya and Isaacson [13] and Faltinsen [14].

Several factors can limit the maximum ice force. When ice moves against a structure, and the structure remains in place, one of three outcomes follows. Figure 4.5 is a series of plan views that illustrate the possibilities schematically, and uses a terminology that has been adopted widely. In mode (a), called "limit stress" the ice fails against the structure, and it is the strength of the ice that determines the force that the ice can apply to the structure: that is the case outlined in the first paragraph of this section. That force must come from somewhere. The ice is driven forward by wind drag on the upper surface and current drag on the lower surface, and if there is not enough force applied to the ice, then the limited driving force determines the maximum force on the structure. In mode (b), "limit force", the force applied to the structure is governed by the driving force and not by the local strength of the ice. In the case sketched in the diagram, the ice sheet contains a thicker and stronger ice ridge, and then the force applied to the structure may be limited by the force that the ice upstream of it can apply to the ridge, but not by the strength of the ridge itself. The third mode (c), "limit momentum", applies when an ice mass drifts down on a structure, strikes it, and slows down in response to the contact force. Eventually, the initial momentum of the ice mass has been used up.

A loose analogy is a hammer driving a nail into a block of wood. One possibility is that the head of the hammer breaks up at the point where it contacts the nail: that corresponds to limit stress. A second possibility is that the hammerhead remains intact but the handle breaks: that corresponds to limit force. A third possibility is that neither the head

Figure 4.4 Molikpaq platform in moving ice in Beaufort Sea, 1986 (Photo by G. Comfort/ M. Metge)

not the handle breaks, but that the head comes to a stop because all its momentum has been absorbed in driving the nail, and that corresponds to limit momentum. The ideas of limit energy and limit momentum are set out in greater detail in chapter 5.

The simplest limit stress case – although not the easiest one to analyse – is a uniformly-thick level ice sheet moving against a structure with vertical or near-vertical sides. This case has been the object of a great amount of research attention, and is examined in section 4.4. Experience shows that ice forces can be reduced if the structure has sloping sides, which tend to break the ice in bending rather than crushing, and that case is discussed in section 4.5. The fragments of ice may clear and be carried downstream around the sides of the structure, but instead (especially on a wide structure) they may pile up in front of it and form a rubble field. The rubble field may extend some distance upstream, and in shallow water the rubble may ground. The ice may break against the upstream face of the rubble field, as in mode (b) in Figure 4.5, and the grounded rubble field itself may transmit a significant fraction of the ice force straight to the seabed beneath, rather than to the structure. Chapter 5 considers the mechanics of ridging and rubbling in more detail.

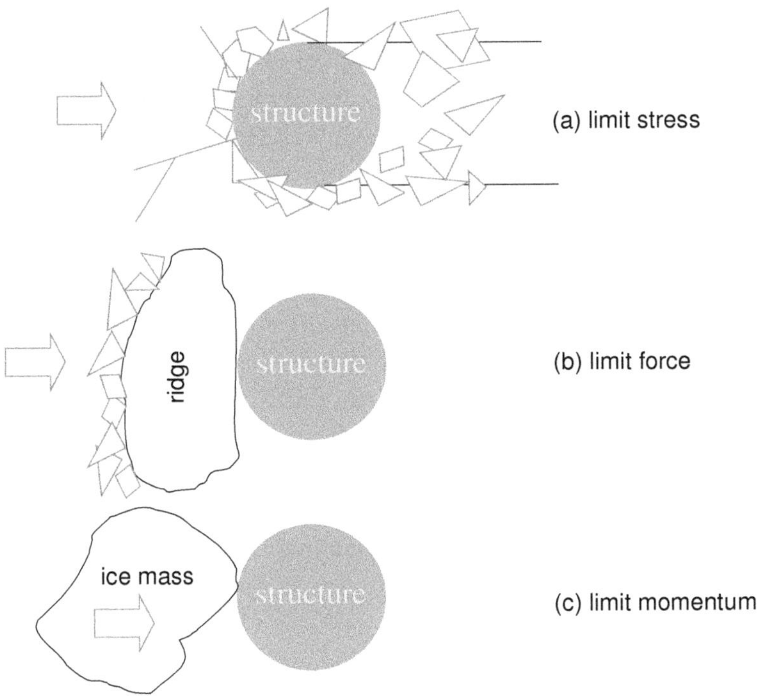

Figure 4.5 Limits to ice loads

Ice mechanics is a primitive and immature subject still in the course of development. Much research has been done, and much more is in progress, but almost every part of the subject remains controversial. Understandably, engineers charged with the design of structures that will operate in ice have not been able to wait until the subject is completely understood, and have had to rely on empirical and semi-empirical calculation methods that reflect the current state of understanding, and are based on a blend of field measurements, small-scale tests, theory and numerical calculation.

Meanwhile, there has been a growing demand for codes. Codes are an attempt to put together a clear summary of the state of knowledge, but like any human product they are imperfect, and often contain mistakes, inconsistencies, and the distortions generated by special interests and multiple egos. They follow the unexamined conventions of code writing [15, 16], and may give an unfounded impression of confidence and reliability. The reader is advised to be cautious, and not to believe everything a code says. The famous words of Otto von Bismarck

> "If you respect the law, or like eating sausage, you haven't watched either being made"

apply forcefully to codes. It is important to avoid the opposite reaction, and not to suppose that something in a code that is not explained or not consistent is thereby necessarily wrong. Moreover, the development of international codes can be very beneficial in encouraging communications between experts and in working towards a consensus.

4.4 Ice Forces on Vertical-sided Structures

4.4.1 *Alternative Modes*

Interaction between level ice and a vertical-sided structure looks straightforward (Figure 4.4) but is not. Much research has been done, and there have been some serious theoretical developments, some full-scale measurements, and some model tests. Controversy and vigorous debate surround all of them, and it is too soon for a consensus to emerge. In the meantime, the practical needs of design have compelled designers to adopt various empirical and semi-empirical methodologies, and for the moment they have been incorporated in codes.

A complication is that the ice can deform in qualitatively different ways. Generally the ice breaks. It may break into quite small fragments, as it does in continuous crushing against the sides of a structure, or the fragments may be larger, if the ice rides up a slope and breaks in bending, or if cracks radiate outward. If it is moving very slowly, though, it deforms in creep, like a slow-moving Alpine glacier.

4.4.2 *Creep*

If the ice is moving very slowly indeed, it deforms in creep, like ice slowly flowing in a valley glacier. This case is unusual and almost invariably short-lived: it happens when ice has previously been moving more rapidly, and comes almost to a stop because the driving force has diminished. Creep loads can also occur in landfast ice which often moves by small amounts due to thermal strains and sometimes due to sustained winds. The ice deformation is governed by the power-law creep equation (3.2.5) described in section 3.2. The force between the ice and the structure is proportional to the $1/n$ power of the relative velocity, where n is the exponent in the power law and is approximately 3. Creep deformation in metals is important for high-temperature applications like gas turbine blades, and there are well-understood methods for analysing it numerically. There are also approximate methods; they have been applied to glaciers [17] and to ice forces on fixed structures [18, 19].

4.4.3 *Buckling*

If the ice is thin, it buckles under the edge loads applied by contact with a structure. If the loading is rapid enough for the deformation to be essentially elastic, the relevant solid mechanics model is a thin elastic plate on a linear Winkler foundation, since the additional vertical force per unit area is proportional to the deflection (provided that the ice sheet does not deflect so far downwards that the top surface floods, or so far upwards that it lifts clear of the water). Sanderson [20] examines this elastic case, summarises the conclusions of Kerr [21], and shows that elastic buckling is likely to be the governing mode only when the ice is rather thin, in his analysis thinner than 0.4 m. He goes on to consider creep buckling, which is often observed in the field when ice moves very slowly against a fixed structure. He develops a simple analysis based on Calladine's linearisation of bending superposed on in-plane compression in power-law creep. He concludes that though creep buckling is interesting "it is generally associated with rather slow loading processes, and is unlikely ever to constitute the design condition for offshore structures"

4.4.4 *Crushing: A Simple but Incorrect Approach*

If the ice is thicker and moves against a vertical structure, the ice crushes. This is the case that generally governs design, and that has had most attention.

A simple analysis follows a minor extension of the scheme outlined at the beginning of section 4.3. The ice is supposed to be characterised by a strength Y, conveniently measured in a uniaxial compression test on a cube or a cylinder, like the specimens used for concrete, and calculated as the maximum force observed in the test divided by the cross-section. The breadth of the contact between the structure and the ice is w. The ice thickness is h. The projected contact area is therefore wh. The maximum ice force F is taken as

$$F = \varphi whY \tag{4.4.1}$$

where ϕ is a dimensionless multiplying factor that depends on the aspect ratio w/h, and accounts for various complicating effects, among them nonsimultaneous failure and vertical constraint if w/h is small. Less charitably, ϕ can be interpreted as an empirical fudge factor that can be used to make the formula fit to measurements. (4.4.1) is often known as Korzhavin's formula.

Equation (4.4.1) is enticingly seductive because of its simplicity, but it is incorrect and deeply misleading. If the formula were correct, the force per unit contact area F/wh would be independent of the contact area wh, and it would depend only on the compressive strength Y and the indentation factor ϕ. In reality, the force per unit contact area is found to depend on the contact area. Ice mechanics customarily calls this a 'size effect' or an 'area effect', though an effect of this kind is to be expected. It would only be surprising if it did not occur. Similar effects are well known in other areas of solid mechanics, such as concrete, ceramics and high-strength steels.

The evidence for the size effect comes from large-scale experiments and from observations on full-scale structures. The evidence is discussed first, and then the different ways in which it has been included in rules and formula that a designer can use.

4.4.5 *Crushing: Evidence from Measurements*

Figure 4.6 is a version of the famous Sanderson pressure-area diagram. It plots observed ice force per unit area against area, in this instance both on logarithmic scales. Each cloud represents a group of measurements. No attempt has been made to separate measurements on ice at different temperatures, or ice of different salinities, or different kinds of structures (except that they are all roughly vertical-sided).

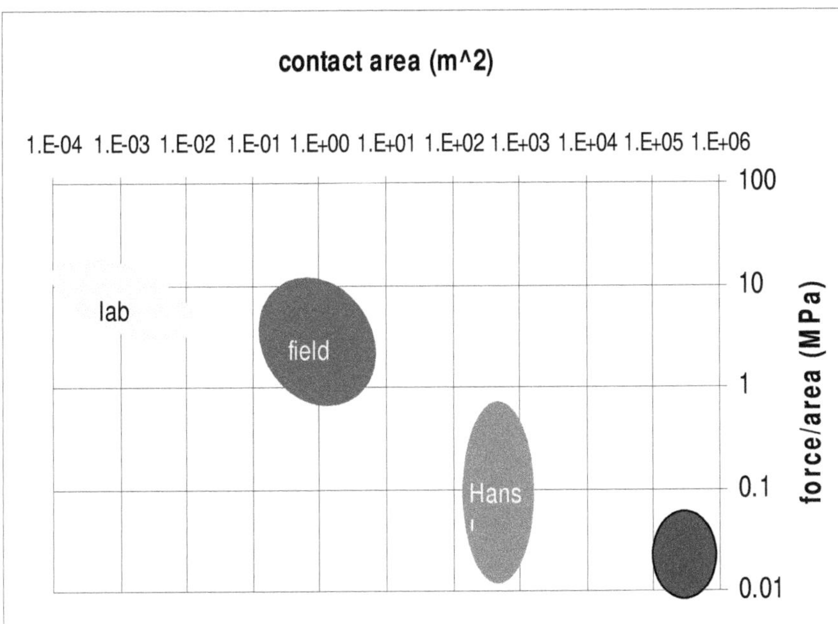

Figure 4.6 Pressure- area data sets

The first tests on ice-structure interaction are represented by the group of points to the left of the diagram, marked 'lab'. They were on laboratory-scale systems, in which sheets of ice were pushed against rigid rectangular and circular indenters. A typical indenter breath was 50 mm. Those tests determined a contact force per unit area, and that force per unit area could be compared with a compressive strength measured in a conventional compression test on a cube or a cylinder.

The next tests ('field') were on a larger scale, in the Arctic and much more difficult and expensive to carry out. Square blocks of ice were cut from floating sea ice, and loaded by platens driven by hydraulic jacks. In a typical test the ice was 1 m thick, the floating block was 5 m square, and the platen was 150 mm wide. In some tests the floating block split in two: that was accounted a test 'failure', though more judiciously it could be said that Nature was trying to tell us something (and that some researchers were not listening). The force per unit area was somewhat lower than in the series 1 tests. That could be attributed to various causes, among them variation of ice properties through the ice thickness, variation of temperature, and eccentric loading: all those influences are indeed present.

In parallel with the above tests (and actually before the field tests), initiatives were underway to investigate systematically the crushing strength of ice in the context of the interaction of Arctic ice with piles (and later with wider structures). One of the authors' entry into ice mechanics commenced in 1969 when he was asked by the drilling department of Imperial Oil in Calgary to help instrument piles that would be driven in the Canadian Mackenzie Delta region to measure ice forces. From a literature review of ice loads on piers, it became clear that it would not be simple to use piles in the weak soils of that region. To get significant ice motion, the piles would have to be placed beyond the landfast ice boundary in water depths of 15 to 20 m. The author suggested another approach – to push the pile through the ice rather than wait for the ice to move against a stationary pile. That approach also had the advantage that strain rates could be controlled and other relevant parameters could be measured (e.g. thickness, index strength, ice temperature, salinity).

The method was adopted, and the first tests were conducted in Tuktoyaktuk harbour in the winter of 1969/70 [22,23]. The apparatus was called the "nutcracker", because it had two long cylinders hinged at their ends below water, and hydraulic rams at the other end above water to push them apart, so that it resembled a giant nutcracker upside-down. This study also became the first project of the newly-formed Arctic Petroleum Operators Association (APOA), an initiative to encourage oil companies to collaborate in joint Arctic research of common interest.

Within the period of intense research by Canadian companies during the 1970s and 1980s, over 200 projects were sponsored. Reports of these studies can generally be obtained through the Arctic Institute of North America in Calgary. The nutcracker device was used in the Arctic during two winters, but it became apparent that logistically it would be much easier to conduct experiments closer to home, and so the same team at Imperial Oil initiated a more ambitious series of tests on Eagle Lake close to Calgary [24]. At the completion of this work several years later, Kry [25] had established a size effect on indentation crushing pressures. The work was also combined with the measurements of ice pressures and observations of ice failure around artificial islands in the Canadian Beaufort to create the concept and quantification of non-simultaneous failure of ice of across wide structures [26]. It was that work that allowed the then-current ice design pressure for Arctic ice of 3.8 MPa (550 psi) that had been advocated by some, to be reduced to 1.72 MPa (250 psi). That value for global ice pressures that has not moved much since then; for example for a typical 50 m wide structure and 2 m of ice, the current ISO formula gives 1.36 MPa.

One concern for the Arctic at that time was that very little was known about how thick multi-year ice would crush against vertical platforms. Could the 1.72 MPa be used? Should the ice pressure be lower because the contact area was larger, or should the ice pressure be higher because the aspect ratio was smaller? Multi-year ice had always feared by mariners, so intuitively was difficult to be confident without some actual data it about which conflicting potential trends to choose. It was also recognised that apart from some measurements of ice pressures in the landfast ice around islands (at low strain rates), all the tests mentioned above were necessarily on a scale much smaller than the interaction between sea ice and a real structure, typically 5 m to 20 m across for a monopod tower, and at least 50 m to 100 m for a caisson-retained island platform.

The next important step forward was made possible by an ice/structure interaction that Nature generously carries out, on a much larger scale than human beings could possibly arrange. Hans Island lies in Kennedy Channel between Greenland and Canada. It is roughly elliptical and about 1700 m long and 1300 m across. In July the sea ice breaks up further north, and large multi-year ice floes, sometimes 5000 m across, drift down the channel. Occasionally an ice floe collides with the island. The force between the floe and the island decelerates the floe. A helicopter can land on the floe before it hits the island and install an accelerometer, and the accelerometer measures the deceleration. The area of the floe can be determined visually, its thickness can be measured, and so the volume and the mass can be estimated. Newton's law of motion then gives the ice force. The contact breadth and thence the contact area can be estimated visually.

These tests, identified here as 'Hans I', produced surprising results [27–29]. The contact area was much larger than in the earlier tests, and the forces per unit area turned out to be much smaller than in tests with smaller contact areas. The discrepancy was too large to be explained by the arguments that had been deployed to explain the differences between series 1 and 2. An argument that there is no size effect could no longer be sustained.

Figure 4.6 includes a fourth group of points, series 4, which have a different and lesser status. Oceanographers construct numerical models of the Arctic Ocean, and find that they need to account for ice/ice interactions. The relevant contact areas are plainly very large. The points in the figure represent to level of ice force that has to be included if the movements of the Arctic ice calculated by the numerical models are to match the movements observed. It should be noted that in this case, the ice forces are being controlled largely by ridge building in which the ice is not crushing but failing in the rubbling mode (see chapter 5)

A further complication is that measurements show that the force between ice and a structure is very far from uniformly distributed. Instead, most of the force is concentrated in small and localised high-pressure zones (HPZs) (Dempsey et al.[30], Sodhi [31,32], Jordaan [33]). At any one time, there are several HPZs, and they occupy only a small fraction of the nominal contact area, generally but not always towards the middle and away from the edges. They form and disappear rapidly, and they jump around from place to place. Figure 4.7 shows one of the force distributions observed in a test by Sodhi, who drove against an ice sheet a platen covered with some 4000 tactile sensors, so that the force on each element of the sensor grid could be followed as it evolved. The plot is a three-dimensional image of the distribution of force per unit area, plotted upwards on a scale from zero to 13 MPa. HPZs are most dramatically seen in a downloadable video of the test results [34].

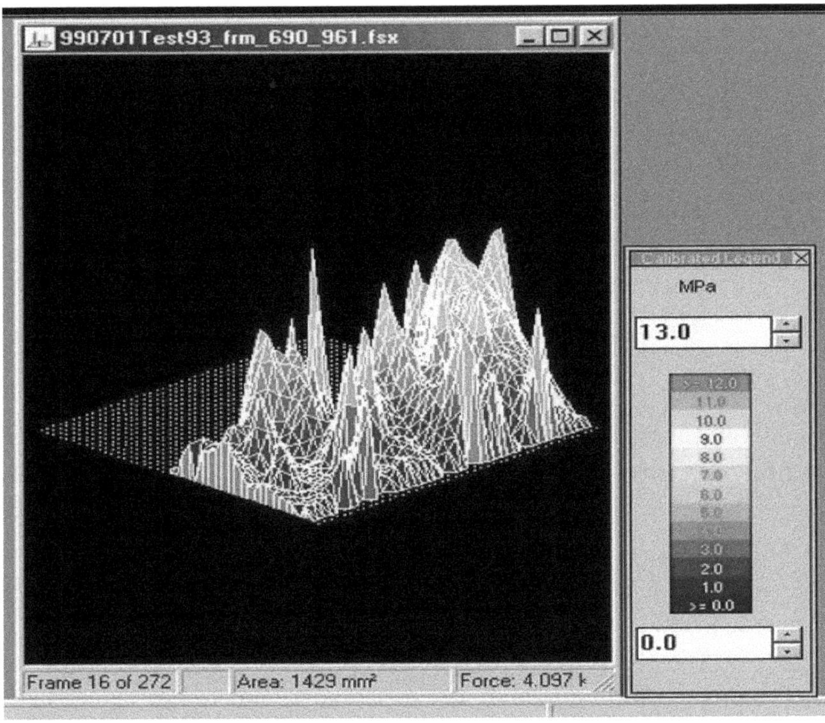

Figure 4.7 Distribution of contact for over contact area observed by Sodhi [34]

HPZs are plainly extremely complicated. As Jordaan emphasised, many processes are occurring simultaneously, among them fracture on many different scales, recrystallisation, plastic and elastic deformation, creep, pressure-melting and slip on grain boundaries. Here we do not attempt to consider all the phenomena, but focus on simple models.

4.4.6 *Crushing: Empirical Representations of the Data*

The reality of the size effect can scarcely be denied. What should be done about it? More than one approach is possible, and there are arguments in favour of all of them.

The simplest approach is in the ISO 19906 code [5]. The global average ice pressure p_G is given by an empirical formula

$$p_G = C_R h^{n-m} w^m \qquad (4.4.2)$$

where

p_G	is the global average ice pressure in MPa,
h	is the thickness of the ice sheet, in m,
w	is the projected width of the structure, in m,
m	is an empirical coefficient, taken as –0.16,
n	is another empirical coefficient, taken as $-0.5+h/5$ if h is less than 1 m and as –0.3 if h is equal to or greater than 1 m, and
C_R	is an ice strength coefficient, in MPa.

The equation applies to rigid structures where the aspect ratio w/h is greater than 2, and where the waterline displacement calculated from the ice action is less than 10 mm.

The ice strength coefficient C_R corresponds to the ice pressure when w and h are both 1 m (through strictly speaking an aspect ratio w/h of 2 is outside the range that (4.4.2) applies to). C_R is extensively discussed in the code. The representative ice actions are determined for the extreme-level ice event (ELIE), which corresponds to an annual exceedance probability of 10^{-2}, and the abnormal-level ice event (ALIE), which corresponds to an annual exceedance probability of 10^{-4}. The code recommends that in a deterministic ELIE analysis C_R be taken as 2.8 MPa, "based on first-year and multi-year data from the Beaufort Sea" and says that that value "could be conservative" because it "potentially" includes some magnification due to the compliance of the structure. It goes on to cite another data set which gives C_R as 1.8 MPa for a stiff structure in the Baltic, in conditions where the ice speed was higher than 0.1 m/s and the maximum waterline displacements were about 0.4 % of the ice thickness. A further discussion suggests a value of 2.4 MPa for sub-Arctic areas.

The code makes no attempt to explain the mechanics reasons for the formula. The user can apply it reasonably confidently without engaging in the ongoing controversy about the reasons for it. It implies an area effect.

A different and stronger size effect is observed for small contact areas less than 10 m^2. Figure 4.8 is Masterson's compilation of observations, from which he suggested that force per unit area p is related to area A by

$$p = 7.4A^{-0.7}\ \text{MPa} \quad A < 10\,\text{m}^2 \tag{4.4.3}$$

That relationship too is incorporated in the ISO code. A tentative mechanical explanation of that formula is outlined later in section 4.4.7.

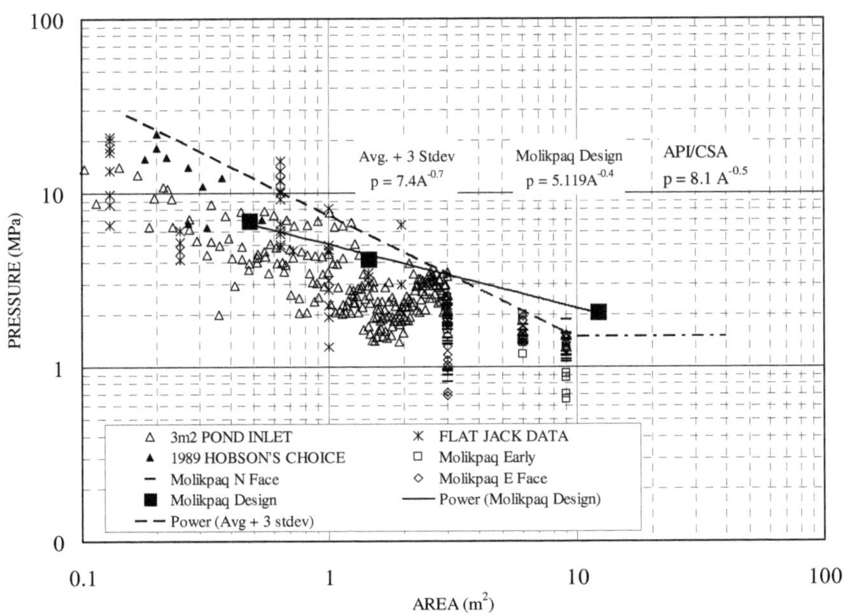

Figure 4.8 Pressure/area data compiled by Masterson

4.4.7 *Crushing: Theory*

Empirical relationships (4.4.2) and (4.4.3) rest on a lot of direct evidence, and can be applied with some confidence. The reader who is not interested in the physical background in applied mechanics can skip the following section.

The model represented by (4.4.1) corresponds to an idealisation of the ice as a perfectly-plastic material. It says that crushing force per unit area is independent of area, and depends only on compressive strength. It fails to work, because it does not lead to a size effect. To understand the size effect, we need to find a better material model.

Ice is an extremely brittle material, and section 3.3 began by describing linear elastic fracture mechanics (LEFM), the simplest version. In LEFM, the only material parameter is the fracture toughness. Imagine then that the ice force F between a structure and the ice depends only on the contact area A and the ice fracture toughness K. We can apply dimensional analysis to find the form of the relationship, and apply the Vaschy-Buckingham Pi theorem [35,36], which says that relationships between physical quantities can always be written as relationships between dimensionless groups. Taking the fundamental dimensions as mass M, length L and time T, the dimensions of F, A and K are

force F	$[MLT^{-2}]$
area A	$[L^2]$
fracture toughness K	$[ML^{-1/2}T^{-2}]$

Only in one way can those three quantities be combined to form a dimension group. The group is $F/A^{3/4}K$. There is nothing else for the group to be a function of, and so it must be a constant, and

$$\frac{F}{A^{3/4}K} = c_1 \tag{4.4.4}$$

where c_1 is an unknown constant, and so

$$\frac{F}{A} = c_1 K A^{-1/4} \tag{4.4.5}$$

and ice force per unit area is proportional to $A^{-1/4}$ and there is an area effect. Comparison with Figure 4.6 shows that this extremely simple result does in fact agree reasonably well with what is observed for large contact areas.

If on the other hand the only material parameter is a yield stress Y with dimension $[ML^{-1}T^{-2}]$, a similar dimensional analysis argument leads to

$$\frac{F}{AY} = c_2 \tag{4.4.6}$$

where c_2 is another unknown constant, and so

$$\frac{F}{A} = c_2 Y \tag{4.4.7}$$

There is then no size effect. Equation (4.4.7) is not consistent with the observations.

Imagine now that the ice force depends on both material parameters Y and K. There are now two independent dimensionless groups, $F/A^{3/4}K$ and $K/YA^{1/4}$. Applying the Pi theorem again, one of the groups must be a function of the other, and so

$$\frac{F}{A^{3/4}K} = f\left(\frac{K}{YA^{1/4}}\right) \tag{4.4.8}$$

That is as far as dimensional analysis alone can take us. $K/YA^{1/4}$ is a Carpinteri brittleness number [37] for brittle-plastic materials, and represents the relative importance of LEFM brittle fracture and plastic yield. Analysis and experiment show that in this instance yield dominates mechanical response at small physical scales, and fracture dominates at large scales. At the very small scale of laboratory tests, there may be no area effect. Atkins' book on cutting [38] provides another dramatic example of that kind of transition. We are accustomed to think of glass as brittle and of mild steel as ductile. If we machine glass in a lathe with a very small depth of cut, measured in microns, we find that the glass responds like a ductile metal, whereas if we could machine mild steel on a very large scale, with a depth of cut measured in m, the steel would respond as a brittle material and would break away in chunks.

This analysis confirms that there is nothing mysterious or unexpected about the observed size effect. It is a natural consequence of the small fracture toughness of ice.

It would not be right to think that fracture is the only factor engaged with ice crushing. When ice breaks against a large structure, it does not break simultaneously across the whole contact area. Instead one heavily-loaded contact breaks first, a fragment comes away, the force is transferred to another contact, perhaps at some distance away, another fragment breaks loose, and so on. Non-simultaneous failure of this kind was identified as important by Kry [26] and Ashby [39] A size effect occurs if fragments of ice break away randomly in different parts of the contact area with a wide structure. .

A difficulty with the application of fracture mechanics to ice has been that fracture mechanics has generally focussed on the growth of a single crack (or occasionally on the growth of a group of identical cracks), whereas ice fracture against a structure creates thousands of cracks on a huge range of scales, some of them tens of metres long and others only fractions of a millimetre long.

One application of fracture mechanics has been to the high-pressure zones (HPZs) described earlier. Dempsey et al. [30] carried out an analysis of the force generated at an isolated HPZ. They idealise an isolated three-dimensional HPZ as a hollow hemisphere, with internal radius a and external radius b. Within the radius a the stress is hydrostatic, and then going outwards come a radially-cracked region of radius, and finally an elastic region with outer radius b. The only material parameter in the idealisation is an apparent fracture toughness K_Q; the apparent fracture toughness is used because of the evidence that measured values of fracture toughness are not independent of scale, because of the influence of the size of the process zone discussed in chapter 3. With this idealised model, the force on the HPZ at which cracks propagate to the outside surface is

$$\pi a^2 p \approx \Omega b^{3/2} K_Q \tag{4.4.9}$$

where p is the pressure on the inner surface of the hemisphere and Ω is a numerical factor that incorporates a correction for the free surface. If we identify the radius to the outer boundary with the half-thickness $h/2$, the force P_{HPZ} on a single HPZ is

$$P_{HPZ} \approx (\Omega/2\sqrt{2})h^{3/2}K_Q \qquad (4.4.10)$$

Now consider a total contact breadth w, large compared to h, and suppose the maximum force to correspond to a line of isolated HPZs along the midsurface. The spacing between one HPZ and the next is λh. All the force is contributed by the HPZs, and none by the contact area in between. The contact area is wh, and the number of HPZs is $w/\lambda h$, and so the force per unit area is

$$p = \frac{w/\lambda h}{wh} P_{HPZ} \approx (\Omega/2\lambda\sqrt{2})\frac{K_Q}{\sqrt{h}} \qquad (4.4.11)$$

and, if K_Q is independent of thickness, that force is therefore inversely proportional to the square root of the ice thickness and independent of breadth [40]. (4.4.11) is dimensionally correct, but gives a significantly stronger dependence on thickness than the empirical ISO 19906 equation (4.4.2).

An analysis based on HPZs can be extended to explain the stronger area effect seen with small contact areas, equation (4.4.3). As another simple model, suppose first

(i) that the maximum force on a contact area A occurs when one and only one HPZ is present within the area;
(ii) that the force transmitted by the HPZ is independent of the area the HPZ forms part of; and
(iii) that the force transmitted across the remainder of the area is negligible by comparison with the HPZ force.

If that were correct, force would be independent of A, and pressure would be proportional to A^{-1}. That is a stronger area effect than the effect observed.

Palmer and Dempsey [41] examined an alternative model. It considers loading on a square on the plane face of an elastic/brittle material, and supposes that one and only one HPZ is randomly located within that square, that the force applied by the HPZ depends on the distance to the nearest edge of the square, and that there is no other force within the square. The measurement area is another random variable. A simulation based on a large number of choices of the random variables then shows that force per unit area is proportional to $A^{-0.8}$, again on the assumption that K_Q is not size-dependent. The agreement with (4.4.3) is encouraging. That equation is only intended to apply to areas smaller than 10 m^2, but over larger areas more than one HPZ will be present at one time.

4.5 Sloping-sided Structures

4.5.1 *Introduction*

The discussion in section 4.4 applies to a structure that is vertical or nearly vertical at the waterline. If the structure has a sloping face, the fracture mechanism changes completely. If the face slopes inward with increasing height, as in Figure 4.2(d), the ice rides up the face and fractures by repeated bending. The ice forces are often significantly smaller than in vertical-sided structures, the fragments are larger, and it is easier for the ice to clear by pushing round the structure. Indeed, it can be argued that structures in ice ought almost always to have sloping sides, and that ice researchers' focus on vertical-sided structures has been mistaken. However, the issue of choice of structure shape is never as simple as that, as will be seen.

This design principle has been applied to many structures, among them the Confederation Bridge between Prince Edward Island and mainland Canada [42], light piers in the Gulf of St. Lawrence, ice barriers in the Kashagan project in the northern Caspian, the Kemi lighthouse in the Gulf of Bothnia, the Prirazlomnoye platform in the Barents Sea, and numerous structures in the Bohai Gulf. Figure 4.9 is a series of photographs and elevation of one of the piers of the

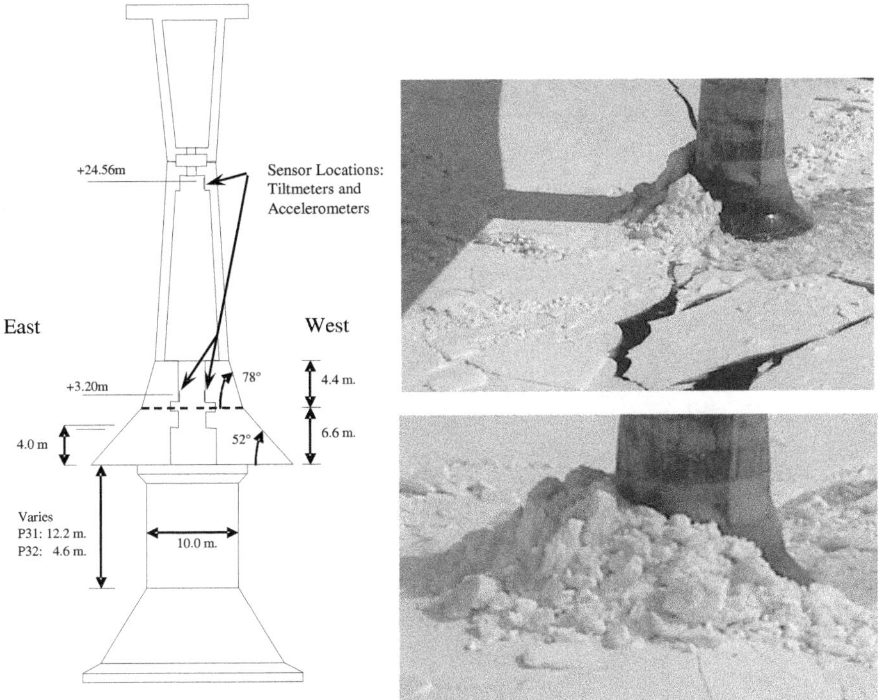

Location of Tiltmeters, P31 and P32

Figure 4.9 Confederation bridge, Canada (with permission T. Brown [42])

Confederation Bridge, where the geometry near the waterline consists of two conical sections, the lower at the water line inclined at 52° to the horizontal and the higher at 78° starting 2.6 m above the highest waterline. Figure 4.10 shows a smaller cone attached to one of the legs of the JZ20-2 MUQ platform in the Bohai Gulf [43]: the inclinations are 60° to the horizontal, upwards from the widest point, and 45° downwards. Similar cones have attached to several other Bohai Gulf platforms. Figure 4.11 is a 100 m long ice barrier in the Caspian Sea [44].

Figure 4.10 Bohai Bay cone [43]

Figure 4.11 Sloping ice barrier in the Caspian Sea [44]

Another possibility is for the structure to slope the other way, so that it becomes narrower with increasing depth below the waterline. Again the ice breaks against the slope, but now the fragments are driven downward below the water. Pushing fragments down against their buoyancy is much easier than lifting them up, because the net buoyancy is only about a tenth of the air weight. However, the fragments may tend to accumulate, particularly if the water is shallow so that the fragments jam between the sloped face and the seabed or the structure base. A geometry in which the structures slopes inward under water was used in the circular drilling vessel *Kulluk.* It can be thought of as an application of the principle applied long ago to ships that might be beset in ice, such as Nansen's *Fram,* which was designed so that ice forces would tend to lift the hull out of the water rather than crushing it.

4.5.2 *Mechanics of Ice Interaction with Sloping-sided Structures*

Ice that moves against a sloping structure breaks primarily in bending rather than crushing, though crushing may also occur locally as part of bending fracture. The ice force has several components, and includes:

(i) the force to break the ice by lifting the advancing edge,
(ii) the force to push the fragments up the sloping face of the structure. (This is generalized to be through rubble which has fallen back onto the advancing ice, as is often seen in nature, but not seen as much in model tests)
(iii) if fragments form a rubble mound that has fallen back onto the advancing surface of the ice, the force to push the ice through the rubble mound,
(iv) the force to lift the ice rubble prior to creating enough vertical force to fail the advancing ice sheet,
(v) the force required to turn the fragments so that they move around the structure and are carried down stream.

Figure 4.12 shows a simple two-dimensional model.

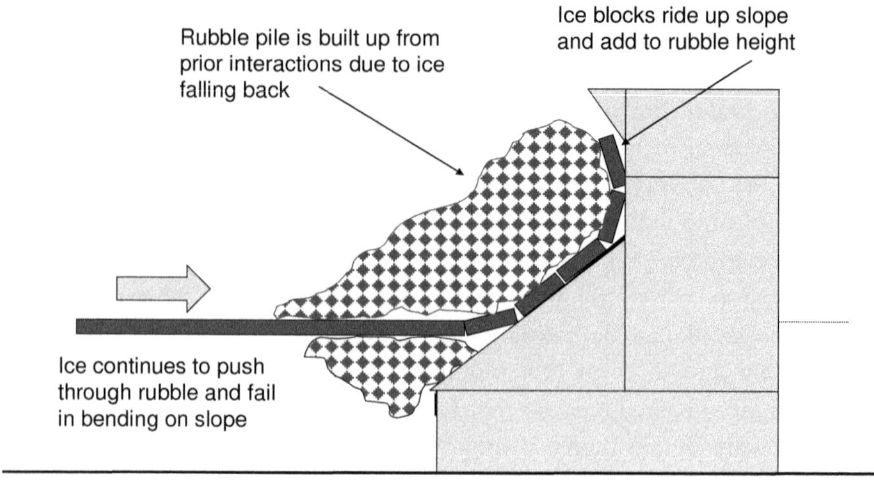

Figure 4.12 Two-dimensional view with typical processes for ice action on a sloping face

This model can be generalised to three dimensions. Several researchers have developed formulas for the more complex three-dimensional situation. The new ISO code [5] presents a formulation developed by Croasdale et al [45,46], which gives the horizontal force F_H on a cone as:

$$F_H = \frac{H_B + H_P + H_R + H_L + H_T}{1 - \dfrac{H_B}{YL_c h}} \tag{4.5.1}$$

in the notation of this book, and in part split up in a different way from the division into five components at the start of this section. Here H_B is the breaking load (component (i) above), H_P is the force required to push the ice through the rubble ((iii) and (iv)), H_R is the force required to push the ice up the slope (ii), H_L is required to lift the ice up on top of the advancing sheet prior to breaking (iv), and H_T turns the ice blocks at the top of the slope (v).

In the following, the notation is

c	is the cohesion of the ice rubble, idealised as a Mohr-Coulomb material,
D	is the waterline diameter of a cone, or the overall width of a sloping structure,
e	is the porosity of the ice,
E	is the elastic modulus of ice,
g	is the gravitational acceleration,
h_r	is the height of the ice rubble
L_c	is the elastic critical length defined by (4.5.4) below,
h	is the thickness of the ice sheet
α	is the inclination to the horizontal of the face of the structure,
ξ	$=\dfrac{\tan\alpha+\mu}{1-\mu\tan\alpha}$ (4.5.2)
θ	is the angle the rubble makes with the horizontal,
μ	is the coefficient of friction between the ice and the structure,
μ_i	is the coefficient of friction between one piece of ice and another,
ν	is Poisson's ratio,
ρ_i	is the density of ice,
ρ_w	is the density of water,
ϕ	is the angle of internal friction of the ice rubble, idealised as a Mohr-Coulomb material,
Y	is the flexural strength of the ice

H_B is then

$$H_B = 0.68\xi Y\left(\frac{\rho g h^5}{E}\right)^{1/4}\left(D+\frac{\pi^2 L_c}{4}\right) \qquad (4.5.3)$$

where L_c is the critical length for elastic plate bending given by

$$L_c = \left(\frac{Eh^3}{12\rho_w g(1-\nu^2)} \right)^{1/4} \tag{4.5.4}$$

H_P is

$$H_P = (1/2)h^2 \mu_i \rho_i g(1-e)\left(1-\frac{\tan\theta}{\tan\alpha}\right)^2 \cot\theta \tag{4.5.5}$$

H_R is

$$H_R = \frac{DP}{\cos\alpha - \mu\sin\alpha} \tag{4.5.6}$$

where

$$P = (1/2)\mu_i(\mu+\mu_i)\rho_i g(1-e)h_r^2\left(1-\frac{\tan\theta}{\tan\alpha}\right)$$
$$\left(\sin\alpha(\cot\theta-\cot\alpha)+\cos\alpha\cot\alpha\right) \tag{4.5.7}$$
$$+\,h_r T\rho_i g(1+\mu\cot\alpha)$$

H_L is

$$H_L = (1/2)\mu_i(\mu+\mu_i)\rho_i g(1-e)h_r^2\xi\left(1-\frac{\tan\theta}{\tan\alpha}\right)$$
$$\left(\cot\theta-\cot\alpha+\left(1-\frac{\tan\theta}{\tan\alpha}\right)\tan\varphi\right)+\xi cDh_r\left(1-\frac{\tan\theta}{\tan\alpha}\right) \tag{4.5.8}$$

and H_T is

$$H_T = \frac{(3/2)Dh^2\rho_i g}{\tan\alpha - \mu} \tag{4.5.9}$$

The horizontal force F_H is then given by (4.5.1), and the vertical force is

$$F_V = \frac{F_H}{\xi} \tag{4.5.10}$$

This model idealises the ice sheet approaching the structure as a elastic plate, idealises the fracture as occurring when the bending stress reaches a critical strength, and applies elementary mechanics to the broken pieces to find the force required to push them up the slope. The denominator in (4.5.1) is to take account of the influence of the horizontal force in the ice on the flexural failure. It can be seen from the formulas that the force is sensitive to the ice-structure friction coefficient μ: in particular, if $\mu \tan\alpha$ approaches 1, ξ becomes very large, and if μ approaches $\tan\alpha$ then H_T becomes large. This shows that it may be beneficial to apply a rugged low-friction coating as well as to avoid steep cones and abrupt changes in slope when merging into a vertical shaft or wall.

The use of the above equations that clearly relate to specific processes which can occur during the interaction is attractive. This is because an experienced user can make adjustments to suit a specific situation (as is done later in section 4.5.5 for very thick ice and in section 4.9 for wide structures in shallow water). The detailed assumptions on which the equations are based given in the papers by Croasdale et al [45, 46].

Users of the equations will also note that the loads are very sensitive to the height of the ride-up and the volume of ice rubble which builds up and which is proportional to the ride-up height as well as the angle of the rubble pile relative to the slope angle of the structure (see Figure 4.13). These parameters must be chosen carefully and ideally based on experience of observations on similar real structures. For rubble angle, experience suggests that a value in the range of 5 to 10 degrees shallower than the structure slope usually matches reality. Note that when the rubble slope is made the same as the structure slope, the model is simulating a single layer of blocks riding up (which may be appropriate for very thick ice).

For ride-up height; on cones with narrow shafts the ice may be expected to be turned sideways as it rides up and clear without extreme ride up. On the other hand, ice riding up a wide flat sloping structure (especially normal to its surface) will not clear easily and tend to ride higher for the same ice thickness. The ride up height in this case may only be limited by ice deflectors turning back the ice or by downwards failure of the oncoming ice due to the weight of the ice rubble. Equations

4.5.3 and 4.5.10 can be used to assess the approximate rubble weight that will fail the oncoming ice downwards and limit the rubble height, although in shallow water, (as will be discussed in section 4.9), rubble created below the ice sheet due to downward failure will ground and may provide support for a much higher ride up. For non-grounded situations, empirical relationships have been developed for limiting ride up height at least for ice up to about 1m thick (Määttänen and Hoikkanen[47] and Mayne and Brown [48,49])

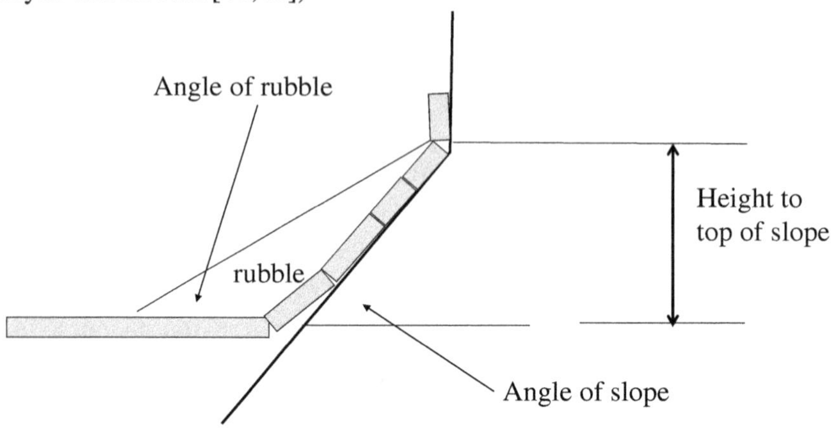

Figure 4.13 Ride-up configuration in the Croasdale model

The geometry of the ride-up configuration used to develop the equations is shown in Figure 4.13. Again this emphasizes that the rubble volume is solely dependent on rubble height, rubble angle and slope angle (and to some extent ice thickness). It is also calculated on a 2D basis using the width of the structure at the water line, which will be appropriate for wide structures but somewhat conservative for cones with narrow shafts. Depending on where the slope ends there can be situations where the ride up is higher than the height of the slope. This is shown in Figure 4.14. Understanding how the equations have been developed allows this situation to be approximated. It is recommended that the height of rubble be that which is expected or has been observed on similar structures and that the location of the rubble toe be the same. The virtual rubble pile is then as shown (with a shallower angle). The calculation is now based on an additional virtual length of ride up, which approximately corrects for the slightly lower volume of ice rubble over the real slope.

Readers may also question how H_T is calculated in this situation where there is additional pressure from the adjacent rubble on the turning blocks and is not accounted for in equation 4.5.9. Whether H_T needs a correction depends on whether the cone is narrow or wide because it is already somewhat overestimated because it is based on the waterline width not the width where the blocks are turned at the top of the slope. This component of load is easily adjusted if need be. An approximate approach could be to simply increase the thickness of the blocks being turned.

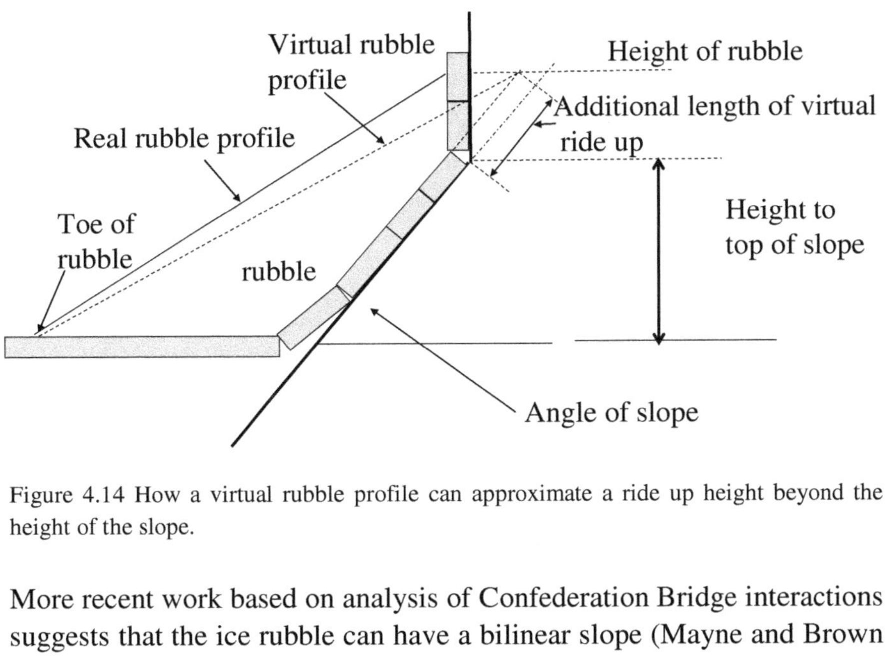

Figure 4.14 How a virtual rubble profile can approximate a ride up height beyond the height of the slope.

More recent work based on analysis of Confederation Bridge interactions suggests that the ice rubble can have a bilinear slope (Mayne and Brown [48]) and Mayne [49]. In that instance it may be a function of the specific cone geometry. In any event, a constant slope for the rubble can still approximate the volume of rubble whatever its actual shape by suitable choice of rubble angle and ride up height. Bonnemaire et al describe such an adaptation [50].

Alternative models for ice force on conical structures have been put forward by Ralston [51] and Nevel [52]. Ralston's model idealises ice as a rigid-plastic material, and is also included in the ISO code [5]. Nevel's method is based on an elastic response of the ice and treats the 3D failure more rigorously than Croasdale. However, these two models do not treat rubble on top of the ride up, except by suggesting the approximate approach of increasing the ride up thickness.

All these models make quite severe idealisations of the mechanical behaviour, and from the standpoint of applied mechanics they are all open to criticism from various directions. However, when the formulas are applied with sensible choices of the many parameters they include, they appear to lead to predictions that agree reasonably well with the measured values discussed below.

The strong size effect seem in crushing leads one to expect that there might be a size effect in ice breaking against a conical structure. If there is such an effect, it has not been observed, though that may well be because it has not been looked for. The ISO code formulas do not include a size effect for sloping structures. Non-simultaneous effects might occur with thick ice and have been discussed (Croasdale [53]); they are reviewed in Section 4.5.5.

4.5.3 *Adfreeze Effects*

A complication is that ice that remains stationary for some time may freeze to the face of a structure, and that before the ice can start to move the adfreeze bonds have to break. Cammaert and Muggeridge [54] provide a simple formula to estimate the corresponding horizontal force H_A

$$H_A = \frac{C_a C_s \pi D h q}{\sin\alpha \cos\alpha} \tag{4.5.11}$$

where the same symbols are as used previously, with the addition of q, the shear strength of the adfreeze bond, and C_a and C_s are factors smaller than 1 that account for uneven bonding and uneven stress distribution.

Using the above equation to estimate adfreeze loads with a value of q of say 0.7 MPa will generally yield large potential loads as shown in Table 4.1. The loads shown are for a range of platform diameters and ice thickness and assuming that the adfreeze situation will be mainly from level ice growing around the platform and freezing to it before an ice motion occurs. The first row in the table is close to the dimensions of Confederation Bridge and suggest an adfreeze load of about 11 MN, which is less than the design load of over 16 MN based on a thick ridge. So even if the adfreeze load occurred it would not govern. In reality, ice load measurements and observations on the piers of the bridge show loads less than the potential adfreeze load in the table. A second example is a possible Arctic platform with an ice-line diameter of 80 m, like the example in the next sub section). The ice thickness is the typical maximum first-year ice of 1.8 m. The calculated adfreeze load is about 103 MN, quite a substantial load. However, the design feature for this platform is a 12 m thick multiyear hummock or a still deeper ridge. Even using the improved methods discussed in the next sub section, these extreme features will lead to loads of over 400 MN, so again adfreeze will not control. A third example in the table might apply to a sloping barrier 100 m wide in the Caspian Sea. The 100-year level ice thickness is assumed to frozen to the slope when the ice moves suddenly. In this case the adfreeze load could be critical for design because the mature flexural failure load is about 30 MN. However, as discussed in section 4.9, for this structure, the design load is governed by the rubbling failure mode, at about the same level as the adfreeze load calculated with 0.7 MPa adfreeze strength.

This brings up the issue reviewed by Croasdale and Metge in 1989 [55], where they proposed that adfreeze bonds will be subject to a gradually increasing load or low strain rate as the ice picks up wind stress before it moves. In their analysis, Glenn's law for creep is assumed to apply. It is shown that the adfreeze bond strength can be weakened by a factor of 10 from measurements with instantaneous loading. The examples are repeated in Table 4.1 with this lower strength and suggest that now adfreeze loads will not be significant. Interestingly, in the discussion following the paper by Croasdale and Metge [55], Mauri Määttänen said; "As the author proposed, adhesion is not a decisive design criterion in conical structures due to strain rate effects. Our in-field data of Kemi 1 test cone, with a 10 m diameter at waterline, confirms the point. Ice, including a significant pile up, was stationary for a week with temperatures well below freezing point, after which a storm made the ice mass move. The initial adhesion break up did not cause the highest loads."

Table 4.1 Potential adfreeze loads; with and without invoking strain rate effects

q	Angle	Sin	Cos	D	h	$C_a.C_s$	H_A
MPa	degrees			m	m		MN
0.70	52.00	0.79	0.62	15.00	1.00	0.50	10.82
0.70	50.00	0.77	0.64	80.00	1.80	0.50	102.35
0.70	45.00	0.71	0.71	100.00	0.80	0.50	56.00
0.07	52.00	0.79	0.62	15.00	1.00	0.50	1.08
0.07	50.00	0.77	0.64	80.00	1.80	0.50	10.24
0.07	45.00	0.71	0.71	100.00	0.80	0.50	5.60

4.5.4 *Experimental and Full Scale Data*

Ice interaction with conical structures has been intensively studied on a model scale. The objections to model tests of crushing explained later in section 4.15 are generally thought to apply less strongly to model tests in situations dominated by bending. Model tests were first carried out on a laboratory scale in Russia and China in the 1970s, and in 1973 Imperial Oil built in Calgary a large 55×30 m outdoor test basin, in which it was possible to test cones 3 m in diameter at the waterline.

There are some full-scale measurements on conical structures. Määttänen [56-58] measured forces on the Kemi 1 lighthouse. Figure 4.15 is a comparison between his measured loads and those calculated with the Croasdale, Ralson and Nevel equations.

It can be seen that the equations predict ice forces quite well, and certainly closely enough for design purposes. The loads are quite modest, and the largest measured load is only 2.9 MN, but it must be kept in mind that Baltic ice is comparatively thin, that it is much less saline than Arctic Ocean ice, and that all of it is first-year. Yue and Bi [59, 60] and Xu [43] measured ice forces on cones attached around the legs of jackets in the Bohai Gulf, but in ice no more than 0.2 m thick.

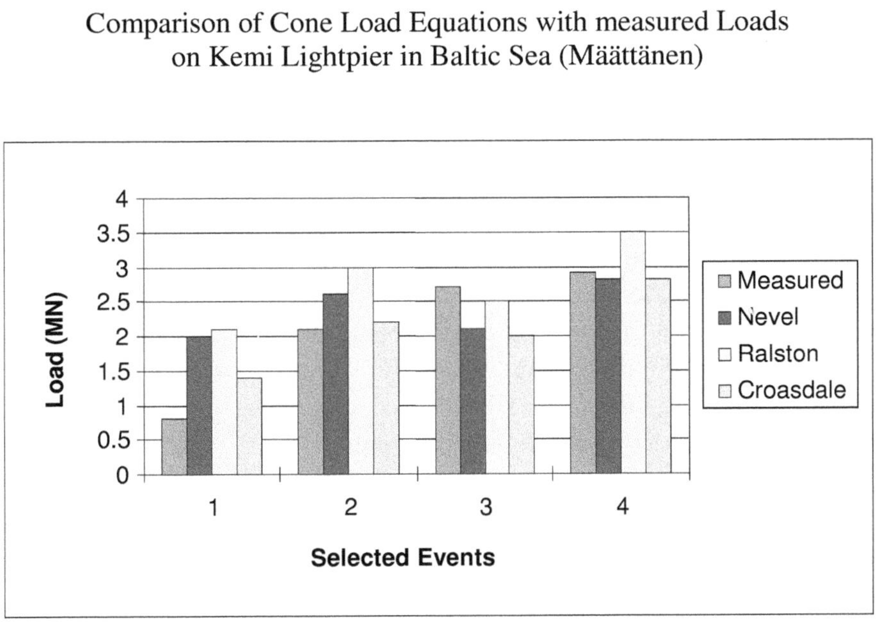

Figure 4.15 Comparison of theoretical models with measured data at the Kemi lightpier (after Määttänen [56-58]).

An ongoing program has been measuring forces on the piers of the Confederation Bridge since it opened in 1997. Numerous papers, theses and reports on the measured ice loads have been published, mostly by the University of Calgary team led by Tom Brown. A summary report for the period 1998–2008 is now available [42]. The highest loads measured on the two instrumented piers to date are between 8 and 9 MN (in 2003 and 2008). In most years the annual maximum load has been below about 5 MN. Note that the 100 year design load is 16.9 MN. The large load events are mostly associated with high continuously sustained rubble piles as the ice moves past. In some cases large loads can also be associated with a stall, implying that there is insufficient driving force to continue to build-up and clear the rubble or to fail a ridge lodged against the pier.

4.5.5 *Modifications for Very Thick Ice*

Croasdale's method described above [46] was originally developed in 1980 and refined for the design of the Confederation Bridge. The bridge piers are conical with a 52 degree slope, an ice line diameter of about 15 m and a shaft diameter of about 8 m (Figure 4.9). It was recognized at the time that certain conservatisms were embedded in the method. One was the assumption that the ice breaking across the width was simultaneous; another was that the ride up components acted as though on a 2D slope with the width at the top being the same as at the ice line. There are possibly other conservatisms (for example that all the five load terms peak at the same time) because comparisons of measured and predicted loads show generally an over-prediction by up to about 50% (Tibbo, [61]).

The two specific conservatisms mentioned above may lead to even higher over-predictions for thick ice. Some simple adjustments can be made to reduce them. Thick ice is important in places with multi-year ice, such as the Beaufort Sea. A potentially challenging design feature could be a large multi-year hummock field with an effective maximum average thickness of 12 m (Johnston et al [62]).

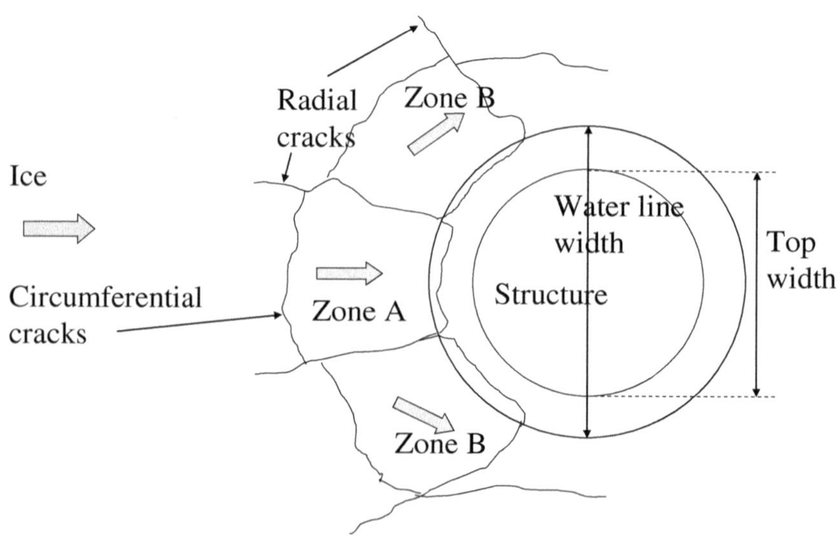

Figure 4.16 Multi-zone model for ice on a sloping structure

Figure 4.16 (from Croasdale [53] shows a possible approach that considers that the ice-breaking term be considered as resulting from the breaking of three separate non-simultaneous zones.

A non-simultaneous factor is introduced in the method, which has a minimum value of 0.33 and a suggested value based on judgment of 0.5. For ride up, it is suggested that the two side zones (B) only ride up half the height of the centre zone (A), and so the average height is 0.67 of the maximum ride up of the centre zone. In addition it is assumed that the H_T term (the force to turn the blocks to the vertical when they meet the shaft) is also only calculated for the centre zone. The blocks riding up from the side wedges are assumed to be cleared sideways without being turned to the vertical. With these simplifications, H_T is calculated based on one third of the width.

Table 4.2 shows the results of these modifications compared to the current ISO formulation for a 80m wide structure. As can be seen, the total global load for 12 m thick ice is reduced to about 336 from 660 MN. Also shown in Table 4.2 is the ice load assuming a vertical face based on ISO (Equation 4.4.2); it is 942 MN. There is a potential significant advantage for the sloping structure with a load, using this new approach, in the range 336 to 400 MN depending on the ride up height.

Table 4.2 Comparison of loads (in MN) using a multi zone model for breaking and ride up

vertical Structure	Sloping structure					
	ride-up height (m)	H_B	H_R	H_T	total H load	ISO method
942	20	46	251	39	336	660
	25	48	314	39	401	763

It can be seen from Table 4.2 that the ride up component of the horizontal load is about five times greater than the breaking term. This implies that we should not fret too much about the large scale strength of multi-year ice in bending (for example: is there a size effect?) but we should be investigating the ride-up process and how it might be limited by structure shape and/or ice failure downwards due to weight of the ice.

The choice of the range 20–25 m in the previous calculation is somewhat arbitrary. No one - as far as we know - has seen multi-year ice 12 m thick or so failing in bending (or in crushing for that matter). Interaction with smaller cones and thinner ice can provide guidance and properly designed model tests can help. These and the work by Määttänen referred to earlier [47] suggest that ride-up for say 1 m thick ice can be up to about 6 to 7 m. For extremely thick ice the ride-up will likely be determined by the height and shape of the structure, and that the clearing processes will limit the height, as in the photos in Figure 4.17.

Figure 4.17 Ice breaking and ride-up in tests in the Esso basin in 1989 [63] (with permission of Imperial Oil, Calgary)

It can be appreciated from Figure 4.17 that overhanging decks could be problematic for thick ice. The design shown in Figure 4.18 is considered to be a good geometry to clear the thick ice, but the drilling and/or producing facilities would need to fit into a smaller footprint than a design with an overhanging deck. Another way to avoid jamming and improve clearing is to have the deck on a high narrow neck (as in the photos in Figure 4.17), but determining the optimum diameter and height of such a neck is subject to uncertainty, and this solution may be more costly. Ride-up may be limited by downward failure of the oncoming ice due to the weight of the ice blocks above. An approximate calculation shows that a 15–20 m ride up can cause downwards failure of 12 m ice, but if the oncoming ice cannot fail downwards, because it is in contact with the structure slope or because there is already grounded rubble below, then 25 m ride up is possible and needs to be considered.

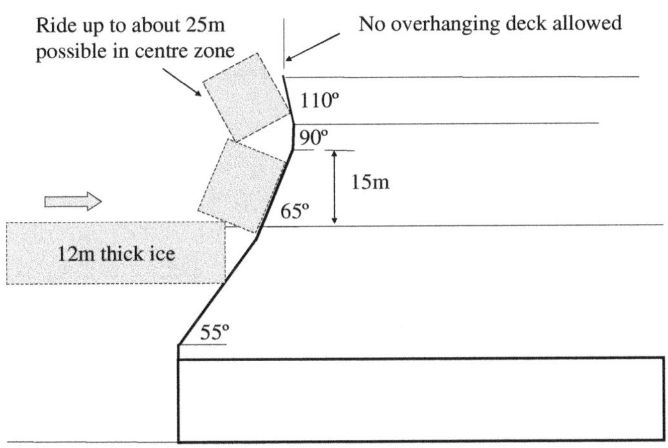

Figure 4.18 Thick ice acting on a sloping structure with expected ride up

The discussion above makes clear that there is very little experience of thick ice interacting with offshore structures, either in crushing or in bending. Full scale data on crushing are extrapolated based on plausible pressure/area trends developed for much thinner ice than 12 m, and 7 m seems to have been the thickest ice in the data sets. There is debate on the interpretation of existing full-scale crushing data [64]. The crushing data on multi-year ice from the Beaufort are higher than those obtained at

Hans Island, even allowing for temperature corrections. In general, it would appear that the crushing loads based on the current ISO Arctic formula may be conservative, but more data on thick ice interacting with vertical structures is needed to be sure of that, and to improve confidence.

For sloping structures, again there are a lot of data for ice less than 2 m thick (e.g. Confederation Bridge; Kemi 1), but no data for thick ice. For sloping structures, data, experience and theory all show that the breaking load is much lower than the equivalent crushing load (and this is without a size effect on flexural strength which may be present but is uncertain). However, it is the clearing process that is critical in keeping loads below the crushing load. With thick ice, broken blocks pushed upwards may jam into overhanging decks and so they must be avoided or positioned much higher than many designs often propose. Although we do not favour model tests as a way to determine ice loads due to ice breaking, if they are devised properly they may help to improve shapes to better cope with ride up and clearing of thick ice. In such tests it is important that ice piece sizes are similar to full scale.

4.5.6 *Velocity Effects*

A potential concern with ice action on sloping structures is whether at higher velocities there are additional loads due to inertia effects. Such effects could be due to the "elastic foundation" becoming effectively more stiff and/or the momentum change as ice blocks are turned from the horizontal to the angle of the slope of the structure. Early work on cones was primarily for the Arctic Ocean and model tests showed negligible velocity effects up to scaled maximum expected ice speeds. Other ice areas can have higher ice speeds, so this initial assessment might not apply. Matskevich in his review paper of 2002 [65] convincingly shows that based on model tests (and some theory) there can be an effect which needs to be considered. His recommended equation was:

$$H_v / H_{0.5} = 1 + 0.5(v - 0.5) \tag{4.5.12}$$

where H_v is the load at velocity v and $H_{0.5}$ is the load calculated assuming no velocity effect. This equation is based on a cone with a 60 degree slope and the paper notes that model tests show a reduced magnification for a 45 degree cone. The load increase is 25% at 1 m/s and 50% at 1.5 m/s.

At the time of the work by Matskevich, data from the Kemi-1 cone were examined and those data showed the possibility of a slight increase in load with velocity. Loads measured on the Hondo bridge pier showed a stronger effect, but this pier and the way ice interacts with it are not typical of larger structures. More recently, data from Confederation Bridge has become available [42]. The data in Figure 4.19 show no load magnification with ice speed and in fact may indicate the reverse. The slope angle of this structure at the ice line is 52 degrees.

Perhaps it is not surprising that real structures show less sensitivity to ice speed when one notes that rubble build up is seen more on full-scale structures than in model tests.

As one can demonstrate using equations 4.5.1 to 4.5.10, with typical rubble present, the breaking term is usually limited to about 20% of the total load on the platform. If this is the only term subject to equation 4.5.12 (speed magnification), then it can be appreciated that the velocity correction would be limited to about 10%, all other parameters being the same. In fact, the observations from Confederation Bridge show that rubble pile height is generally less at higher speeds. The reduced rubble pile height more than compensates for an increase in the breaking term due to inertia Taking the pier slope 52° and a tangent correction, the magnification on the breaking term would be 36% rather than 50% for a speed of 1.5 m/s, reducing the potential effect on total load to about 7% (again assuming all other parameters and contributions are the same).

The trends and uncertainties discussed above suggest that this is an area worthy of more research. Model tests should be used with caution and require confidence that the model tests properly replicate full-scale processes such as the significant contribution of ride up and rubble build up on the slope. Moreover, the extensive full scale data from Confederation Bridge appear to show no velocity effect in the range of speeds experienced (to 1.5 m/s)

Ice-induced vibrations in sloping structures are considered separately in section 4.9.

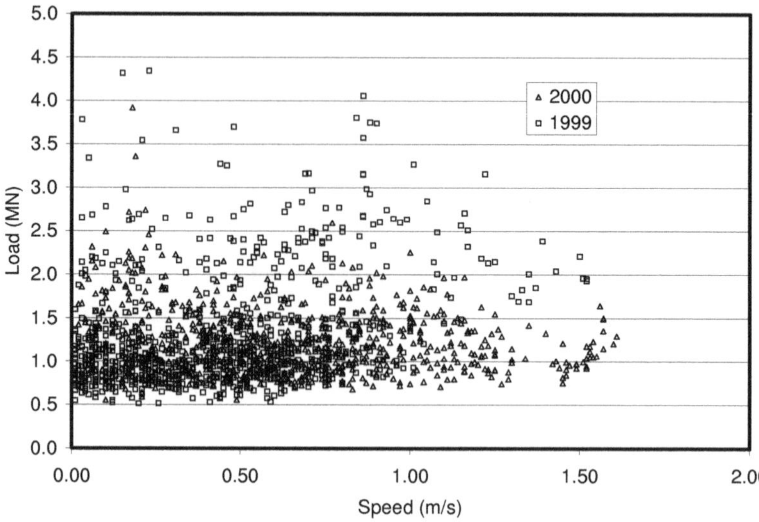

Figure 4.19 Ice speed effects on measured loads on Confederation Bridge (by courtesy of Tom Brown [42])

4.6 Local Ice Pressures

Choosing ice pressures for the design of the plates and stiffeners (or concrete thickness) of a structure is a different problem than that of global ice loads. One can choose to make this a rather simple exercise or one which is terribly complex. Complex methods, although perhaps moving closer to the real physics, have the potential to be applied incorrectly, so we will attempt to recommend simple approaches.

Some background is relevant. Designers and operators of ships in ice were the first to recognize that local ice pressures can be high enough to fail steel plating. Bad experiences such as the sinking of the SS Titanic in 1912 after striking an iceberg off Newfoundland, with the loss of over 1500 passengers and crew, provided ample evidence of the pressure of ice against steel. Consequently, designers of ice breaking vessels developed empirical methods for the ice pressures to be used in plating design, and those rules clearly accepted that pressures could be high over small areas.

The tests discussed earlier on small ice samples in the laboratory indicated that ice can easily attain crushing strengths well beyond the

1 MPa to 2 MPa now recommended for global ice loads (section 4.4.6). Indentation experiments began with "nutcracker" experiments on Arctic ice in 1970 [22,23] and continued with several years of testing by Imperial Oil (Esso) on lake ice (Eagle Lake) near Calgary. They were clearly showing high ice pressures on small areas and a pressure area and aspect ratio effect. In the main series of these tests, the indenters were powered by two rams each of 10 MN (1000 tonnes) force capacity. The indenters were up to 4 m wide. These tests were described by Taylor [24], and Kry [25]. Sample data from the Eagle Lake tests as reported by Taylor are shown in Figure 4.20.

The data show clearly that on areas below 1 m^2 pressures up to 9 MPa could occur, between 1 and 1.5 m^2 pressures were about 4 MPa, and beyond that about 2 MPa at 3 m^2. These particular data sets from the first indentation tests were not included on the generalized pressure area curves which Sanderson and Masterson et al compiled in the late 1980s (see Figures 4.6 and 4.8 of section 4.4), but it can be seen that the two sets of data would fit quite well.

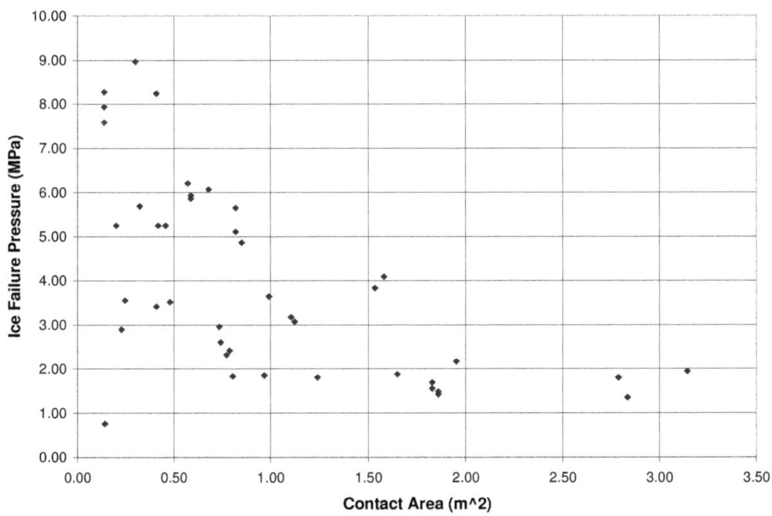

Figure 4.20 Ice crushing pressure versus area obtained from the Imperial Oil Eagle Lake indentation tests [24]

One of local pressure-area curves recommended by ISO 19906 is the one already given in section 4.4 for ice crushing:

$$p = 7.4A^{-0.7} \qquad \text{(4.6.1, repeat of equation 4.4.3)}$$

This relationship is based on the mean of the data plus three standard deviations. ISO suggests that it be used when the ice is thicker than 1.5 m and the area is less than 10 m^2, and that above 10 m^2 the local pressure be constant at 1.48 MPa (presumably up to the area that gives the same value for global pressure, based on the ISO method equation 4.4.2 in section 4.4).

If we look at the data sets from which the above equation is derived, we see most of the highest values are from "confined tests" such as Pond Inlet and Hobsons Choice. In these tests, indenters were pushed into the walls of tunnels in icebergs and deep trenches in ice islands and thick multi-year ice. Such data may indeed be appropriate for small areas within a larger area of interaction which can occur with large icebergs, ice islands and thick multi-year floes. However, with thinner ice and with sloping structures and thick ice, the use of equation 4.4.3 may be very conservative. Besides, since the data from the above tests were in cold ice with virtually no salinity, one must ask: do they apply to more temperate regions?

Each of these issues will now be reviewed. First, with respect to temperature and salinity, it is relevant to point out that ISO allows a recognition of the effects of basic ice strength on its global crushing pressure. The coefficient of C_r equal to 2.8 in equation 4.4.2 is for the Arctic. ISO suggests values of 2.4 and 1.8 for sub Arctic and Baltic regions respectively. The authors have some concerns that these potential reductions may be overstated because, for example, the Baltic tests from which the value of 1.8 was derived were with much thinner ice, which has more tendency to spall and flake to the free surfaces because it is less confined. Nevertheless, perhaps some reduction on the coefficient of 7.4 in equation 4.4.3 may be possible. In the Canadian (CSA) code, this is recommended. The value of the coefficient in the CSA ice crushing equation for low aspect ratios is 8.5 (with A to the power –0.5) for the Arctic, but reduces to 4.2 (with A to the power –0.4) for sub-Arctic and is 1.8 for temperate area. The authors recommend a combination of the CSA and ISO approaches for local ice pressures for first-year ice. This is

$$p = 4.8A^{-0.5} \tag{4.6.2}$$

The above equation also takes care of the problem of confinement effects, which will be less for first-year ice (which will range in thickness from say 1 to 2 m) than for the confined areas of thick multi-year ice, icebergs and ice islands. In other words it is recommended for level ice thickness up to about 2 m thick. In the opinion of the authors it can also be applied to local ice pressures in consolidated layers of first-year ridges, because, although they may be thicker than 2 m, they have greater porosity and are generally weaker than level ice, especially in the lower half of the layer.

The matter of local ice pressures when ice acts on a sloping structure also deserves some review. There is actually nothing in the literature or codes on this issue. Figure 4.21 shows how ice acting on a sloping structure will initially crush along the line of contact. The crushing "thickness" will increase until the vertical component of the crushing load breaks the oncoming ice in bending. The equations for sloping structures (from section 4.5.2) can be used to estimate this crushing "thickness" on the slope for a given level ice thickness and rubble height relevant to the interaction. Table 4.3 shows the line loads associated with some typical first-year ice scenarios. The vertical force at the ice line is estimated from the vertical components of H_B, H_L and H_R (but only using half of H_R as the other 50% is assumed distributed up the slope of the structure). The vertical force at the ice line is resolved back into N, the normal force at the ice line. Also shown are the crushing thicknesses that result based on a crushing strength of 1.5 MPa. Those "thicknesses" are quite narrow (up to about 0.5 m). Therefore there will be no structural design significance in specifying ice pressures greater than the 1.5 MPa within these strips. Essentially both plate thickness and vertical frames can be based on the line loads.

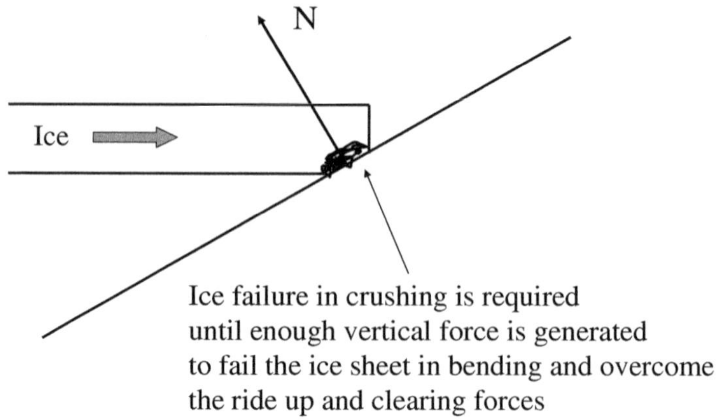

Figure 4.21 Scheme for estimating local ice pressures on a sloping structure

Table 4.3 Thickness of the crushing zone during typical ice interaction on a sloping structure

Ice thickness	Rubble height	N	Crushing thickness
m	m	kN/m	m
1	0	96.25	0.06
1	7	196.18	0.13
1.5	0	205.13	0.14
1.5	10	413.43	0.28
2	0	353.84	0.24
2	15	776.55	0.52

Beyond the ice belt, the slope should be assumed to be loaded by 50% of the ride up force (H_R); H_T should be applied where the slope changes to a vertical wall or shaft higher up.

Where a structure is subject to the action of ice rubble (e.g. from a ridge keel) it is considered sufficient to design for the passive pressure from the ice rubble using maximum expected shear strengths for ice rubble (as reviewed in section 3.6). This is about 0.5 MPa. With ice rubble, it is certainly possible to have local crushing of ice blocks at higher pressures than 0.5 MPa, but over such small areas as to have no significance in the design of the plating.

The approach to local ice pressures discussed so far is "deterministic" but is based on the mean of the data plus three standard deviations. This could imply an ELIE level (or even an ALIE), and it really depends on how many ice interactions are expected per annum (exposure). ISO 19906 does have a clause on the treatment of local ice pressures based on exposure. It is based on the statistics of ice pressures measured primarily from ship rams (Jordaan et al., [66]). It offers a useful approach where a platform may only rarely experience ice or when it may become protected from direct ice crushing by grounded ice rubble etc.

4.7 Ice Encroachment

Ice encroachment is the term given to ice moving onto the surface of a platform. The risk of ice encroachment is inversely proportional to the freeboard of the platform. There are two kinds of ice encroachment; ice over-ride and ice pile-up.

Ice over-ride is depicted in Figures 4.22 and 4.23. It is rare, but can occur with continuous ice, smooth surfaces, low freeboards and gentle slopes [67]. An example of ice over-ride in the Caspian Sea is shown in Figure 4.22. During an ice movement event, ice about 0.5 m thick started to rubble directly against the quay. After a few minutes, advancing ice climbed over the ice rubble and rode across the perimeter of the island. It was stopped by some equipment which was not damaged but gave enough resistance to trigger instabilities in the ice blocks at the island perimeter.

A concern with ice over-ride is that the ice advances onto the surface at the speed of the ice movement. Therefore for manned structures and those with equipment that can be damaged by ice, ice over-ride must be designed against. Methods of avoiding ice over-ride include high freeboards, steep slopes, rough surfaces and other means to cause ice jamming and to trigger ice pile up (see Croasdale et al. [68]). They can include various beach geometries, massive concrete blocks, sheet pile walls and steps in the island cross-section.

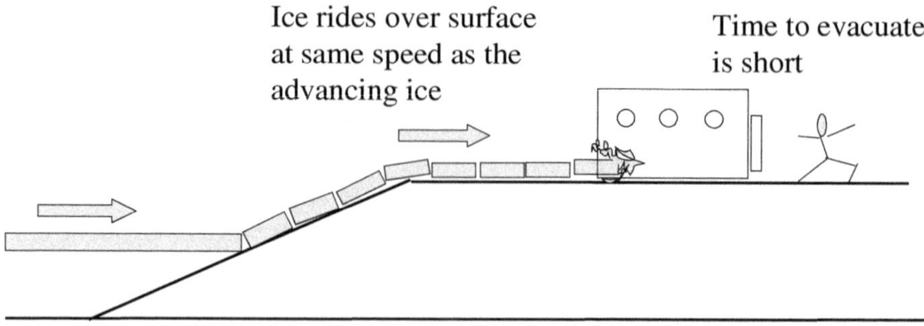

Figure 4.22 Ice ride-up on low freeboard structure

As discussed by McKenna et al. [69], steeper sloped islands (and those constructed from rock) favour ice pile up rather than ride up. In this case, an ice encroachment perimeter may be a reasonable solution. No permanent facilities are placed in this perimeter allowance, but it could be a perimeter road. The width of this perimeter depends on an assessment of risk but in the Caspian Sea a width of 15 m is normally used. Figure 4.24 shows the principles of how the width of the perimeter should be calculated for various island freeboards and slopes. An alternative to the perimeter strip is to raise the surface (or deck) above the height of natural pile-ups seen in the region, or the top of an ice deflector should be at this height. This approach can be very costly. A perimeter design which provides a "management" strip for ice pile up and a step to trigger instabilities during ice over-ride events to avoid ice encroachment onto the working surface is shown in Figure 4.24.

The equation for the width of the ice encroachment management strip is based on simple geometry. Using the parameters defined in Figure 4.24 the width of the perimeter strip (w_p) is given by.

$$w_p = \frac{h_p - y}{\tan\theta} - w_t \tag{4.7.1}$$

Figure 4.23 Ice over-ride incident in the Caspian Sea (from McKenna et al.[69])

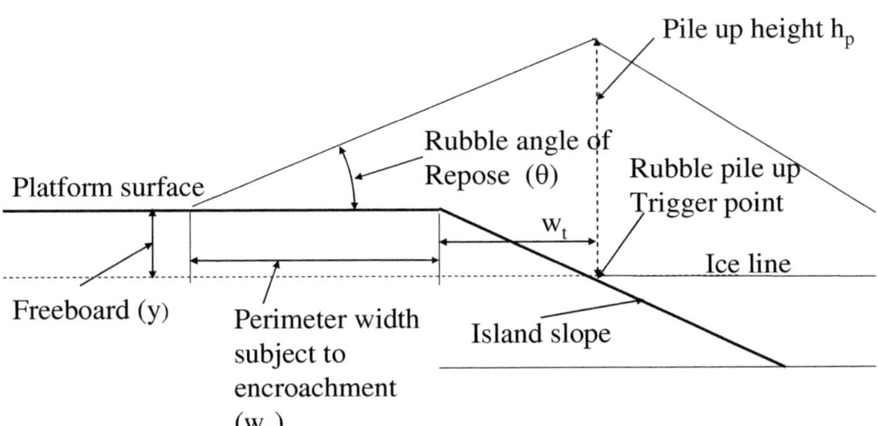

Figure 4.24 Scheme to specify edge perimeter allowance for ice encroachment due to ice pile-up

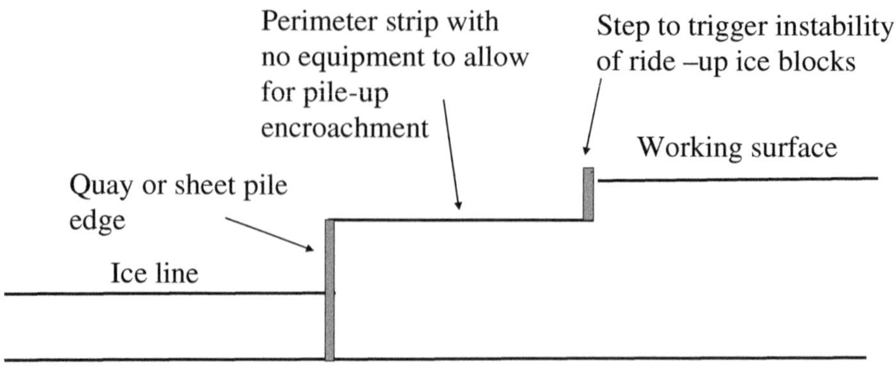

Figure 4.25 Perimeter design to reduce consequences of potential ice encroachment

4.8 Model Tests

The mechanical interaction between ice and a fixed structure is complicated, and this has led engineers to think that the analytical and theoretical difficulties might be avoided by carrying out tests on a physical model.

Models appeal to subliminal childhood memories of playing with dolls' houses and model aircraft. A scale model offers a miniature world that is perfectly controlled by its creator, free from the untidiness and compromise of the wider real world. A model almost always "works", in the sense that some results almost always emerge, whether or not the results are actually relevant to the intended prototype. A modeller is highly motivated to believe that his results can be scaled up and will apply to a full-scale system.

Models can also be deeply misleading: see, for example, Svendsen [70] and Palmer [71]. Svendsen's wise remarks apply to many laboratory experiments on ice. The conditions for similarity between the small scale of the test and the large scale of the application have often been ignored.

The earliest model tests were on icebreaking ships. They followed conventional towing tank practice, which invariably adopts Froude scaling and seeks to make Froude number $U/\sqrt{(g\ell)}$ the same in the model as in the prototype: U is velocity, g is gravitational acceleration, and ℓ is a representative length. It follows that in a 1/10 scale model at 1 g velocities are scaled in the ratio $1/\sqrt{10}$. An interpretation is that Froude scaling maintains the ratio between forces associated with inertia (and that scale with $\rho U^2\ell^2$, where in addition ρ is density) and forces associated with gravity (and that scale with $\rho g\ell^3$).

Froude scaling may be appropriate for model ships in ice, where wave-making resistance is important and where the interaction between inertia and gravity is significant for waves, in dynamics in ice cover and in ramming. In many ice-structure interactions, however, the velocities are quite small and there are no waves and no significant effects associated with the interaction between gravity and inertia. Froude scaling can then be dropped. A rough calculation that idealises the ice as a dense fluid confirms this. If an ice sheet 1 m thick is moving at 1 m/s against a 20 m diameter fixed structure, the inertia component of the total force on the structure is 0.01 MN (taking the drag coefficient as 1), and therefore negligibly small by comparison with other components. This is consistent with the conclusions of Schwarz [72], described in Palmer and Dempsey [73].

Cauchy scaling makes Cauchy number $\rho U^2/S$ the same in the model as in the prototype: S is a strength with the dimensions of stress. For fluid-structure interactions or high-speed impacts, Cauchy similarity maintains the ratio between forces associated with inertia (again scaling with $\rho U^2\ell^2$) and forces associated with strength (scaling with $S\ell^2$). In a 1-g model that has both Froude and Cauchy scaling, S must then be scaled with U^2, and in a 1/10 scale model at 1 g the strength S must be 1/10 of the strength at full scale.

Many ice modellers have chosen this option, and much effort has been given to the development of various kinds of artificially weak and flexible ice, through the addition of contaminants such as additional salt, urea and EG/AD/S. Timco [74] reviewed the scaling issues in detail, and further work was carried out by Schwarz [72], Cammaert and Muggeridge [54], Tatinclaux [75], Hirayama [76], Zufelt and Ettema [77], Lau et al. [78] and others.

The supposed requirement for weak ice is not a consequence of Cauchy scaling by itself, but results from the combination of Froude scaling and Cauchy scaling. Palmer and Dempsey [73] argue that the search for weakened ice was mistaken, and that Froude scaling is unnecessary, because the objective of Froude scaling is to maintain the ratio between forces associated with inertia and forces related to gravity. At least for ice moving slowly against fixed structures, inertia forces are negligibly small by comparison with the other forces present, and so it becomes unnecessary to include inertia forces in the scaling parameters.

A complication is that we would like the characteristic length for elastic bending of floating ice to scale in the same way as other linear dimensions, because some interaction processes appear to be governed by fracture in bending: this applies particularly to sloping structures and to icebreaking ships. Characteristic length L in elastic bending is given by equation (4.4.4)); $\rho_w g$ is the unit weight of water, h is ice thickness and ν Poisson's ratio. If ρg and ν are unchanged in a model, and if in addition ice thickness scales with other lengths ℓ, E has to scale with ℓ. If in addition Cauchy similarity applies, E/S ought to be the same in the model as at full scale. Many people have worried about this, and have chosen to prefer versions of weakened 'ice' that have roughly the same E/S as at full scale. If on the other hand we have Cauchy similarity without Froude similarity, and the model ice has the same E as the prototype, then characteristic length scales as $h^{3/4}$, and we can maintain the ratio of characteristic length to other lengths by a modest increase in h. Many small-scale laboratory experiments have in fact been on undoped freshwater ice rather than 'model' ice.

Rather than characterise ice by a failure stress, a more realistic alternative is linear elastic fracture mechanics (LEFM). Atkins and Caddell [79, 80] wrote that

> "Scaling laws for ice tank work are ill understood at present, and the techniques used to translate from model to prototype are quasi-empirical"

They discussed similarity conditions for systems made of brittle solids, and went on to develop an ice number

$$\mathrm{At} = \frac{U^2 \rho \ell^{1/2}}{K_{1c}} \tag{4.8.1}$$

where K_{1c} is a critical stress intensity (fracture toughness), the subscript 1 stands for mode I fracture, and the other symbols are defined earlier. The number At can be interpreted as a ratio between a force $\rho U^2\ell^2$ associated with inertia and a force $K_{1c}\ell^{3/2}$ associated with fracture. This kind of similarity brings together both Froude similarity and LEFM similarity. Applied to models of icebreaking ships, the ratio between the model velocity and prototype velocity is

$$\frac{U_m}{U_p} = \left(\frac{\rho_p K_m}{\rho_m K_p}\right)^{1/2}\left(\frac{\ell_p}{\ell_m}\right)^{1/4} \tag{4.8.2}$$

and so if the model and prototype ice have the same density and fracture toughness, a 1/10 scale model ought to be tested at $(1/10)^{1/4} = 0.56$ prototype speed, which is clearly perfectly practicable.

Atkins and Caddell's work attracted notice in the early years of ice tank testing, but appears to have been largely ignored since then. Some experiments that purport to measure fracture toughness are open to criticism, because the process zone is not small by comparison with the apparent crack length [81–83]. It can also be argued that the critical parameter is the weakening rate in the fracture process zone: see Palmer and Dempsey [73] and Uenishi and Rice [84]. Research on the fracture toughness of artificially weakened ice has found that the fracture toughness does not scale down as much as the flexural strength does, and Lau [78] agreed that "the fracture toughness of most model ices is somewhat greater than that required of model ice". Loosely, model ice is too ductile. Palmer and Dempsey concluded that ice is best modelled by real ice.

Modelling of ice-induced structural vibrations is discussed in section 4.9 below.

The role of model tests remains contentious, and the controversy is likely to continue. Croasdale has put forward a moderate and balanced opinion: in his view model tests are likely to be useful for sloping-sided structures where the interaction is dominated by bending and clearing, but they are less useful for vertical-sided structures.

4.9 Ice-induced Vibrations

Structures in ice often vibrate when ice is pushed past them. The effect has been observed both in models and in full-scale structures. Vibration destroyed the Kemi 1 lighthouse in the northern Baltic, which had an unusually slender steel column (in order to minimise the breadth at the waterline), a low natural frequency, and very low structural damping [47,85] Less dramatic oscillations have been observed in bridge piers (Montgomery et al.[86]), platforms (Jefferies [87]) and other lighthouses (Bjerkås, [88,89]). In addition, vibrations have often been observed in physical models (Sodhi [90]), but because the mechanics of ice-induced vibration were not understood, it was unclear how to transfer information between the model scale and the full scale. If one researcher observes vibrations in a model test in 10 mm thick urea ice moving at 80 mm/s, and another researcher observes vibrations of a full-scale offshore structure in 1500 mm thick sea ice moving at 100 mm/s, is the first test in some sense a model of the second, or does it represent something completely different? In particular, can it be assumed that vibrations will always occur when the ice velocity reaches 80 mm/s or 100 mm/s, whatever the size of the structure and whatever the thickness of the ice?

If the ice moves very slowly against a structure, the ice force builds up until the ice fractures in front of the structure. The force falls back almost to zero as fragments clear around the sides, and then builds again when the unfractured 'solid' ice reaches the structure again. If the ice moves faster, the structure deflects under the ice load, and then springs back when the ice load temporarily falls. It can be set into oscillation, and the oscillation can either be random (if the ice is moving very fast) or regular (if the ice is moving at an intermediate speed). Yue and Guo [91–93] and Palmer et al. [94] describe the modes in more detail.

There is a loose analogy with vortex-excited oscillations in marine structures, which have been the source of much trouble, have been extensively researched, and are much better understood. If water flows across a vertical cylindrical structure such as a platform pile or a riser, vortices are shed alternately on one side and the other. As the vortices leave the cylinder, they exert small oscillatory forces on the cylinder, and if the frequency matches a natural frequency, the cylinder can go into oscillation

Imagine now the heavily oversimplified picture of ice interaction with a structure shown in Figure 4.27.

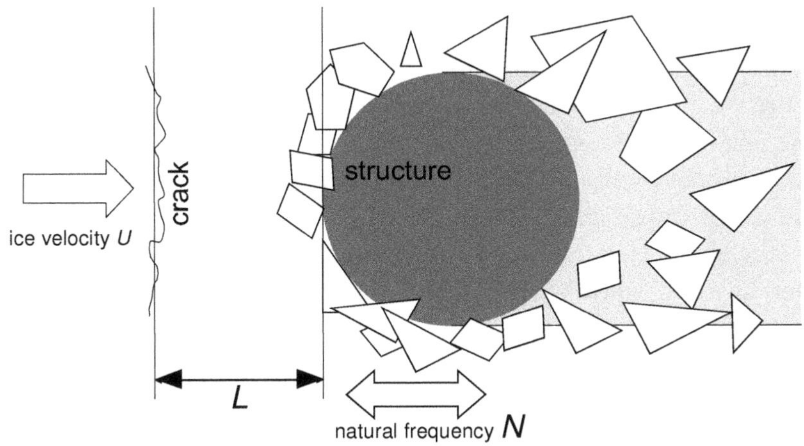

Figure 4.27 Simplified model of ice-induced vibration

The ice advances with velocity *U*. Transverse fractures across the whole breadth of the structure break the ice into fragments. The distance between fractures is *L*, measured in the direction of the ice velocity. The frequency of the alternating force on the structure is *U/L*, and we can expect resonant oscillations if *U/L* is close to one of the structure's natural frequencies *N*. A dimensionless group *U/NL* ought then to be illuminating.

L is ill-defined and highly variable, but ice thickness *h* is reasonably well known in most practical situations, though the ice thickness will often vary and will be punctuated by thicker ridges and thinner refrozen leads. It is then unclear whether the length parameter that controls the effective spacing *L*, and thence the excitation frequency, ought to be related to *D* or to *h* or to some mixture of *D* and *h*, or to something else.

Palmer, Yue and Guo [94] investigated this difficulty. They focussed on the mode of the vibrations that occur, and looked at eleven instances of ice-induced vibration for which the parameters *U h D* and *N* were available, five of the eleven in full-scale structures and six in small-scale models. All the data came from structures with vertical or near-vertical sides, to avoid the additional complexities of sloping sides and cones. If the structure diameter is the controlling parameter, the mode will depend on *U/ND*. If the ice thickness is the controlling parameter, the mode will depend on *U/Nh*. The results showed that the mode depends on *U/Nh*.

This is consistent with other research. Research into the high-pressure zone model discussed in section 4.4 indicates that the mode of failure against wide structures, where h is small compared to D, is unstable propagation of a fracture to the upper or lower surface of the ice. Loosely, the upper and lower surfaces are closer to the high-pressure zone at which the ice forces are concentrated, whereas the sides are much further away. Unsurprisingly, the cracks propagate to the closer free surface.

These results also indicate the scaling rule to be applied to models of structures in ice. If we are to model ice-induced vibrations correctly, the parameter *U/Nh* ought to be the same in the model as in the prototype. Consider a prototype structure that is 10 m in diameter and has a lowest natural frequency of 0.8 Hz, and that will operate in ice 1.5 m thick moving at 0.1 m/s (0.2 knots). We wish to model it at 1/20 scale with the same kind of ice, and to keep constant the ratio of the ice thickness to the structure diameter. Then the model diameter is 10 m/20 = 0.5 m, the model ice thickness is 1.5 m/20 = 75 mm and since

$$\left(\frac{U}{Nh}\right)_{model} = \left(\frac{U}{Nh}\right)_{prototypel} \tag{4.9.1}$$

it follows that

$$\left(\frac{U}{N}\right)_{\mathrm{mod}\,el} = \left(\frac{0.1}{(0.8)(1.5)}\right)(0.075) = 0.00625\,\mathrm{m} \tag{4.9.2}$$

and so if the model ice velocity is 0.01 m/s (10 mm/s), the model natural frequency ought to be (0.01 m/s)/(0.00625 m) = 1.6 Hz.

Yap [95] made an exhaustive review of research on ice-induced vibrations, and re-examined Engelbrektson's observation [96] that in steady-state oscillation the maximum velocity of the structure matches the maximum velocity of the ice. This provides a useful way of quantifying the amplitude of oscillations, if one knows the ice velocity and the structure frequency and knows that steady state oscillation is indeed occurring.

Sloping-sided structures are generally agreed to be less prone to ice-induced vibrations than vertical-sided structures, but they can nevertheless occur. Xu [43] and Yue [59,60] report ice-induced vibration on platforms in the Bohai Gulf. Each platform leg carries a double cone that slopes 60° up and 45° down: that configuration was chosen because of the 4 m tidal range. The accelerations reached 1 m/s^2 (0.1 g) on the JZ20-MSW platform and 0.4 m/s^2 on the JZ20-MUQ platform, enough to create damage to pipework, structural fatigue, and reduction in efficiency of people working.

Xu [43] has made a particularly useful comparison. A 20-tonne monopod column 1.06 m in diameter at the waterline was added in 2007 to an existing jacket platform, JZ9-3E in the Bohai Gulf (Figure 4.10). Severe vibrations occurred in the following winter, and the acceleration reached a maximum of 2.5 m/s^2 and exceeded 1.5 m/s^2 on 11 days. An up-down double cone was then retrofitted to the column. The cone is 3.55 mm in diameter, 4 m high, and weighs 6 tonnes, and both the upward and downward cones are at 60° to the horizontal. A comparison was made between the ice-induced accelerations on 18 January 2008, before the double cone was retrofitted, with the accelerations on 17 January 2009, after the retrofit. The character of the vibrations is different. Before the cone was fitted, the vibrations were steady-state and continuous. After the cone was fitted, the general level of vibration was much lower, though there were brief high accelerations at intervals of about 3 s, presumably linked to bending fracture of the ice sheet. The peak accelerations with the cone were only 17 percent smaller than the steady-state maximum accelerations without the cone, but the great reduction in the number of high-stress load cycles implies that the increase in fatigue life is substantial. An approximate fatigue analysis based on the measured accelerations before and after the retrofit, and applying S-N curves and the Palmgren-Miner rule indicated that the fatigue life would be increased by a factor of 20.

4.10 Ice Load Measurements on Platforms

Much of the material in this book emphasises the difficulties in using ice strength parameters from small samples to predict large scale processes and in particular ice loads on platforms. For this reason, starting in the 1960s, engineers have pursued initiatives to measure loads on the platforms themselves. Those load measurements are also a normal part of the verification process that is part of engineering practice and are much to be encouraged.

Ice load measurement methods for structures include the following approaches:

- Structure response using strain gauges
- Foundation response using soils instruments
- Structure response using extensometers
- Structure response using accelerometers
- Structure/foundation response using tiltmeters
- Ice load cells and panels at the interface between ice and the structure
- Instrumenting the surrounding ice for strains or pressures
- Estimating ice loads from the deceleration of free floating floes impacting the structure.

Table 4.4 summarises some of the key ice load measuring initiatives starting in the late 1960s. The table builds on earlier ones [97,98], and cites references.

The work by Peyton and Blenkarn was an early initiative by the oil industry during the design of the Cook Inlet platforms in Alaska. Much of the early work concentrated on bridge piers and from this work Neill concluded that the 2.76 MPa (400 psi) figure for ice pressure in the Canadian bridge code was probably too conservative. As soon as artificial islands for offshore drilling started to be used in the Beaufort Sea, Imperial Oil mounted a significant initiative to measure ice pressures on them as well as their geotechnical response. It was this initiative that led to a lot of work on ice pressure panels, which were initially used inserted in the ice surrounding the islands, which was generally landfast for most of the winter. Occasionally, a significant motion of several tens of metres occurred and gave ice pressures, as interpreted by the sensors, up to about 1.0 MPa averaged across the island width.

In Europe there were also significant initiatives. Schwarz instrumented the Eider bridge by using small load cells at the ice pier interface; Määttänen instrumented the Kemi 1 lightpier in the Baltic, and at the same time early work began on the Nordstrømsgrund lighthouse.

When drilling caissons started to be used in the Beaufort Sea, they were generally well instrumented with strain gauges and ice pressure panels. The technology used for the in-situ ice panels was adapted for use in panels attached to the structure at the ice line. In addition to the bottom-founded caissons, drillships with some ice capability were employed, notably the Kulluk. This had instrumented mooring lines, and as reviewed in Chapter 6, these measurements have proved very valuable in assessing loads on floaters in managed ice. Starting in 1997, the newly constructed Confederation bridge was instrumented and has provided considerable data on ice loads on conical piers.

At about the same time, a brand new initiative was undertaken at the Nordstromsgrund lighthouse. It was instrumented with ice pressure panels which were continuous around half the perimeter at the ice line; this commenced as the LOLEIF project and continued as the STRICE project. Results from the latter provided the full scale data that was used to derive the ice strength coefficient for the Baltic in equation (4.4.2).

In the late 1990s, the Molikpaq caisson platform was moved from the Canadian Arctic and deployed off Sakhalin as an early production platform. Its instrumentation was refurbished specifically with improved ice panels that did not have some of the drawbacks of the Medof panels used in the Canadian Beaufort. Some data from this initiative have been published in the book by Vershinin, Truskov and Kuzmichev [130] which gives valuable insights on offshore Sakhalin.

Table 4.4 A list of examples of ice load measurements on structures (after Croasdale & Frederking [97], and Blanchet [98])

Site	Initiation Date	Company or Agency	Method, Size and Instruments	Reference
Cook Inlet - Pile	1963	U of Alaska and Amoco	Beam/pile 0.91m dia. Ice to 0.3m	Peyton [99], Blenkarn [7]
Cook Inlet Platform	1964	Amoco	Strain gauges on structural members of 4 leg platform	Blenkarn [7]
Hondo Bridge, Alberta	1967	Alberta Research Council & Alberta Highways	Nose section hinged at bottom, load cell at top. 2.3m dia. 23° from vertical. Ice thickness to 1m.	Neill ([100]. Lipsett and Gerard [101]
Eider River Estuary	1967	University of Hannover	Array of 50 load cells covering 0.6m width and 1.5m high. Ice to 0.4m	Schwarz [102].
Norstromsgrund lightpier, Sweden	1972, 1978	VBB Consulting	4 Accelerometers initially but inclinometers and panels installed later.	Engelbrekston [8,94]
Adgo P-25 Artificial Island. 2m water. Mackenzie Delta	1974	Imperial Oil	U of Alaska small cylindrical in-situ sensor and Esso thin wide sensors in surrounding ice	APOA project. 104. Metge [103]: Nelson and Sackinger (1976) [104]
Netserk South B-44 Island. 4.5m water. Beaufort Sea	1975	Imperial Oil	Twenty in situ ice panels	APOA 104: Metge [103]
Kemi 1 lightpier, Finland	1975	University of Oulu	Deformation of structure. 20cm pressure plates and accelerometers	Määttänen [56]
Yamichiche lightpier, Quebec	1975 and 1985	Transport Canada and NRC	Load panels supported on load cells. Also strain gauged load panel. Ice to 0.5m	Danys [105] Frederking and Sayed [106]
Saroma Lagoon	1976	Mitsui	2.5m cantilevered cylinder. Strain gauged	Oshima et al [107]
Netserk North F-40. 7m of water. Beaufort Sea	1976	Imperial Oil	13 in situ ice panels	APOA Project 105: Strilchuk [108]
Kannerk G-42 and Arnak L-30 islands in 8 and 8.5m water. Beaufort Sea	1977	Imperial Oil	In situ ice panels	APOA 122. Semeniuk [109]
Tarsiut N-44 (Caisson retained island) in 22m water. Beaufort Sea	1981 - 83	Gulf, Dome, BP	Flat jack panels on caisson. Shear bar plates. Strain gauges. In situ panels in surrounding ice. Soils instruments	APOA 197. Weaver and Berzins [110] Pilkington et al [111]
Ottauquechee River, Vermont	1982	CRREL	Four 0.56m x 1.22m panels covering 2.5m vertical height of V shaped pier	Sodhi et al [112]
SSDC at Uviluk (31m water). Beaufort Sea	1982 - 83	Dome/Canmar	Medof panels on structure and in surrounding ice. Soils instruments	Blanchet (1990)[98]
SSDC at Kogyuk (28m water) Beaufort Sea	1983 - 84	Dome/Canmar	Medof panels on structure. Soils instruments	Blanchet (1990) [98]

Esso Casson Retained Island at Kadluk (14.5m water) Beaufort Sea	1983 - 84	Imperial	Strain gauges and panels on caisson. In situ ice panels in surrounding ice. Soils instruments	Johnson et al.[113]. Hawkins et al [114]
Esso CRI at Amerk 0-99 (26m water) Beaufort Sea	1984 - 85	Imperial	Strain gauges and panels on caisson. In situ ice panels in surrounding ice. Soils instruments	Croasdale [115]. Sayed et al [116]
Kemi 1 lightpier, Finland	1984-1987. Data analyzed 1994 onwards	JIP managed by Helsinki University of Technology	Conical section supported on load cells.	Maattanen et al, [117]
CIDS off Alaska	1984	Exxon: Global Marine	In situ ice pressure panels. Strain gauges. Soils instruments	Wetmore (1984) [118]
Molikpaq (water depth range 11m to 32m)	1984 onwards	Gulf Canada	Ice pressure panels. Strain gauges. Extensometers. Soils instruments	Jefferies et al, [119] Hewitt [120]
Kulluk (a floating round moored drillship) Beaufort Sea	1983 onwards	Gulf Canada	Instrumented mooring lines	Pilkington & Haverson [121] Wright [122]
SSDC at Phoenix, Aurora, Fireweed and Cabot. (16 to 21m water) US Beaufort Sea	1986 - 1992	Canmar	Medof panels on structure and in ice. Soils instruments.	Blanchet & Kennedy [123]
JZ 20 Platform Leg – Bohai Bay, China	1989/90	HSVA	Load panels on 2.3m dia. Leg . Ice to 0.55	Yue et al [91]
Iceberg Impacts at Grappling Island: Labrador	1995 - 1996	C-CORE: KRCA	Icebergs towed into an instrumented structure fixed to a cliff face. Structure was instrumented with 36 triangular panels.	Crocker et al, [124] Ralph et al. [125]
Piers of the Confederation Bridge. Canada. Water depth to 30m	1997 on	Public Works Canada. Strait Crossing. Industry JIP	Inclinometers. Load panels. Upward looking sonar. Video cameras.	Brown, [126]. Tibbo et al, [127]
Nordstromsgrund lightpier, Sweden	1997 - 2003	EU funded project – LOLEIF followed by STRICE	9 ice force panels (1.65m x 1.25m) Upward looking sonar	Schwarz and Jochmann [128]
Molikpaq on Piltun Astokhskoye field off Sakhalin Island Russia in 31m water	1998 on	Sakhalin Energy Investment Co.	Flat jack panels on structure (18). Over 80 strain gauges. Soils instruments	Weiss et al [129] Vershinen et al (2006) [130]

Instrumenting the surrounding ice

As already mentioned, the first offshore wells in the Beaufort Sea were drilled from artificial islands in shallow water. At these locations, the ice soon became landfast, but continued to grow to about 2 m thick. It was also known from prior ice measurement programs, that the landfast ice could move several metres under wind stress and thermal expansion, sometimes a ice major motion could occur due to pack ice pressures from the offshore pack [108].

Although most islands were capable of taking significant ice loads, there were concerns about some that had low freeboards or were on weak foundations or used weak fill. Therefore it was imperative for operating safety, that a method was devised to measure ice loads against them. It was also recognized that there was a need to understand ice loading in order to optimize structures for the deeper water locations planned later.

Measuring ice loads directly at the edges of the islands was not easily accomplished as they often had sandbags or other slope protection. The interior of the islands were instrumented with slope indicators. However, to measure ice loads it was recognized that devices to measure ice pressures inserted in the stable ice around the islands was perhaps the only viable approach.

The first use of such devices was around the Imperial Oil Adgo P-25 island in 2m of water in the Mackenzie Delta, Canada. Two devices were used, the University of Alaska's small cylindrical sensor [104] and a thin panel sensor developed by Imperial Oil, [131]. By this time it was recognized that sensors inserted in ice might give erroneous readings of ice pressure unless they had a particular geometry and stiffness. The theory of inclusions suggested that wide thin panels with stiffness beyond a certain threshold would lead to an inclusion factor of 1.0. Also the inclusion factor would remain close to 1.0 for a range of ice stiffness. This is important because the effective "modulus" of ice changes with time due to ice creep. With an inclusion factor of 1.0, the pressure experienced by the panel is that prevailing in the ice.

Even knowing that the sensor is measuring the internal stress in the ice, the prediction of the load on the structure requires the additional step of converting stresses in the ice sheet to the global load. Wang and Ralston [132], examined the radial stress distribution in an ice sheet pressing against a circular cylinder. The results showed that a simple elastic analysis gave a good approximation of radial stress decay. It should be noted that variations in ice sheet thickness and the presence of cracks may seriously affect the prediction of global loads on a structure from in-situ sensor readings.

In 1984, a series of tests were performed in a large outdoor ice basin [133] in order to develop confidence in the use of in-situ sensors, and to confirm the theories of inclusions for them. In these tests, a variety of insitu sensors and strain meters were installed in an ice sheet that could be loaded with a known force against a structure. The ice sheet was 30 × 55m in size with an ice thickness of about 0.7 m. Twenty two separate tests were conducted at a variety of load levels. Pressures up to about 800 kPa were applied to the structure. Overall, the average percentage difference between between the actual load on the structure and that predicted from the sensor readings was less than 30%.

Measuring ice deceleration

An elegant method to determine ice forces caused by an isolated floe impacting a structure is to deduce ice forces from the ice floe motions during the impact. Such an approach was pioneered at Hans Island in tests conducted 1980, and repeated in 1981 and 1983. At Hans Island, the decelerations of multi-year floes were measured, which when combined with the estimated mass of the floes, yielded the force between ice and island. (See Metge & Danielewicz, [134]; Danielewicz & Blanchet, [135]).

Foundation response

It has been normal in many installations to instrument the foundations of platforms, or in the case of artificial islands the interior fill. The aim being to monitor both short term deformations due to ice, waves (or earthquakes) and longer term settlements due to gravity. Typical instruments have included inclinometers, pressure cells, and piezometers. The disadvantages of using soils instruments for ice loads are:

1) For all but the highest ice loads, soil responses are small and the resolution of ice load measurements will be quite coarse.
2) The soil responses often depend on the soil properties. These may not be well known and could vary with time.
3) A given soil response may be due to a range of ice loads applied at different levels and locations. Therefore, back calculating an ice load is subject to error.

4) A load peak applied for a short time will likely not be transmitted into the foundation due to the inertia and damping in the system.

Despite these problems, soil responses have been used to back calculate ice loads (see, for example, Hewitt [136]). If the soils instruments are present, it may be a useful reality check to assess whether extreme ice loads predicted by another system have likely occurred. As well, if a prime concern in an area is the ability of the foundations to resist ice loads, then monitoring the foundation soils makes perfect sense even if the accuracy of predicting peak ice loads is uncertain.

Structure response using strain gauges

Strain gauging technology is well proven. Furthermore, modern data acquisition and computing systems allow a large number of channels to be monitored and the data stored. By definition, the strain gauges measure strain response to a load (or other effects such as temperature change). To back-calculate an applied ice load from strain can be subject to uncertainty for the following reasons:

1) Structures usually have multiple load paths (i.e. they are described as redundant – not "statically determinate"). It is important to appreciate that only non-redundant or "statically determinate" structures have load paths that are independent of the stiffness of the elements and deflections of the supports. In a redundant structure, the load paths calculated by analysis can be subject to error due to incorrect stiffness allocations (especially in the foundation). Also, in a redundant structure, a strain gauged element can respond with the same strain for a variety of different loads applied at different locations. Sophisticated finite element analyses are often performed to achieve calibrations of strain gauged elements. However, they are still subject to the previously described difficulties as well as idealization of the structural elements.
2) The load application points of ice loads are not always certain. Strain gauge calibrations (whether by analysis or by physical testing), have to input a load application point. This is particularly

difficult on a sloping structure and/or when a first-year ridge acts on a structure.

3) Cyclic loads can lead to dynamic excitation of the structure. In which case, measured strains are not a true indicator of the applied loads.

Despite the above difficulties, strain gauged elements may be the only practical approach. They can be very suitable for simple structures such as a monopod –where only one load path exists into the foundation. They can also be useful to measure local plate loads. For steel structures, they have the advantage that they can usually be applied internally after structure installation. This is not likely the case with a concrete structure. Strain gauges can be subject to drift and temperature effects and failure in a wet environment, but well proven techniques are now available to overcome these difficulties. Strain gauges are cheap but proper installation is fairly labour intensive.

Structure response using extensometers

Extensometers also measure strain or deflection. They are therefore subject to similar drawbacks and difficulties as strain gauge methods. They can be a useful stand-alone check on loads especially if they can be used to measure structure deflection relative to a deep pile (e.g. the well casing) and can be physically calibrated. However, applying a sizable load to achieve calibration is not easy.

Structure response using accelerometers

If the structure can be idealized as a mass spring system with known damping and stiffness, then theoretically a measurement of acceleration can be integrated twice to get the forcing function. In general, this is neither simple nor reliable because structures usually do not behave as simple mass spring systems and allocating damping and stiffness is subject to errors. Accelerometers can be a useful ancillary measurement in assessing structural vibrations which might lead to confusion in load interpretation from other devices (and to assess whether structural vibration is affecting the ice loads).

Structure/foundation response using tiltmeters

Tiltmeters use a pendulum system or some other method that references the earth's gravitational field. Tiltmeters can be placed at different heights in the structure to not only measure global tilt, but to also give angular deflection of one point relative to another. They have been used with some success on the piers of the Confederation Bridge between mainland Canada and Prince Edward Island.

Tiltmeters are relatively simple to install and can be conventionally linked to sophisticated data acquisition and storage systems. They can be installed after commissioning of the structure. They work best on tall slender structures such as bridge piers and mono-leg type structures. Their accuracy is usually dependent on a large-scale calibration test (as was achieved for the Confederation Bridge).

Their disadvantages can bea lack of adequate resolution for stiff, short structures, the dynamic response of the tiltmeter itself to vibrations and cyclic loads, and tilt caused by other loads such as winds. Tiltmeters can only give global loads. The point of application of the load should be known.

Ice load cells and panels at the interface between ice and structure

Ice load cells and panels, placed at the ice line overcome many of the problems ascribed to the devices already discussed. A major advantage is that if properly constructed and installed, they are capable of measuring the actual load imposed by the ice at that point on the structure. They can measure loads over small areas to give local loads. As well, adjacent panels can be integrated to give global loads over whatever width and area they are deployed.

Ice load panels are much favoured, and the data from them have heavily influenced current ice design criteria. Table 4.5. summarises the various types of ice load panels with their attributes.

Table 4.5 Summary of ice load panel types

Load Panel Type	Example	Advantages	Disadvantages
Plate on 3 load cells	C-CORE iceberg impact panel element. 2m x 1m triangle; about 0.25m thick	Unambiguous interpretation of global load regardless of deformation of supporting structure. Load cells are proven technology High frequency response	Relatively thick and heavy. Needs to be in a waterproof "box". Triangular shape not always convenient for mounting. Generally only able to give normal load; shear loads may give errors.
Plate on four load cells	LOLEIF – STRICE panel	Good interpretation of global load on panel. Load cells are proven technology High frequency response	Heavy and potentially thick. Needs to be in a waterproof "box". May give erroneous signals if supporting structure deforms (especially from adjacent loads). Relatively expensive. Generally only able to give normal load; shear loads may give errors.
Panel made up of strain gauged plates and beams	Fleet Shear Bar Sensor. Proposed C-Core panel. Weir Jones Microcell.	Can be customized for the specific application. OK for small areas Strain gauged elements are fairly reliable. May be designed to measure shear loads.	May be lighter than load cell sensors. Redundant load paths may lead to errors of interpretation. Difficult to "thin down"
Thin panel with distributed support elements. Some elements strain gauged	Weir Jones "Ideal Panel" Fleet "Hexpack"	Can be thin and relatively light if many internal support elements used. Generally cheaper than load cell panels. Signals from individual elements can be added electronically	Depending on how many elements are not instrumented, significant errors may occur in global load.
Fluid filled panel with internal elements providing strength & stiffness	Medof Panel as used on Molikpaq in Beaufort.	Can be thin and relatively light One signal output (change in fluid volume - height in vertical tube).	Internal elements, if elastomer, can exhibit non linearity and hysteresis. Temperature sensitive. Some concerns about loss of stiffness with repeated loads. Depending on fluid pathways, may have limited frequency response.
"Flat Jack" supported plates	Sandwell Panel used at Tarsiut and on Molikpaq off Sakhalin	Robust and proven. Depending on front plate thickness, can be thinner than load cell panels. Can use several flatjacks to help avoid "bottoming" Lower cost than load cell panels	Thicker and heavier than distributed support panels. Not clear if support structure deformations can lead to errors. Required edge support leads to some errors,
Optical Panel	IMD Panel as used on Terry Fox. 5ft x 8ft	Can measure both global and local loads. Minimal wiring.	Not yet well proven. High cost. Durability unknown
Pressure Sensitive Films	Tekscan film used on indentation tests	Relatively cheap.	Low durability. Not clear that absolute pressures can be measured.

References

1 Kvenvolden, K.A. (1998).A primer on the geological occurrence of gas hydrate. Gas hydrates: relevance to world margin stability and climate change. Geological Society of London Special Publication 137, pp. 9-30

2 Soga, K., Lee, S.L., Ng, M.Y.A. and Klar, A. (2007) Characteristics and engineering properties of methane hydrate soils. *Characteristics and engineering properties of natural soils.* Taylor & Francis, London, 4, pp.2591-2642.

3 Sloan, E.D. and Koh, C.A. (2008) *Clathrate hydrates of natural gases.* CRC Press, Boca Raton, FL.

4 Gravesen, H. and Kärna, T. (2009) Ice loads for offshore wind turbines in the southern Baltic Sea. *Proceedings, Twentieth Conference on Port and Ocean Engineering under Arctic Conditions*, Luleå, POAC09-3

5 International Standards Organisation. (2010) *Petroleum and natural gas industries — Arctic offshore structures.* ISO 19906:2010.

6 Menglan, D., Huachan, F. and Ruheng, F. (1994). Investigation of the pushover of the Bohai2 platform by ice. *Oil Field Equipment*, 23, pp. 1-4.

7 Blenkarn, K.A. (1970). Measurement and analysis of ice forces on Cook Inlet structures. *Proceedings, Offshore Technology Conference*, 2, pp. 365-378, paper OTC161.

8 Engelbrektson, A. and Janson, J.E. (1985) Field observations of the ice action on concrete structures in the Baltic Sea. *Concrete International*, 7, pp. 48-52.

9 Marshall, P., Palmer, A.C., Liew, J.Y.R., Wang, T. and Ma, K.W.T. (2010) Bond enhancement in sandwich shell ice wall. *Proceedings, IceTech conference*, Anchorage, AK.

10 Baudais, D.J., Watts, J.S. and Masterson, D.M. (1976) A system for offshore drilling in the Canadian Arctic Islands, *Proceedings, Eighth Annual Offshore Technology Conference*, Houston, 3, pp. 31-44, paper OTC2622.

11 Palmer, A.C., Baudais, D.J. and Masterson, D.M. (1979) Design and installation of an offshore flowline for the Canadian Arctic Islands, *Proceedings, Eleventh Annual Offshore Technology Conference*, Houston, 2, 765-772.

12 Palmer, A.C. and Loth, W.D. (1987) A hybrid drilling system for deep water in the Arctic. *Journal of the Society for Underwater Technology*, London, 13 (2), pp. 3-5.

13 Sarpkaya, T. and Isaacson, M. (1981) *Mechanics of wave forces on offshore structures.* Van Nostrand Reinhold, New York, NY.

14 Faltinsen, O.M. (1990) *Sea loads on ships and offshore structures.* Cambridge University Press, Cambridge.

15 Palmer, A.C. (2007) Pipeline codes or pipeline cookbooks? *Journal of Pipeline Engineering*, 6, pp. 69-74.

16 Palmer, A.C. (2011) Moving on from ISO 19906. *Proceedings, Twenty-Second conference on Port and Ocean Engineering under Arctic Conditions*, Montréal POAC 11-60.

17 Palmer, A.C., Goodman, D.J., Ashby, M.F., Evans, A.G., Hutchinson, J.W. and Ponter, A.R.S. (1983) Fracture and its role in determining ice forces on offshore structures, *Annals of Glaciology*, 4, pp. 216-221.

18 Palmer, A.C. (1967) Creep velocity bounds and glacier flow problems, *Journal of Glaciology*, 6, pp. 479-488.

19 Ponter, A.R.S, Palmer, A.C., Goodman, D.J., Ashby, M.F., Evans, A.G. and Hutchinson, J.W. (1983) The force exerted by a moving ice sheet on an offshore structure: I: the creep mode, *Cold Regions Science and Technology*, 8, pp. 109-118.

20 Sanderson, T.J.O. (1988) *Ice mechanics: risks to offshore structures.* Graham & Trotman, London.

21 Kerr, A.D. (1980) On the buckling force of floating ice plates. *Proceedings, IUTAM (International Union of Theoretical and Applied Mechanics) Symposium on physics and mechanics of ice,* Copenhagen, pp. 163-178.

22 Croasdale, K. R. (1970) The nutcracker ice strength tester and its operation in the Beaufort Sea. *First International Symposium on Ice*, IAHR, Reykjavik Iceland.

23 Croasdale, K.R., Morgenstern, N. R. and Nuttall, J. B. Indentation tests to investigate ice pressures on vertical piers. Journal of Glaciology, l 19 (81) (1977).

24 Taylor T. P. (1981) An experimental investigation of the crushing strength of ice. *Proceedings, Sixth International Conference on Port and Ocean Engineering under Arctic Conditions*, Quebec City, Canada.

25 Kry, P. R., Lucente, R. F. and Hedley, R. E. (1978) Continuous crushing of ice. Esso Resources Canada Ltd. Report IPRT-28ME-78, APOA Project 106.

26 Kry, P.R., (1980) Ice forces on wide structures. *Canadian Geotechnical Journal*, Vol. 17, No, 1, pp. 97-113.

27 Danielewicz, B.W. and Metge, M. (1981) *Ice forces on Hans Island*, August 1980. APOA project no. 180.

28 Danielewicz, B.W. and Metge, M. (1982) *Ice forces on Hans Island*, 1981. APOA project no. 181.

29 Danielewicz, B.W. and Cornett, S. (1984) *Ice forces on Hans Island*, 1983. APOA project no. 202

30 Dempsey, J.P, Palmer, A.C. and Sodhi, D.S. (2001) High pressure zone formation during compressive ice failure. *Engineering Fracture Mechanics*, 68, pp. 1961-1974.

31 Sodhi, D.S. Crushing failure during ice-structure interaction. *Engineering Fracture Mechanics*, 68, pp. 1889-1921.

32 Sodhi, D.S.. (1998) Non-simultaneous crushing during edge indentation of freshwater ice sheets. *Cold Regions Science and Technology*, 27, pp. 179-195.

33 Jordaan, I.J. (2001) Mechanics of ice-structure interaction. *Engineering Fracture Mechanics*, 68, pp. 1923-1960.

34 Sodhi, D.S. Downloadable video of forces on high-pressure zones.

35 Vaschy, E. (1892) Sur les lois de similitude en physique. *Annales télégraphiques*, 19, pp. 25-28.

36 Palmer, A.C. (2008) *Dimensional analysis and intelligent experimentation.* World Scientific, Singapore (2008).

37 Carpinteri, A. (1982) Notch sensitivity in fracture testing of aggregative materials. *Engineering Fracture Mechanics.* 16, pp. 467-481.

38 Atkins, A.J. (2009) *The science and engineering of cutting.* Butterworth-Heinemann, London.

39 Ashby, M.F., Palmer, A.C., Thouless, M., Howard, M., Hallam, S.D., Murrell, S.A.F., Jones N., Sanderson T.J.O and Ponter, A.R.S. (1986) Nonsimultaneous failure and ice loads on Arctic structures, *Proceedings, Eighteenth Annual Offshore Technology Conference*, Houston, OTC 5127, 1, pp. 399-404.

40 Palmer, A.C. and Dempsey, J.P. (2002) Models of large-scale crushing and spalling related to high-pressure zones. *Proceedings, IAHR Conference*, Dunedin, NZ.

41 Palmer, A.C., Dempsey, J.P. and Masterson, D.M. (2009) A revised pressure-area curve and a fracture mechanics explanation. *Cold Regions Science and Technology*, 56, pp. 73-76.

42 Brown, T.G., Tibbo, S., Shrestha, N,. Tripathi, D, Obert, K., Bruce, J.R., Maes, M.A. (2009) Analysis of ice interactions and ice loads on Confederation Bridge. University of Calgary report to Strait Crossing, 2010.

43 Xu, N., Yue, Q., Shi, Z, Guo, F. and Qu, Y. (2007) Ice load and structure vibration when ice acts on the up-down cone. *Proceedings, Nineteenth International Conference on Port and Ocean Engineering under Arctic Conditions*, Dalian, 1, pp. 328-336.

44 Croasdale, K., Jordaan, I. and Verlaan, P. (2011) Offshore platforms and deterministic ice actions: Kashagan phase 2 development: North Caspian Sea. *Proceedings, Twenty-Second International Conference on Port and Ocean Engineering under Arctic Conditions*, Montréal, POAC 11-117.

45 Croasdale, K.R., and Cammaert, A.B. (1993) An improved method for the calculation of ice loads on sloping structures in first year ice. *Proceedings of the First International Conference on Development of the Russian Arctic Offshore.* St Petersburg, Russia. pp. 161-168.

46 Croasdale, K.R., Cammaert, A.B. and Metge, M. (1994) A method for the calculation of sheet ice loads on sloping structures. Proceedings of the Twelfth International Symposium on Ice of IAHR. Trondheim, Norway.

47 Määttänen, M and Hoikkanen, J., (1990) The effects of ice pile up on the ice force of a conical structure. *Proceedings, Tenth International Symposium on Ice (IAHR)*, Espoo Finland, pp.1010-1021.

48 Mayne, D.C. and Brown, T.G. (2000) Rubble pile observations. *Proceedings, Tenth International Offshore and Polar Engineering Conference (ISOPE)*, Seattle, USA, pp. 596-599.

49 Mayne, D.C.. (2007) *Level ice and rubble actions on offshore conical and sloping structures.* PhD Thesis, Schulich School of Engineering, University of Calgary, Calgary, Alberta.

50 Bonnemaire, B., Lundamo,T., Serre, N. and Frederiksen. A. (2011) Numerical simulations of moored structures in ice: Influence of varying ice parameters. *Proceedings, Twenty-First International Conference on Port and Ocean Engineering under Arctic Conditions,* Montréal, POAC11-176.

51 Ralston, T.D. (1980) Plastic limit analysis of sheet ice loads on conical structures. *Proceedings, IUTAM (International Union of Theoretical and Applied Mechanics) Symposium on physics and mechanics of ice,* Copenhagen, pp. 289-308.

52 Nevel, D.E.. (1992) Ice forces on cones from floes. *Proceedings, Eleventh IAHR International Symposium on Ice*, Banff, Canada, 3, pp. 1391-1404.

53 Croasdale, K. R. (2011) Platform shape and ice interaction: a review. *Proceedings, Twenty-First International Conference on Port and Ocean Engineering under Arctic Conditions*, Montréal, POAC11-029.

54 Cammaert, A.B. and Muggeridge, D.B. (1988) *Ice interaction with offshore structures*, Van Nostrand Reinhold, ISBN-0-442-21652-1, (1988).

55 Croasdale, K. R. and Metge, M. (1991). Structure geometry and ice interaction. *Proceeding,IUTAM-IAHR Symposium on Ice-Structure Interaction.* St John's , Canada.

56 Määttänen, M. (1975) Experiences of ice forces on a steel lighthouse mounted on the seabed, and proposed constructional refinements. *Proceedings, Ninth International Conference on Port and Ocean Engineering under Arctic Conditions,* Fairbanks, AK.

57 Määttänen, M. (1977) Ice force measurements at the Gulf of Bothnia by the instrumented Kemi-1 lighthouse. *Proceedings, Fourth International Conference on Port and Ocean Engineering under Arctic Conditions*, St.John's, 2, pp. 684-694.

58 Määttänen, M. (1978) On conditions for the rise of self-excited ice-induced vibrations. *Styrelsen för vintersjöfartsforskning* (Winter Navigation Research Board), research report 25.

59 Yue, Q. and Bi, X. (2000) Ice induced jacket structure vibration. *Journal of Cold Regions Engineering,* 14, pp. 81-92 (2000).

60 Yue, Q. and Bi, X. (1998) Full-scale test and analysis of dynamic interaction between ice sheet and conical structure. *Proceedings, Fourteenth International Association for Hydraulic Research Symposium on Ice.*

61 Tibbo, J. S., (2010) *Flexural Failure of sea ice - analysis of interactions at the Confederation Bridge and assessment of selected prediction techniques.* PhD thesis, University of Calgary, AB.

62 Johnston, M., Masterson, D., and Wright, B., (2009) Multi-year ice thickness: knowns and unknowns. *Proceedings, Twentieth International Conference on Port and Ocean Engineering under Arctic Conditions,* Luleå, Sweden. POAC09-120.

63 Metge, M. and Tucker, J.R. (1990) *Multifaceted cone tests: Year two: 1989 -1990.* Esso Resources Report ERCL.RS.90.11. Part of Joint Industry/Government Project.

64 Frederking, R., Hewitt, K., Jordaan, I., Sudom, D., Bruce, J., Fuglem, M., Taylor, R. Overview of the Molikpaq multi-year ice load re-analysis 2007 JIP. *Proceedings, Twenty-second International Conference on Port and Ocean Engineering under Arctic Conditions*, Montréal, *Canada* POAC 11-165

65 Matskevich, D. G. (2002) Velocity effects on conical structure ice loads. *Proceedings, Twenty-first International Conference on Offshore Mechanics and Arctic Engineering (OMAE)*, Oslo.

66 Jordaan, I.J., Maes, M.A., Brown, P., and Hermans, I. (1983) Probabilistic analysis of local ice pressures, *Journal of Offshore Mechanics & Arctic Engineering*, 115, pp. 83-89.

67 Croasdale, K. R, Allyn, N. Roggensack, W. (1988). Arctic Slope Protection: Considerations for Ice. In ASCE Monograph - Arctic Coastal Processes and Slope Protection Design. Editors Chen, A.T. and Leidersdorf, C. B. ASCE, New York, 1988.

68 Croasdale, K. R., Metge, M. and Verity, P.H. (1978). Factors governing ice ride up on sloping beaches. Proceedings, IAHR Ice Symposium, Lulea, Sweden, (1978).

69 McKenna, R., Stuckey, P., Fuglem, M., Crocker, G,. McGonigal, D., Croasdale, K., Verlaan P., and Abuova A. (2011) Ice encroachment in the North Caspian Sea. *Proceedings, Twenty-first International Conference on Port and Ocean Engineering under Arctic Conditions*, Montréal, Canada, POAC11-002.

70 Svendsen, I.A. (1984) Physical modelling of water waves. in *Physical Modelling in Coastal Engineering*, pp. 13-47, AA Balkema, Rotterdam.

71 Palmer, A.C. (2008) *Dimensional analysis and intelligent experimentation.* World Scientific, Singapore.

72 Schwarz, J. (1977) New developments in modeling ice problems. *Proceedings, Fourth International Conference on Port and Ocean Engineering under Arctic Conditions*, St. Johns, pp. 45-61.

73 Palmer, A.C. and Dempsey, J. (2009) Model tests in ice. *Proceedings, Twentieth International Conference on Port and Ocean Engineering under Arctic Conditions*, Luleå, paper POAC09-40.

74 Timco, G.W. (1984) *Ice forces on structures: physical modelling techniques.* Second IAHR State-of-the-art report on ice forces on structures..

75 Tatinclaux, J.P. (1983) *Ship model testing in level ice: an overview.* US Army Cold Regions Research and Engineering Laboratory, CRREL Report 88-15.

76 Hirayama, K. (1993) *Properties of urea-doped ice in the CRREL test basin.* US Army Cold Regions Research and Engineering Laboratory, CRREL Report 83-8 .

77 Zufelt, J.E. and Ettema, R. (1996) *Model ice properties.* US Army Cold Regions Research and Engineering Laboratory, CRREL Report 96-1.

78 Lau, M, Wang, J. and Lee, C. (2007) Review of ice modeling methodology, *Proceedings, Nineteenth International Conference on Port and Ocean Engineering under Arctic Conditions, Dalian*, 1, pp. 350-361.

79 Atkins, A.G. (1974) Icebreaking modeling. *Journal of Ship Research*, 18, pp. 40-43.

80 Atkins, A.G. and Caddell, R.M. (1974) The laws of similitude and crack propagation. *International Journal of Mechanical Sciences*, 16, pp. 541-548.

81 Dempsey, J.P. (1991) The fracture toughness of ice. in: *Ice-Structure Interaction, IUTAM-IAHR Proceedings* (Ed. S.J. Jones, R.F. McKenna, J. Tillotson and I.J. Jordaan), Springer-Verlag, Berlin, pp. 109-145.

82 Dempsey, J.P., Adamson, R.M. and Mulmule, S.V. (1999) Scale effects on the insitu tensile strength and fracture of ice. Part II: First-year sea ice at Resolute, NWT. *International Journal of Fracture*, 95, pp. 347-366.

83 Mulmule, S.V. and Dempsey, J.P. (1999) Scale effects on sea ice fracture. *Mechanics of cohesive-frictional materials*. 4, pp. 505-524.

84 Uenishi, K. and Rice, J.R. (2003) Universal nucleation length for slip-weakening rupture instability under nonuniform fault loading. *Journal of Geophysical Research,* 108 (B1), 2042, doi:10.1029/2001JB001681.

85 Määtänen, M. Stability of self-excited structural vibrations. (1977) *Proceedings, Fourth International Conference on Port and Ocean Engineering under Arctic Conditions Conference,* St. John's, 2, pp. 684-694.

86 Montgomery, C.J., Gerard, R. and Lipsett, A.W. (1980) Dynamic response of bridge piers to ice forces. *Canadian Journal of Civil Engineering*, 7, pp. 345-356.

87 Jefferies, M.G. and Wright, W.H. (1988) Dynamic response of Molikpaq to ice-structure interaction. *Proceedings, Seventh International Conference on Offshore Mechanics and Arctic Engineering Symposium*, Houston, pp. 201-220.

88 Bjerkås, M. and Skiple, A. (2005) Occurrence of continuous and intermediate crushing during ice-structure interaction. *Proceedings, Eighteenth International Conference on Port and Ocean Engineering under Arctic Conditions*, Potsdam, NY.

89 Bjerkås, M. (2006) *Ice actions on offshore structures*. PhD thesis. Norwegian University of Science and Technology, Trondheim.

90 Sodhi, D.S. (1989) Interaction forces during vertical penetration of floating ice sheets with cylindrical indentors. *Proceedings, Eighth International Conference of Offshore Mechanics and Arctic Engineering*, pp. 377-382.

91 Yue, Q., Zhang, X., Bi, X. and Shi, Z. (2001) Measurements and analysis of ice induced steady vibration. *Proceedings, Sixteenth International Conference on Port and Ocean Engineering under Arctic Conditions*, Ottawa.

92 Yue, Q., Guo, F. and Kärnä, T. (2009) Dynamic ice force of slender vertical structures due to ice crushing. *Cold Regions Science and Technology* MORE

93 Guo, F., and Yue, Q. (2009) Model test of ice-structure interaction. *Proceedings, ASME Twenty-eighth International Conference on Ocean, Offshore and Arctic Engineering*, Hawaii, OMAE 2009-79780 (2009).

94 Palmer, A.C., Yue, Q. and Guo, F. (2010) Ice-induced vibrations and scaling. *Cold Regions Science and Technology* 60, pp. 189-192.

95 Yap, K.T. *Level ice-vertical structure interaction: steady-state self-excited vibration of structures.* Unpublished PhD dissertation, National University of Singapore, Singapore (2011).

96 Engelbrektson, A. (1977) Dynamic ice loads on a lighthouse structure. *Proceedings, Fourth International Conference on Port and Ocean Engineering under Arctic Conditions,* St John's.

97 Croasdale, K. R. and Frederking, R. (1987). *Field Techniques for Ice Force Measurements. Working Group on Ice Forces,* 3rd IAHR State of the Art Report. CRREL Special Report 87-17.

98 Blanchet, D. (1990) Design criteria for wide Arctic offshore structures. Canadian *Geotechnical Journal* 27. (6) pp. 701-725.

99 Peyton, H.R. (1968) *Sea ice forces.* National Research Council of Canada, Tech. Memo. 92. Ottawa.

100 Neill, C. R. (1970) *Ice pressure on bridge piers in Alberta, Canada. IAHR Ice Symposium,* Reykjavik, Iceland.

101 Lipsett, A. W., Gerard, R. (1980) *Field measurements of ice forces on bridge piers 1973 – 1979.* Alberta Research Council Report SWE 80-3..

102 Schwarz, J. (1970) The pressure of floating ice-fields on piles. *IAHR Ice Symposium,* Reykjavik, Iceland.

103 Metge, M. (1976) *Ice conditions and ice defence at Netserk B-44 and Adgo P-25 during the winter of 1974-75.* APOA Project 104. (available through the Arctic Institute of N. America, University of Calgary).

104 Nelson, R. D. and Sackinger W. M. (1975) *Ice stress measurements at Adgo and Netserk Islands, 1974 - 75.* APOA project No. 104 -3 (available through Arctic Inst. of North America, University of Calgary) (1975).

105 Danys, J. V.. (1975) Offshore installations to measure ice forces on the lightpier in Lac St. Pierre. *Proceedings, Ninth Intl. conf. on lighthouses and other aids to navigation,* Ottawa.

106 Frederking R., Sayed, M., Hodgson, T. and Berthelet, W. (1985) Ice force results from the modified Yamachiche Bend Lightpier, winter 1983 -84. Proceedings, Canadian. Coastal Conf. 1985, St. John's NF).

107 Oshima, M., Narita, H., Yashima, N.and Tabuchi, H. (1980) Model and field experiments for development of ice resistant offshore structures. *Twelfth Annual Offshore Technology Conference,* Houston, paper OTC3885.

108 Strilchuk, A. R. (1977)*Ice Pressure Measurements Netserk F-40, 1975-76.* Imperial Oil Ltd., Report No. IPRT-14ME –77. APOA Project 105.

109 Semeniuk, A. (1977) *Ice pressure measurements at Arnak L-30 and Kannerk G-42.* APOA Project No. 122-1.

110 Weaver, J. S., Berzins, W. (1983) The Tarsiut island monitoring program. *Fifteenth Annual Offshore Technology Conference*, Houston, paper OTC 4519.

111 Pilkington, G. R., Blanchet, D. and Metge, M. Full-scale measurements of ice forces on an artificial island. (1983) *Proceedings, Seventh International Conference on Port and Ocean Engineering under Arctic Conditions*, Helsinki.

112 Sohdi, D. S., Kato, K., and Haynes, F. D. (1983) *Ice force measurements on a bridge pier in the Otauquechee River*, Vermont. Cold Regions Research and Engineering Laboratory, Hanover, NH, Report 83-20.

113 Johnson, J. B., Cox, G. F. N., Tucker 111, W. B. Kadluk ice stress measurement program. (1985) *Proceedings, Eighth International Conference on Port and Ocean Engineering under Arctic Conditions*, Narssarssuaq, Greenland, (1985).

114 Hawkins, J. R., James, D. A., and Der, C. Y. (1983) Design, construction and installation of a system to measure environmental forces on a caisson retained island. *Proceedings, Seventh International Conference on Port and Ocean Engineering under Arctic Conditions*, Helsinki. Vol.3.

115 Croasdale, K. R. (1985) *Ice Investigations at a Beaufort Sea Caisson*, 1985. A report prepared for the National Research Council, Ottawa and the US Dept of the Interior. (1985).

116 Sayed, M., Croasdale, K. R., Frederking, R.. (1986) Ice stress measurements in a rubble field surrounding a caisson retained island. *International conference. on ice technology*, MIT, Cambridge, MA.

117 Määttänen, M., Hoikkanen, A. and Avis, J. (1986) Ice failure and ice loads on a conical structure : Kemi-1 cone full scale ice force measurement data analysis. *IAHR Symposium on Ice*, Beijing, 1.

118 Wetmore, S. B. (1984) Concrete island drilling system: Super series (Super CIDS). *Sixteenth Annual Offshore Technology Conference*, Houston.

119 Jefferies, M., Rogers, B., Hardy, M. and Wright, B. (2011) Ice load measurement on Molikpaq: Methodology and accuracy. *Proceedings, Twenty-Second International Conference on Port and Ocean Engineering under Arctic Conditions*, Montréal, POAC 11-189.

120 Hewitt, K.J. (2011) Ice loads on the Molikpaq at Amauligak I-65 based on geotechnical analyses and responses. *Proceedings, Twenty-Second International Conference on Port and Ocean Engineering under Arctic Conditions*, Montréal, POAC 11-134.

121 Pilkington, R. and Haverson P. (1985) Drilling in ice from the conical drillship - Kulluk. *Arctic Offshore Technology Conference.* Anchorage, AK.

122 Wright, B. (2001) Ice loads on the Kulluk in managed ice conditions. *Proceedings, Sixteenth International Conference on Port and Ocean Engineering under Arctic Conditions*, Ottawa, Vol 2. pp 553 – 566.

123 Blanchet, D. and Kennedy, K. (1996) Global first year ice load measurements in the Arctic. *Proceedings, Offshore Mechanics and Arctic Engineering Conference.*

124 Crocker, G, Croasdale, K, Mckenna, R., Guzzwell, J. and Bruneau, S. (1996) *C-CORE Iceberg impact experiment Phase 2 final report.* C-CORE Publication 96-C16, St John's, NF.

125 Ralph, F., McKenna, R., Crocker, G. and Croasdale, K.. (2004) Pressure/area measurements from the Grappling Island iceberg impact experiment". *Seventeenth IAHR Symposium on Ice,* St. Petersburg, Russia.

126 Brown, T. G. (2001) Four years of ice force observations on the Confederation Bridge. *Proceedings, Sixteenth International Conference on Port and Ocean Engineering under Arctic Conditions*, Ottawa, Canada. Vol 1, pp 285-298.

127 Tibbo, S., Obert K.M., Shrestha, N., Tripathi, D., Mayne, D., and Brown, T. G. (2009) Year twelve of the Confederation Bridge ice monitoring program. *Proceedings, Twentieth International Conference on Port and Ocean Engineering under Arctic Conditions*, Luleå, Sweden. POAC09-49.

128 Schwarz., J. and Jochmann, P. Ice forces affected by temperature and thickness of the ice. (2009) *Proceedings, Twentieth International Conference on Port and Ocean Engineering under Arctic Conditions,* Luleå, Sweden. POAC09-41

129 Weiss, R. T., Wright, B. and Rogers, B. In-ice performance of the Molikpaq off Sakhalin Island. (2001) *Proceedings, Sixteenth International Conference on Port and Ocean Engineering under Arctic Conditions*, Ottawa.

130 Vershinin, S. A., Truskov, P. A., Kuzmichev, K. V. (2006) *Sakhalin Island – offshore platform structures – the impact and influence of ice.* Institute Giprostroymost, Moscow.

131 Metge, M., Strilchuk, A. and Trofimenkoff, P. N. (1975) On recording stresses in ice. *Proceedings, IAHR Ice Symposium,* Hanover, NH.

132 Wang, Y.S. and Ralston, T. D. (1983) Elastic-plastic stress and strain distributions in an ice sheet moving against a circular structure. *Proceedings , Seventh International Conference on Port and Ocean Engineering under Arctic Conditions,* Helsinki.

133 Croasdale K. R., Graham, B. W., Comfort, G., and Marcellus, R. (1986) Evaluation of ice pressure sensors and strain meters in a large test basin. *Proceedings, IAHR Ice Symposium*, Iowa.

134 Metge, M., Danielewicz, B. and Hoare, R. On measuring large scale ice forces; Hans Island, 1980. (1981) *Proceedings, Sixth International Conference on Port and Ocean Engineering under Arctic Conditions*, Quebec City, Canada.

135 Danielewicz, B. W. and Blanchet, D. Multi year ice loads on Hans island during 1980 and 1981. (1987) *Proceedings, Ninth International Conference on Port and Ocean Engineering under Arctic Conditions*, Fairbanks, Alaska.

136 Hewitt, K. J. (1994) Molikpaq ice interactions: predicted and actual performance. *Proceedings, IAHR Ice Symposium*, Trondheim, Norway.

Chapter Five

Broken Ice, Pressure Ridges and Ice Rubble

5.1 Introduction

Flying over a region of mobile ice and looking down, it can be seen that Nature's preferred form of ice failure is the creation of pressure ridges. Furthermore, observations of ice acting against obstacles such as shore lines, stamukhi, and wide structures show that ice prefers to fail out-of-plane, not by crushing. This process is often referred to as rubbling, because the oncoming ice is acting against ice rubble and creating additional ice rubble. The processes of ridge-building and rubbling are similar from an ice mechanics perspective and of high relevance to ice engineering, but they have not been studied to the same level as other ice mechanics topics. This chapter is concerned with different forms of broken ice. Ice can form a rubble field when it builds up on the upstream face of a structure. The process can control the limit force ice loads when thin ice builds ridges on the upstream side of thick features such as multi-year ridges., multi-year hummocks and even ice islands.

5.2 Formation of Ridges

In 1980 one of the authors was involved in designing platforms for the Canadian Beaufort Sea. Ice islands had been observed. The management of the company charged him with developing ice loads and designs to withstand these features. Ice islands are rare, but they can be up to 50 m thick. Limit stress loads are of course extremely high. Rough calculations at the time indicated that typically on a 100 m wide platform using 2 MPa crushing strength, the load would be 10,000 MN. Even an assumption of failure in the 2 m thinner ice on the back of the ice island gave high loads, if crushing was assumed to be the governing mechanism. For a 3 km wide ice island and 1.5 MPa crushing stress in the 2 m ice behind it, the load would still be about 9,000 MN. The author then recalled that the ice circulation modellers were using ridge-building loads that were much lower than the crushing loads, and he advanced the proposition that the ice on the back of the ice island (or close to it) would fail in ridge building rather than in crushing (as nature prefers). The "limit force" scenario was coined for this situation, and it led to deliberations on what ridge-building loads to use. A review at that time gave values reproduced here in Table 5.1.

Table 5.1 Typical ranges of ridge building forces published by circulation modelers

Investigators	Method	Range of values for ridge building force (kN/m)
Parmerter & Coon [1]	Ridge building math model (energy based)	10 to 30
Hibler [2]	Large scale ice motion modelling	10 to 100
Rothrock [3]	As above	40 to 100

Using 100 kN/m in the last example of a 3 km ice island gives a load of only 300 MN; that is still substantial, but it is a lot less than the 9,000 to 10,000 MN obtained from the other methods.

This finding, and the proposal that limit force methods needed to be investigated were incorporated into papers at IAHR [4], and OTC [5]). At that time, the topic of ridging forces had been addressed mostly by the circulation modellers using theoretical approaches. It seemed that some field data were needed, but it was also recognized that measuring in-situ stresses in ice was not easy. However, there had been some progress in developing in-situ ice pressure panels in order to measure ice loads around some of the earliest Beaufort islands situated in landfast ice (Metge and Strilchuk [6]). In 1983 there was a JIP which tested and compared various in-situ stress sensors and strain meters in the large Calgary Esso basin, where known stresses could be induced into a large ice sheet. This work showed that if designed correctly in-situ sensors could reasonably accurately measure the internal stress in an ice sheet [7]. Proposals to measure internal pack ice stresses across a wide front were made to government and industry. It was a slow sell, because there was a high risk of not getting any useful data, (even though this should be the essence of research). However, there were a few with vision enough to see the potential (e.g. Lynn Lewis), and a pilot project was conducted in 1986, supported by the Canadian government, in the Beaufort Sea [8]. This was followed by two more years of measurements in the Canadian Beaufort by Comfort and Ritch. In addition, Coon measured pack ice pressures in the Eastern Arctic in 1989; as did Perovich and Tucker in 1992. Later, the US Navy supported more measurements in 1993/94 in the Arctic Ocean by Coon and others. (See the summary in [9]).

The measured data are sparse but have been subject to extensive review and analysis. In 1992, the data to that time were reviewed (Croasdale et al, [10]). A more detailed background to the treatment of the data is given in Croasdale [11] including the equation currently in ISO 19906 and given below for the ridge building line load.

$$w = Rh^{1.25}D^{-0.54}\ \text{kN/m} \quad 100 < D < 1500\ \text{m} \qquad (5.2.1)$$

where,

- w is the force per unit length (in kN/m), averaged over the breadth D over which the rubble builds (measured in m),
- h is the ice thickness upstream of the upstream edge of the rubble (measured in m),
- R is a value based on fitting a curve to the data

A value of R of 2000 gives a best fit to the data. However based on the discussion in Croasdale et al. [8,10,11] , where measured data are reported and analyzed (and also on other work by Comfort [12]), an uncertainty factor of about 2 was recognized. So R could be from say 1000 to say 4000. Equation (5.2.1) with a range of R values between 1000 and 4000 gives values of ridge-building line loads for 1 m of ice over a 1500 m width of 20 to 80 kN/m. This range is not dissimilar to the ranges of values used by the circulation modellers given in table 5.1. ISO 19906 [13] suggests a conservative upper bound of R equal to 10,000, especially for narrower widths. This is partly to account for uncertainties, but also for the possibility of a frozen-in condition around the blocking floe, and also because for narrow widths the application might be action against grounded rubble in front of a platform rather than floating rubble. As will be discussed later, typical grounded rubble heights observed imply higher forces than those involved in the build-up of floating ridges.

Recent work [9] has further addressed ridging and rubbling forces by making use of the algorithms for ice action on sloping structures as presented in Section 4.5.2. The following material is based on the paper referenced above, but there are some changes and some additional refinements have been made.

The new method for ridging forces is based on considering three distinct bounding ice sheet/rubble interaction cases and then combining them across the width of interaction. The three cases are based on the limiting conditions when a ridge is approaching its maximum height and depth. They are:

(1) *Ice penetration of the existing rubble and failure downwards.* The ice penetrates a short distance then fails in flexure downwards and ramps down either within or along the outside surface of the keel. The general case is shown in Figure 5.1.
(2) *Ice penetration and failure upwards.* The configuration is the same as Figure 5.1 but turned upside down. In this case the ice is ramping up within or on the outside slope of the sail. A photo of ice ramping up a rubble slope is shown in Figure 5.2.
(3) *Ice interacts with ice blocks steeply oriented.* Flexural failure is inhibited, but the rubble fails in the "footing failure mode". This is based on soil mechanics. It will usually require less force than crushing which is not usually seen in rubble-building.

Case 1: Ice fails and ramps downward

As shown in Figure 5.1, the ice penetrates the newly forming ridge. At some point, the ice meets inclined rubble blocks and it fails in downwards flexure. It continues to ramp down internally within the rubble exiting at the bottom of the keel. It could also ramp down the outside of the keel, but this requires less force and is a special case of the general analysis presented.

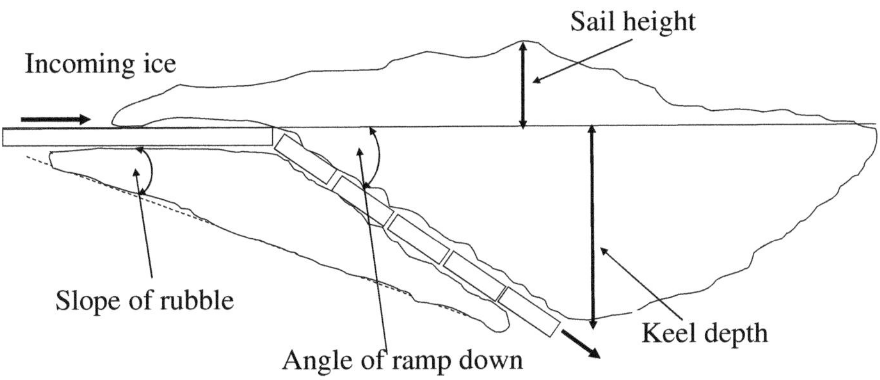

Figure 5.1 Ridge building Case 1: ice ramping down

The ramping ice increases the keel depth, which increases the required force to continue this specific interaction. With increasing force, the oncoming ice either initiates this process elsewhere, starting with a smaller keel or switches to one of the other limit cases above.

Case 1 is analyzed using existing algorithms for ice loads on sloping structures as already presented in chapter 4 but with one significant change. The change is to account for downward motion of the ice so that all gravity forces are now buoyancy forces. The bending term is unaffected, but in calculating the other load components, the unit weight of ice (say 8.9 kN/m^3) is replaced by the buoyant unit weight (typically about 1.18 kN/m^3). In addition in estimating line loads for ridging per linear length, the 3D correction is omitted to give the following expression for the horizontal component of the bending failure load (H_B):

$$H_B = 0.68\varepsilon Y b\left(\frac{\rho_w h^5}{E}\right)^{0.25} \tag{5.2.2}$$

Definitions of the parameters are given in chapter 4, Section 4.5.2. Although the width of interaction b is included, the ridging line load is not sensitive to b (all other parameters being unchanged).

The bending term (H_B) is only one of four components of the load which need to be considered in this case. The other three are (a) H_R the load associated with ride down of the broken ice pieces; (b) H_P the load to push the oncoming ice through the rubble before it starts to ride down; and (c) H_L the load to lift and shear the rubble below the incoming ice before it can break in bending downwards. The equations for these components are given previously in chapter 4 (but as already mentioned, in downward ramping the buoyant unit weight replaces ice unit weight).

In conducting the calculation for a range of ice thicknesses, we will first correlate maximum keel depths with ice thickness using empirical relationships developed from field measurements. Based on the literature (e.g. the review by Timco et al, [14]), the correlation between ice thickness (h) and average maximum sail height (h_s) is taken as

$$h_s = 3.7h^{-0.5} \tag{5.2.3}$$

Table 5.2 Case 2: Ice ride-down in the keel (from Croasdale [9])

Down Ridging		0.5m ice	1.0m ice	1.5m ice	2.0m ice
Initial Data:		h_k correlated	h_k correlated	h_k correlated	h_k correlated
Flexural strength of ice	(kPa)	500	500	500	500
Specific weight of ice	(kN/m^3)	8.90	8.90	8.90	8.90
Specific weight of water	(kN/m^3)	10.08	10.08	10.08	10.08
Buoyant Weight	(kN/m^3	1.18	1.18	1.18	1.18
Young's modulus	(kPa)	5.00E+06	5.00E+06	5.00E+06	5.00E+06
Poisson's ratio		0.3	0.3	0.3	0.3
Slope Angle of interaction	(deg)	45	45	45	45
Rubble angle of repose	(deg)	35	35	35	35
Keel depth	(m)	11.8	16.7	20.4	23.6
Zone width of ride down	(m)	25	25	25	25
Ice-slope friction		0.1	0.1	0.1	0.1
Ice thickness	(m)	0.5	1	1.5	2
Rubble porosity		0.3	0.3	0.3	0.3
Shear strength of rubble	(kPa)	5	5	5	5
Results					
Horizontal Load	(MN)	1.37	2.73	4.22	5.86
Vertical Load	(MN)	1.11	2.20	3.41	4.73
HB	(MN)	0.18	0.43	0.70	0.99
HP	(MN)	0.02	0.04	0.06	0.07
HR	(MN)	0.40	1.05	1.86	2.81
HL	(MN)	0.77	1.22	1.61	1.98
Total	(MN)	1.37	2.73	4.22	5.86
Total line load	**(kN/m)**	**54.77**	**109.27**	**168.87**	**234.32**

The correlation between sail height (h_s) and keel depth (h_k) is taken as

$$h_k = 4.5h_s \qquad (5.2.4)$$

Typical results for Case 1 are shown in table 5.2 for ranges of ice thickness up to 1.5 m, correlated keel depths and other key inputs.

Case 2: Ice fails and rides upwards

Figure 5.2 shows this happening with grounded ice rubble. A floating ridge would have a much lower sail. This case is analyzed using the same equations as for Case 1 but reverting back to the unit weight density of ice. Results for upwards failure and ride-up are given in detail in Croasdale [9]. For 1 m of ice and the sail height given by equation 5.2.3, the line load is 112 kN/m, very similar to the ride down load of 109 kN/m derived for Case 1.

Figure 5.2 Field example of an ice ride-up into grounded rubble with partial penetration [Photo by Ron Ritch and Greg Crocker with permission]

Case 3: Ice rubble failure

This failure may be a transient situation occurring just before ice ramps up or down. In fact this slip plane failure may encourage the oncoming ice to ramp up or down. A very simplistic approach is taken and is described in Croasdale [9]. The method for this case gives a line load of 90 kN/m for 1 m of ice.

The three cases can now be compared and averaged. Using the correlations between thickness and sail heights and keel depths, combined with base-case values for other parameters, the ridging line loads are given in table 5.3 as a function of thickness.

Table 5.3 Ridging and rubbling loads versus ice thickness (Croasdale [9])

Ice thickness	Sail height	Keel depth	Line loads kN/m)			
(m)	(m)	(m)	Footing	Ride up	Ride down	Average
0.50	2.62	11.77	45.00	48.36	54.77	49.38
1.00	3.70	16.65	90.00	112.22	109.27	103.83
1.50	4.53	20.39	135.00	187.67	168.87	163.85
2.00	5.23	23.55	180.00	272.43	234.32	228.91

Because the analyses are 2-dimensional, the line loads for each case will not vary with width, if all other parameters also do not vary. In nature, it is not very likely that sail heights and keel depths will be uniform with width. The correlations for these parameters based on field measurements will tend to upper bounds (because field measurements are often biased to the highest sails and deepest keels). Therefore, it is suggested that in nature there will likely be a reduction of ridge-building forces with width because average sail heights and keel depths will reduce with width. An approximate scheme for addressing this issue follows later in this section.

It is also noted that to use an average line load assumes an equal distribution of interaction cases (footing failure, ride up, ride down). But it can be seen in table 5.3, that the line loads for the various modes are quite similar. Even if one was more heavily weighted in occurrence, the average would not change very much.

A comparison with ISO (equation 5.2.1) for a range of thickness is shown in Figure 5.3. This is done for a width of 500 m for Equation 5.2.1

(which has a width term), whereas as discussed above the values derived in this work for floating ridges are not width dependent. On the other hand, the results of this work show similar trends to ISO with thickness. The results from this work lie closer to the R = 2000 curve than the higher R = 4000. A best fit to the results from this work gives:

$$w = 105.3h^{1.104} \tag{5.2.5}$$

where w is the ridging line load in kN/m and h is the ice thickness in m. Interestingly Comfort et al [12] developed a relationship of $w = 100\ h^{1.25}$ after analysis of field data. The power of 1.104 is less than that derived from previous work of 1.25. The previous work raised h to the power 1.25 to normalize various measured data for different thicknesses because the bending term is proportional to the power 1.25 (see equation 5.2.2). The models used in this work indicate that bending is less important than the frictional, gravity and buoyancy terms (which are approximately proportional to thickness to the power 1.0).

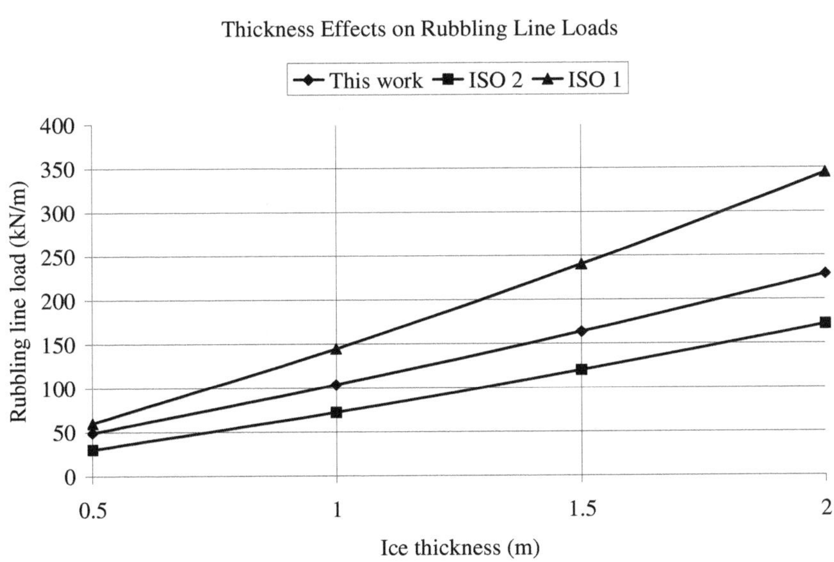

Figure 5.3 Ridging loads vs. ice thickness (floating ridges) (from Croasdale [9])

So far, the application of the three-zone model has been to apply rubbling mechanisms to ridge-building line loads. This approach can be used in limit force calculations. For platforms in shallow water, the scenario with grounded rubble in front of a platform is also of interest.

The main difference from floating ridges is that the rubble is grounded on either the slope of the structure or on the sea floor in front of the structure. This means that the "sail" height can be significantly higher than for a floating ridge. Observations in the Caspian Sea show that a 15 m pile-up height on the slope of a structure is possible (with 0.5 to 1 m ice thickness) (McKenna [15]). Observations and measurements also show that the average maximum pile up height reduces with rubble area (or width). In this scenario, it is the ride up process (Case 2) which will control. To demonstrate how this effect is manifested, table 5.4 shows results for various widths and rubble heights for 1 m ice thickness using the ride-up equations discussed earlier in this section. This is done for widths from 50 to 500 m with average maximum rubble heights diminishing from 15 m at 50 m width to 6 m at 500 m width. At width of 1000 m it is assumed that the rubble is no longer grounded and the calculation is performed assuming the ridge is floating with a keel depth based on equations 5.2.3 and 5.2.4. Now the rubbling line load is estimated using the ride-down method discussed earlier. In recognition

Table 5.4 Rubbling line loads as a function of width and assumptions on rubble height and keel depths

Width (m)	Thickness (m)	Rubble height over width (m)	Ride-up Rubbling line load (kN/m)	Keel Depth over width (m)	Ride-down Rubbling line load (kN/m)	Composite Rubbling line load (kN/m)
50	1.00	15.00	778.93	Grounded	NA	678.82
100	1.00	12.00	550.71	Grounded	NA	480.00
200	1.00	8.00	303.53	Grounded	NA	339.41
500	1.00	6.00	205.72	Grounded	NA	214.66
1,000	1.00	4.50	143.40	16.65	NA	151.79
1,500	1.00	Floating	NA	15.61	124.9	123.94
2000	1.00	Floating	NA	14.57	116.04	107.33
3000	1.00	Floating	NA	12.49	107.54	87.64
4000	1.00	Floating	NA	10.41	91.25	75.89
5000	1.00	Floating	NA	8.33	75.97	67.88
10000	1.00	Floating	NA	8.33	61.85	48.00

that the maximum keel depth for a given thickness has probably been based on spot measurements at selected high sail points, it is assumed that as width increases beyond 1000 m the average maximum keel depth reduces. It is assumed that at 5000 m it is half the maximum and remains at that for wider widths. This approach may seem somewhat contrived, but it is simply a way of creating a possible plausible relationship between ridging/rubbling forces and width of action, which, as discussed in [10] is apparent in the limited measured data on rubbling forces.

A best fit relationship to the calculated points in columns 4 and 6 for 1m thick ice is given by;

$$w = 4800D^{-0.5} \tag{5.2.6}$$

(The values in the column on the right of table 5.3 are based on this relationship)

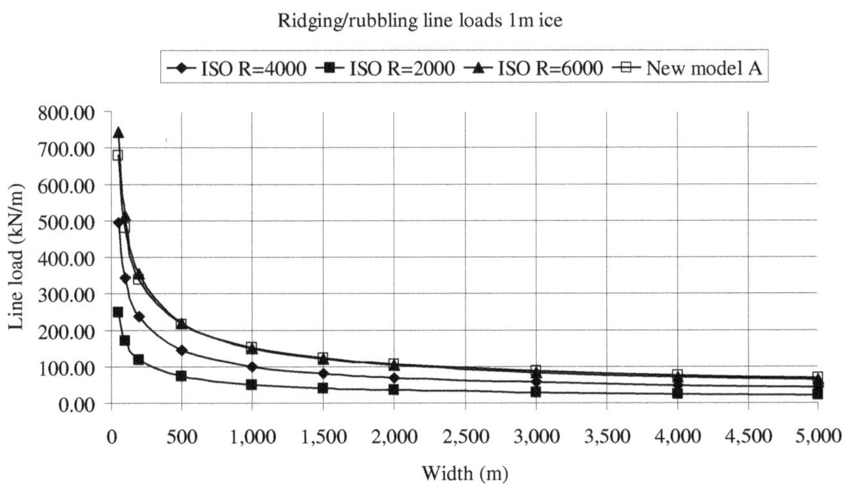

Figure 5.4 Rubbling loads vs. width of interaction for 1 m thick ice (rubble heights for the narrower widths are based on grounded rubble – New Model A)

If we use the power relationship for thickness derived earlier, the revised relationship including thickness for this scenario of rubbling/ridging is given by;

$$w = 4800D^{-0.5}h^{1.1} \tag{5.2.7}$$

This new relationship, referred to as "New Model A" is similar in form to the ISO equation and is compared with ISO for 1m thick ice in Figure 5.4. The best fit is when a value of R=6000 is used in the ISO equation. But remember that this relationship has been developed assuming the rubble is grounded for widths less than 500m width.

It is of interest to develop a relationship assuming the rubble is not grounded over all widths (e.g. for floating ridges). For this we will develop the rubbling load solely on the ride-down case with keel depths varying with width. The relationship for keel depth with width is somewhat arbitrary. For the width of 1000 m, the keel depth is that given by equation 5.2.4 (i.e. keel depth is 4.5 times the sail height which in turn is given by Equation 5.2.5, sail height = 3.7 $t^{0.5}$. For greater widths, the keel depth is assumed to reduce linearly to half the maximum at a width of 5000 m (as for the previous floating portion). For widths less than 1000 m, it is assumed that extreme keel depth at 50 m width can be twice that at 1000 m width with a linear interpolation between.

Using this approach (and the ride-down model), the fitted relationship becomes,

$$w = 1360D^{-0.34}h^{1.1} \tag{5.2.8}$$

This is referred to as "New Model B". Figure 5.5 is a comparison of Models A and B. Based on the assumptions and theory used in these models, we recommend that Model A be used for rubbling forces when it is expected that ice rubble can become grounded over the width of interest. For floating ridges, Model B is recommended and would be the one to use in limit force calculations. Beyond about 1000 m width the models are essentially the same. If a practitioner wishes to use a single relationship and be slightly conservative for floating ridges over widths less than 1000 m, then Model A could be used.

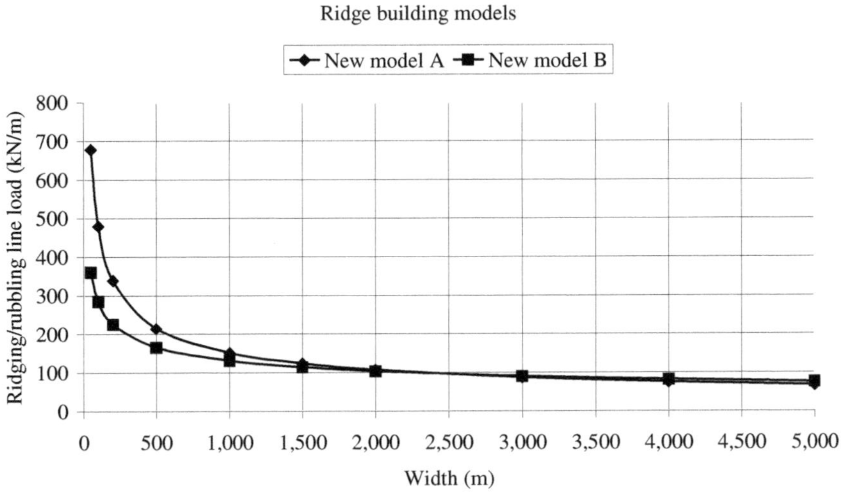

Figure 5.5 Comparison of Models A and B for rubble/ridge building line loads (1 m thick ice)

If the some or all of the rubble is grounded on the seabed, shear between the rubble base and the seabed takes some of the ice force, so that not all of it is transmitted to the structure downstream. The force that can be transmitted can be limited by one of the following mechanisms:

(i) the shear strength of the seabed itself: if the bed is cohesive, that strength can be estimated by multiply the plane area of the rubble by the undrained shear strength of the soil, whereas if the bed is cohesionless (and if there is time for pore pressures to diffuse away), the strength can be estimated from the vertical load applied to the seabed multiplied by the tangent of the angle of internal friction;

(ii) the shear strength of the interface between the ice rubble and the seabed, which can be estimated by multiplying the vertical load applied to the seabed by a 'coefficient of friction', determined experimentally or by interpretation of field measurements;

(iii) collapse of the rubble: the corresponding force can be determined from field measurements, or by idealising the rubble as a Mohr-Coulomb material with parameters determined experimentally.

5.3 Limit Force Calculations

With the topic of ridge building forces fresh in our minds this is an appropriate moment to look at typical limit force calculations as might be done for a range of scenarios in platform design. We will consider three examples ranging from the Caspian Sea to the Arctic Ocean.

The North Caspian Sea is a relatively mild ice region with annual ice reaching a nominal 100 year thickness of about 0.8m (Verlaan et al. [16])). Pressure ridges and rafting can occur. The keel depths of ridges are limited by the water depth, but even so, they have to be considered. In this example we choose a water depth of 9 m so the maximum ridge depth is set at 10 m (allowing for a 1 m water level surge). In most winters the consolidated layers of ridges in the Caspian are not very thick and there is not much data on extremes. In this example we shall use 1.2 m as reported in Verlaan et al.[16]. The structure width is assumed to be 100 m and it is vertical-sided.

Limit stress methods for first-year ridge loads are covered in detail in Section 5.5. As reviewed there, this example gives a potential limit stress design load in the range of about 140 to 195 MN. This range exists because of the uncertainty surrounding the failure mode of ridge keels when they are just clearing the sea floor, and passive failure downwards may be inhibited. This is discussed further in Section 5.5. Regardless, the limit force load is independent of that specific uncertainty, so we can proceed.

The limit force load in this case requires the ridge length and the ice thickness driving it. In this example we will assume that the ridge length to combine with the 100 year ice thickness of 0.8 m is 300 m. (This is based on qualitative observations by one of the authors during many helicopter flights over Caspian Sea ice. The ridging force behind the ridge is determined using equation 5.2.7 or 5.2.8. On a 300 m long ridge the limit force load ranges from 46 to 65 MN for the two equations. This compares to the limit stress load for the consolidated ridge, even at its lowest value of 140 MN and so for a ridge it appears that the limit force condition will control. Note however that for this structure in this location, the crushing of rafted ice (assuming a 1.15 m ice thickness) gives a load of about 97 MN which will therefore be the deterministic design load This is derived from the ISO method (equation 4.4.2), but with a *Cr* value of 1.8 based on the Baltic which is considered to be a region reasonably analogous to the Caspian. The limit force approach might also be applied to the rafted ice scenario if the size of the rafted ice zone was known with some certainty and was surrounded by thinner ice. Nevertheless, in this regional scenario, the application of the limit force method has potentially eliminated the extreme ridge load of up to 195 MN by reducing it to the range of 46 to 65 MN.

Offshore the North East Coast of Sakhalin is chosen as the second example. In this case the location is in 100 m water depth and a bottom-founded structure is being considered. The design ice feature is a first-year pressure ridge. In this water depth, ridge keels are not limited by the water depth, and data obtained over several years by drilling and upward looking sonars give a nominal 100 year design ridge depth of about 25 m. Field programmes have also given good data on the extreme thicknesses of the consolidated layer. This work has also shown that the consolidated layer thickness is variable across an area, so peak values should not be used. For a 75 m diameter structure we will assume 3.5 m of solid ice for the consolidated layer thickness. If considering a range of diameters from say 50 to 150 m, this thickness will vary slightly

Again, using the currently accepted limit stress method for such a feature, with plausible inputs (Section 5.5), the limit stress loads will be in the range of 334 MN to 369 MN. This calculation assumes that *Cr* in equation 4.4.2 is equal to 2.4 for sub-Arctic regions as recommended by ISO.

For limit force loads, a ridge length range of 400 to 1500 m is used in order to be conservative, noting that in Sudom et al. [17], ridge lengths averaged 80 m with a maximum of 500 m. In any real design work ridge lengths should be based on data from surveys. The ambient ice thickness behind the ridge is taken as 1.4 m, a conservative 100-year value for this region [13].

Using the Model B ridging equation (5.2.8) with 1.4 m ice gives the limit force loads as a function of ridge length shown in table 5.5. For a vertical structure with a typical waterline width of 75 m, the upper limit force load is 245 MN compared to 369 MN for limit stress. If both loads were predicted with the same level of confidence, the limit force load could be chosen for design. The level of confidence is affected by a number of factors that would be considered by a seasoned ice specialist and would include the quality and comprehensiveness of the input data as well as the degree of verification of the ice load models themselves.

Table 5.5 A comparison of typical limit stress and limit force loads for a deepwater Sakhalin caisson (vertical-sided)

Limit stress loads for first-year ridges (MN)				Limit force loads for various ridge lengths (KRCA method B) (MN)			
	Keel load	Consolidated layer load	Total	Ridge lengths (m)			
				600	800	1000	1500
Narrow keel	95	239	334	134	162	188	245
Wide keel	130	239	369	134	162	188	245

In regions where the ice could be moving with significant velocity, for instance off Sakhalin as ISO 19906 [13] indicates, the effects of momentum also need to be considered. This is because, by definition, in the limit force scenario, the ice feature has stopped in front of the platform. Its momentum and kinetic energy have to be dissipated before it stops, and this can only be accomplished by a build-up in the limit stress ice force at the platform, until the work done by that force is equal to the initial kinetic energy of the ice feature. This issue is reviewed in more detail in the next example of a multi-year floe as well as in Section 5.8 later.

A different perspective on this comparison is obtained if we look at a more promising candidate for a platform in this environment and water depth. This would be a platform with a sloping section through the ice line, which could be either up or down. In this example we will first look at an upward breaking structure similar to that shown in Figure 5.13 in Section 5.5. With such a structure, only the consolidated layer load is reduced by bending failure, the keel load being estimated as though it was failing on the vertical shaft. Now we have loads as shown in table 5.6. The limit force loads are unchanged, but the limit stress loads are now reduced. For this type of structure in this environment, it might now be argued that the need to refine the limit force approach is low and the emphasis should be on confirming the inputs and components of the limit stress load. It can be seen from the table that the calculated keel load is still significant. The estimation of keel loads and their uncertainties are covered in Section 5.5.

In Section 5.5 on first-year ridge loads, a new approach is proposed for first-year ridges interacting with down-breaking slopes, which could also be used at this site because of the deep water. As will be shown, the limit stress ridge loads would then be much lower than the limit force loads.

Another interesting aspect of this example is that the potential ranges of limit stress and limit force loads are similar. This is encouraging in any analysis because it implies that if one of these loads was completely wrong, the potential effect would not be catastrophic.

Table 5.6 First-year ridge loads (limit stress) on a sloping structure compared to typical limit force loads (depending on ridge length)

Limit stress loads for first-year ridges (MN)				Limit force loads for various ridge lengths (KRCA method B) (MN)			
	keel load	consolidated layer load (from table 5.11)	total	ridge lengths (m)			
				600	800	1000	1500
narrow keel	95	120	205	134	162	188	245
wide keel	130	140 (higher ride up)	270	134	162	188	245

A final example of the limit force approach is a thick multi-year ice interaction in the Arctic. This scenario is the design case previously examined (in chapter 4) of a thick multi-year floe 12 m thick acting on a 80 m diameter structure. In the previous limit stress approach, the floe size did not matter, but in the limit force method we need the floe size and the surrounding ice thickness. The thickness will be taken as 1.6 m and several floe sizes will be considered in the range 1000 to 5000 m. Table 5.7 shows the limit stress and limit force loads for these parameters. The limit force load is calculated using the equation for floating ridge building (model B). Also shown is the limit stress load for a 25m thick ridge failing in crushing.

Table 5.7 Vertical platform; 80 m dia.; limit force loads compared to limit stress loads

Floe dia.	Static limit force	Limit stress load (12m)	Ridge load (25m)
m	MN	MN	MN
1000	218	942	1770
3000	450	942	1770
5000	630	942	1770

The results in the above table suggest that the limit force loads will control because there is not enough driving force from the surrounding pack ice to generate the large limit stress loads, especially for the thick ridge.

However, the picture is not complete. As already mentioned in the previous scenario, the thick floe has to be stopped before the "static" limit force load is manifested. In the case of these thick large floes, we need to look at the limit momentum (or energy) forces which can be generated in bringing these floes to rest. The theory and methods for limit momentum (energy) calculations are covered in Section 5.8, so they will not be repeated here. However, some results are shown in table 5.8 for two floe speeds.

Table 5.8 Limit energy; limit stress and "static limit force loads"

Floes in 1.6 m pack ice					
Floe dia.	Impact load (0.25ms-1)	Impact load (0.5ms-1)	Limit stress load (12m)	Ridge load (25m)	Static limit force
m	MN	MN	MN	MN	MN
1000	420	589	942	1770	218
3000	773	933	942	1770	450
5000	935	942	942	1770	630

It can be seen that in this example floes greater than about 3000 m travelling at 0.5 m/s will generate impact forces close to or the same as the limit stress loads for a 12 m thick sheet. What this means is that the floe fully envelops the structure before it is brought to rest. Even though the limit force load is lower, it cannot govern unless the initial kinetic energy of the floe is dissipated in some other means than by ice crushing.

Possibilities for such are discussed in Section 5.8 which also includes an approach to energy dissipation on a sloping platform.

5.4 Multi-Year Ridges

5.4.1 *Introduction*

Based on field surveys, extreme multi-year ridges in the Arctic Ocean may have keels up to about 25 m thick (and more) [18]. Owing to the processes of creation and ageing over several years, multi-year ridges have been found to consist of mostly solid ice with a lower salinity than normal sea ice. Multi-year ice has always been respected by mariners who will generally avoid it.

In the previous section some potential loads from thick multi-year ice on a vertical structure were discussed (in the context of limit force loads). It was seen that without the relief of a limiting driving force, the potential loads in crushing are very large (e.g. 1770 MN). It is for this reason that for thick multi-year ice and ridges, sloping structures which will fail the ice in bending are often considered.

The equations for level ice on sloping structures were developed in chapter 4 where thick multi-year ice (up to 12m) was considered as an extreme "level ice" case. However, as noted above ridges will be thicker than the 12 m attributed to multi-year hummock fields, but once they are broken may clear better on a sloping structure because they have an approximate beam-like geometry.

In the past, the breaking of such a feature has been analyzed using various methods ranging from simple structural mechanics to complex finite element formulations. Here, we will err on the side of simplicity and then provide some discussion of limitations and potential improvements.

5.4.2 *Ridge Breaking Analysis*

As a first approximation, solid ridges can be considered as floating beams infinitely long. Failure loads can then be calculated using the theory of beams on elastic foundations (Hetenyi, 1946 [19], Croasdale, 1975 [20]). Observations of ridge interaction with sloping structures in model and basin tests indicate that the first failure of the ridge is a centre crack at the point of contact with the slope (as the ridge is lifted by the slope). The vertical load to cause this failure is given by

$$V_1 = \frac{4I\sigma_f}{yL_b} \tag{5.4.1}$$

where, I is the second moment of area of the beam, σ_f is the bending strength of the ice (upper surface in tension), y is the distance from the neutral axis of the beam to the upper surface and L_b is the characteristic length for the beam given by,

$$L_b = \left(\frac{4EI}{\rho_w gb}\right)^{0.25} \tag{5.4.2}$$

where, b is the width of the beam (ridge) through the water line, E is the ice elastic modulus, ρ_w is the density of water and g is the gravitational constant.

The ridge has been fractured by the centre crack but it cannot clear until each of the two side pieces have been fractured again, creating what are called hinge cracks (also seen in basin tests). To be conservative, it is assumed that these two cracks occur simultaneously and the load is twice the load to fail a semi-infinite beam on an elastic foundation. The load is given by,

$$V_2 = \frac{6.17I\sigma_f}{yL_b} \tag{5.4.3}$$

In this case y is the distance from the neutral axis to the bottom of the ridge and σ_f is the bending strength based on tension in the lower surface.

It can be seen that it is most likely that the controlling load is V_2, but note that this depends on simultaneous creation of the hinge cracks.

As in the case for level ice, the corresponding horizontal load (H_2) is given by,

$$H_2 = \xi V_2 \tag{5.4.4}$$

where, ξ is the transformation factor relating horizontal and vertical loads on slopes and depends on slope angle and friction as given by

Section 4.5.2 in chapter 4. The above loads are breaking loads only. It is generally assumed that the thinner ice surrounding the ridge will not be capable of pushing these very thick fragments up the slope to create a larger load than the ridge breaking load. This could be checked using the ride up equations in chapter 4; also see the later comments on results from the Esso basin tests.

An issue raised by Ralston [21] relates to short ridges. He pointed out that short ridges on elastic foundations require more vertical load to fail them in bending than infinite or semi-infinite beams. This is theoretically true and should certainly be considered. The analysis by Ralston indicates that the first crack load will be doubled if the ridge length is comparable to the characteristic length. The hinge crack load is doubled if the ridge is twice the characteristic length and even higher for shorter ridges. However if the short ridge is part of a larger, relatively thick, multi-year floe the short ridge effect is much reduced. Furthermore, if the ridge length is limited by the floe size, then the load may be limited by the driving force of the pack ice surrounding the floe (see Section 5.3); also, as shown in the example below, when looking at design loads the ridge load may not create a controlling design situation.

As an example, we will estimate typical loads for a 25 m deep multi-year ridge using the above method. In this simple example we will assume a rectangular ridge 80 m in width. The base-case bending strength is taken as 350 kPa, the modulus as 5000 MPa, the cone angle as 45 degrees and the friction coefficient as 0.15; with these inputs

$$H_1 = 64 \text{ MN} \quad \text{and } H_2 = 98 \text{ MN}$$

These loads are for long ridges and table 5.9 shows the various inputs and results, including some sensitivities; also shown are the curves for ridge length effect. For the assumed ridge lengths shown, the short ridge factor for H_2 is in the range of 1.3 – 1.9. The revised ridge breaking load accounting for a ridge length of 500m is about 186 MN.

These loads are lower than the loads from a 12 m thick multi- year floe shown previously in table 4.2 (typically 400 MN with the improved method and 763 MN with the current ISO method).

Table 5.9 Parameters and results for multi-year ridge loads

Multi Year Ridge	(Rectangular)		
Bending strength	kPa	350	500
Modulus	kPa	5.00E+06	5.00E+06
Water weight density	KN/m^3	10.2	10.2
Ridge depth	m	25	25
Ridge width	m	80	80
Ridge length (L)	m	500	800
I	m^4	104,166.67	104,166.67
y	m	12.5	12.5
Characteristic length (lc)	m	224.78	224.78
V1	MN	51.90	74.14
V2	MN	80.06	114.37
Slope angle	degrees	45.00	45.00
Slope angle	radians	0.79	0.79
friction		0.10	0.10
e		1.22	1.22
H1	MN	63.43	90.61
H2	MN	97.84	139.77
L/lc		2.22	3.56
Ralston correction on H2		1.9	1.3
(From figure below)			
H2 corrected		185.89	181.70

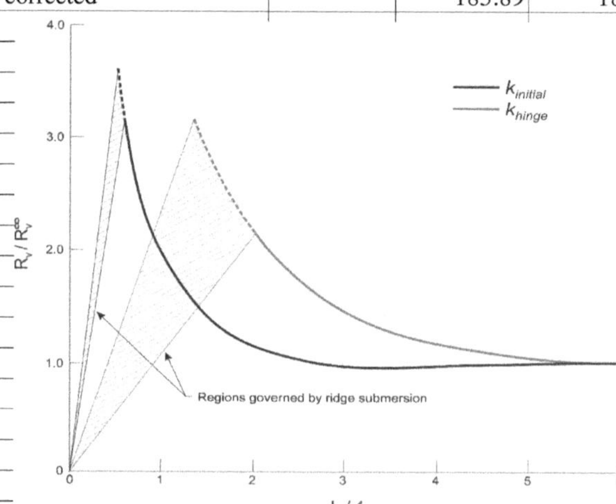

Graph From C-CORE (After Ralston 1977)

Upward-looking sonar data obtained from submarines (Wadhams [65]) show that multi-year ridges are by no means uniform in cross-section along their length, and much of a ridge may have keel depths less than the reported maximum depth. The method of simple beam bending described above can be modified to recognise a varying keel depth with length. A first approximation might be to average the keel depth over the ridge length.

(a)

(b)

Figure 5.6 Simulation of thick multi-year ice ridge acting on a multi-faceted cone on the Esso basin in 1989 [24] (Imperial Oil, Calgary, with permission).

Model tests on ridge interactions were conducted in a saline ice basin 2.4 m wide and are reported by Lewis and Croasdale [23]. The equivalent full-scale ridges were up to a maximum depth of only 5.8m based on the targeted 1/50 scaling. Comparison with the above theory gave good agreement for the deepest ridges, when the short ridge factor of about 2.2 was applied.

Ridge tests were also performed in the Esso basin in 1989 and 1990 [24]. Photographs of a typical interaction are shown in Figure 5.6. The first picture shows the formation of the centre and hinge cracks; the second shows that the broken ridge on one side has collapsed and would rotate around as the ice motion continued (the other side could also be expected to do the same). The results have not been fully reported and analyzed, and a new initiative is underway to do this. However, at the time of the work, the results were incorporated into a joint industry – university – government project and Z. Wang (Memorial University) [25] evaluated the results as well as model tests (at two scales and in two basins) representing the tests. A paper describing the overall initiative was published by Muggeridge and Croasdale in 1993 [26]. This work is worthy of further evaluation. In terms of a results overview, the beam on elastic foundation model gives reasonable results for horizontal loads for the higher load situations when compared to measured values see Figure 5.7. It should be noted that the method developed by Y. S. Wang [27] for ridges using plasticity theory gives higher loads than the simple elastic beam approach.

Z. Wang also noted that for the thickest ridges tested in the Esso basin the load due to the hinge cracks was within about 85 - 90% of the total load measured, which suggests that there is about an additional 10 – 15% loading due to ride up and clearing. In one test this was as high as 50%, but this higher ride up percentage was associated with a lower hinge crack load than in other tests. Nevertheless, the results do show that a ride up and clearing term after breaking may be necessary in order to avoid being non-conservative. Based on the Esso basin tests, it is suggested that a 1.2 factor would cover the ride up term (in lieu of a more authentic model).

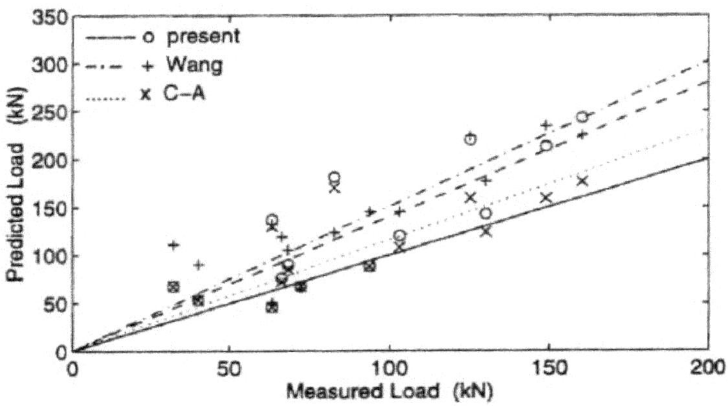

From Z.Wang's thesis **Horizontal Loads from Three Models for the ERC Tests**

- O present is Z Wang's model; + is Y.S. Wang's plasticity model [66]; C-A is the beam on elastic foundation model (Croasdale – Abdelnour) with appropriate correction for finite length effect

Figure 5.7 Comparison of ridge load models with measured loads from the Esso basin tests (Wang's thesis [25])

The work noted above relates to "faceted cones". A "faceted cone" is defined as one made up of flat faces (facets) instead of a circular cone. In the Esso tests, the cone had six facets. One goal of the tests was to assess whether the equations and methods developed for circular cones could be applied to such geometries. In the comparisons shown in Figure 5.7 the best fit (the C-A method) was unmodified from the equations for a circular cone, so perhaps no correction is needed. Nevertheless, during his research Z. Wang developed theoretical corrections for the centre crack and hinge crack loads when the ridge acts on a flat face. These corrections themselves depend on assumptions for the load distribution across the facet.

For example, the coefficient of 6.17 in the equation for the hinge cracks (equation 5.4.3) is replaced by the coefficient F, given by:

$$F = 6.17 + 2.7\left[\frac{2l}{L_b}\right] + 2.0\left[\frac{2l}{L_b}\right]^2 \qquad (5.4.5)$$

where, l is half the width of the facet and L_b is the characteristic length as defined by equation 5.4.2.

In the example of a 25 m deep ridge previously reviewed, the characteristic length is about 220 m. If this ridge was acting on a facet 60 m across, the factor F is calculated as 7.3. This compares to 6.17 for a round cone, about a 20% increase.

Thick multi-year ice ridges acting on vertical structures will need to be considered if that is the favoured platform shape. As discussed earlier, the failure of 25 m thick ice in crushing is outside any experience. If the ISO crushing relationship is used (Equation 4.4.2), then the load on a 80 m wide structure is estimated at about 1770 MN, a very challenging load indeed! In this scenario, multi-year ridge profile data should also be examined very carefully to see if an average thickness over the width less than the extreme depth can be justified. Penetration between a circular platform and a trapezoidal shaped keel (say) may also justify a lower thickness than the extreme thickness across the full diameter. Finally it is with such very thick features that limit force failure in thinner ice behind the thick feature may give lower loads (as reviewed in Section 5.3).

In summary, multi-year ridges and hummocked floes present formidable challenges in the design of offshore platforms for the Arctic. Using current approaches for ice crushing, the limit stress loads on vertical platforms are extremely large. There is also uncertainty about the effect of cyclic forces during the crushing of such very thick ice features. Sloping faces undoubtedly give lower breaking loads, but the ride up and clearing processes are uncertain and to avoid large ride-up loads very careful design of how the deck is placed and protected is very important and will add to the platform costs.

Ice islands are another issue. They can be 30 to 50 m thick and several square km in area. For ice islands, it is likely that either protective underwater berms will be needed, or the platforms abandoned rather than be designed for these rare features.

5.5 Loads due to First-year Ridges

5.5.1 *Introduction*

Figure 5.8 Photograph of an area off Sakhalin with pressure ridges. (Keel testing underway in 1998 [28] - photo by Ken Croasdale)

In the previous section, the interaction of multi-year ridges with platforms has been reviewed. Multi-year floes and ridges will usually govern design loads in regions where they occur, but the first-year ridge will govern design loads in most other ice-covered regions, and so developing a reliable method for first-year ridge loads is very important. Ridges form when one ice sheet fails against another. A typical first-year ridged area, as seen from the surface is shown in Figure 5.8. Only sails are visible, and these represents about 10 per cent of the total floating ridge volume. Figure 5.9 shows a general ridge cross-section. It indicates that the keel represents the largest component of the ice volume, and is therefore a significant potential contributor to the ice load when a ridge interacts with a platform.

Initially when a ridge forms, the whole volume is made up of ice blocks in the sail and in the keel (and sometimes level and rafted ice which has pushed through the rubble at the ice line). Depending on the temperature of the ice sheet forming the ridge, there may be some slight freezing between blocks in the keel due to negative sensible heat being released as the ice blocks attain the water temperature (which will generally be at the freezing point). More importantly, with time, heat is lost from the surface to the cold Arctic air and the upper layer of the ridge refreezes (and is called the consolidated layer). Because the keel has a porosity of only about 30%, the freeze front progresses much faster than it does in the growth of level ice. This generally results in the consolidated layer being thicker than the surrounding level ice (See Croasdale et al [29]).

The values and relationships for ridge morphological parameters have been studied in various regions during numerous field investigations. Very good reviews have been written by Timco et al. [14] and more recently Sudom et al. [17]. On Figure 5.9, some simple typical relationships are shown which can be used for preliminary analyses. Designs for specific regions should use field data from that region if they are available.

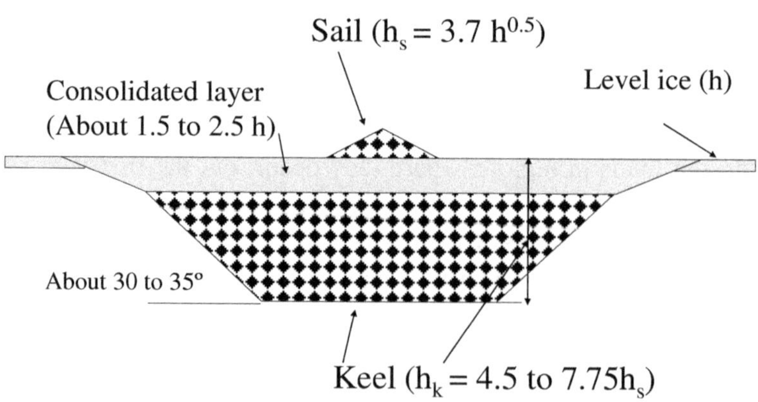

Figure 5.9 Typical first-year ridge cross section

Using the simple relationships given in Figure 5.9, table 5.10 shows how typical ridge parameters vary with maximum level ice thickness for various regions. But note also that ridge keels can be limited in shallow water by the sea floor.

Table 5.10 Range of ridge parameters relative to level ice (after Timco et al. [14])

Level ice h	hcl	hcl2	hs	hk	hk1	hk2
(m)	(m)	(m)	(m)	(m)	(m)	(m)
0.8	1.6	2	3.31	14.89	21.01	27.14
1	2	2.5	3.70	16.65	23.50	30.34
1.4	2.8	3.5	4.38	19.70	27.80	35.90
1.7	3.4	4.25	4.82	21.71	30.63	39.56
2	4	5	5.23	23.55	33.23	42.91

On this table: *hcl* (the typical refrozen layer maximum thickness is based on $2h$; *hcl2* (a typical extreme)) is $2.5h$; the sail height *hs* is $3.6\ h^{0.5}$; *hk* the average maximum keel depth is 4.5*hs*; *hk1* the extreme keel depth is taken as (4.5 + 1 standard deviation)*hs*; *hk2,* the abnormal keel depth is taken as (4.5 + 2 standard deviations)*hs*.

5.5.2 *Ridge Interaction with Vertical Structures*

When a first-year pressure ridge moves against a vertical platform, various ice failure modes can occur, and they are shown in Figure 5.10. The ridge will either fail locally against the platform (limit stress) or the thin ice will fail behind the ridge (limit force). The lower of these two loads will govern. The limit force load is dependent on the ambient ice thickness, the ridge length and the rubbling load in the thin ice behind the ridge, and has already been addressed in Section 5.3. In this section we will focus on the limit stress failure and corresponding loads.

On a vertical structure, a simplified approach is to estimate the load due to crushing of the consolidated layer using the general crushing formula (equation. 4.4.2) and then add to it the load to fail the keel. In this approach it is reasonably assumed that the presence of the keel does not change the crushing failure of the consolidated layer. It is further assumed that the keel will not fail upwards and break through the consolidated layer because it takes less force to fail it in passive failure downwards, as depicted in Figure 5.10.

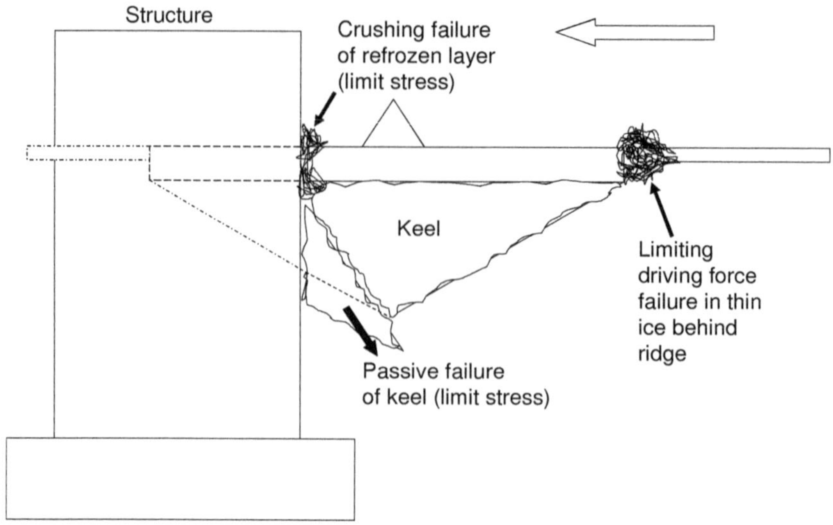

Figure 5.10 Limit stress and limit force failures relating to a first-year ridge acting against a vertical structure

In the crushing load assessment, it may be appropriate to have a reduction factor because of potentially higher porosity in the consolidated layer compared to level ice. There is very little data on how a consolidated layer fails in crushing. In the authors' experience this reduction factor could be in the range 0.85 to 1.0. As a first approximation we recommend 0.9. Furthermore, evidence from actual structures suggests that consolidated layers often fail out of plane at lower loads than crushing, possibly due to their lack of continuity due to variability in thickness; so a crushing assumption, even with a reduction factor is likely to be conservative.

In the keel failure assessment, a refinement over the passive failure calculation is to recognize that as the structure penetrates the keel, it may also induce the horizontal plug failure shown in Figure 5.11.

An approach that combines the passive and plug failures was developed for the design of the Confederation Bridge [30]. It is based on the Dolgopolov et al [31] model for local (passive) failure of the keel rubble, together with a shear plug model for keel failure, based on the original development by Croasdale [32].

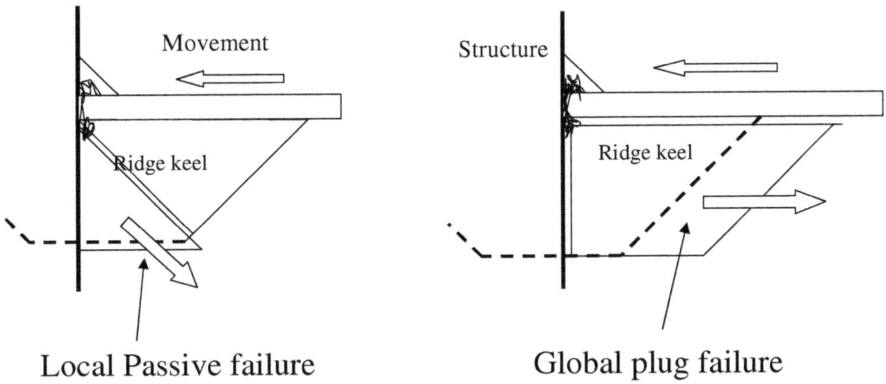

Figure 5.11 Two bounding keel failure modes on a vertical structure

Dolgopolov's model for passive local failure of rubble in the keel is based on experimental studies of the behaviour of ice blocks in tests and also recognizes soil mechanics theory. It is the algorithm published in ISO 19906 [13] as the sole method to estimate keel loads (with a slight modification). It is

$$F_{pa} = K_p h_k D\left(1+\frac{h_k K_p \gamma_e}{2}+2c\right)\left(1+\frac{h_k}{6D}\right) \tag{5.5.1}$$

where D is the structure width, h_k is the effective keel depth which includes the effect of any surcharge, c is the cohesion in the keel, K_p is the passive pressure coefficient of the material ($\approx \tan[45° + \varphi/2]$), φ is the internal friction angle and γ_e is the buoyant unit weight of the keel rubble (accounting for keel porosity), given by

$$\gamma_e = (1-e)(\rho_w - \rho_i)g \tag{5.5.2}$$

where e is the porosity of the keel rubble, ρ_w is the density of water, ρ_i is the density of ice.

Dolgopolov recommended that the keel depth be increased by a surcharge factor (s) to account for the surcharge created at the bottom of the keel due to non-clearing rubble; that is

$$h_k = h_{ko} + sD \tag{5.5.3}$$

where h_{ko} is the original keel depth and D is the structure width. For wide structures, the authors recommend the surcharge factor be limited to about 0.1. Model tests can be useful in determining surcharge factors for various widths.

The shear plug model based on the original development by Croasdale [32], improvements in [30], and refined by Cammaert et al [33] leads to an expression for the force to create the plug as

$$F_{pl} = \left[\gamma_e \tan(\phi) \int h_k dA + cA\right] + \left[\gamma_e \tan(\phi) K_p \int \frac{h_k^{\ 2}}{2} dx + c \int h_k dx\right] / \cos(\omega) \tag{5.5.4}$$

in which the vertical shear planes are assumed to deviate at an angle ω to the direction of motion (as observed in experiments by Bruneau [34]). The first bracket is the shear force on the underside of the consolidated layer over the area A. The second term models the forces on the two vertical failure planes through the depth of the keel. The integrals depend on the nature of the variation of the keel depth across the ridge.

One analytical approach proceeds by stepping through the ridge in convenient increments and determining the two loads (Dolgopolov and shear plug) at each increment. The actual keel load at each increment is the lower of the two. The controlling load is the Dolgopolov load or the shear plug load if the two curves cross prior to penetration to the bottom of the keel (including any surcharge)

To illustrate the method, a typical extreme pressure ridge off the North Coast of Sakhalin is examined (as previously discussed under Section 5.3 on the limit force method). The values of consolidated layer and total keel depth are taken from Table 5.10 for the level ice thickness of 1.4 m, a consolidated layer thickness of 3.5 m and a total ridge depth of 27.8 m. The keel rubble depth entered in the calculation is therefore 24.3 m, and the keel width is about 61 m (assuming it is triangular). Other input parameters for a typical calculation are shown in Figure 5.12, which also shows the results from using the approach described above.

Some of the other inputs should be reviewed. The unit weights for ice and water are typical. The porosity of 0.3 is also typical (see Timco et al [14]). As reviewed in chapter 3 the rubble strength can be represented by a friction angle and a cohesion. However, in-situ tests conducted by one of the authors [35] show that the cohesion represents frozen bonds between the ice blocks which need to be broken before friction between blocks is mobilised. In addition, tests have shown that in general the force to fail the cohesive bonds along a typical failure plane is greater than the residual friction force. Therefore in this calculation, the friction angle is set to zero when cohesion is specified. The cohesion value specified in this calculation is 24 kPa at the top of the keel reducing to half this value at the bottom. These values represent typical high values and this selection is discussed further in chapter 3. In keeping with zero friction angle, the angle omega is also set to zero. The other important input is the surcharge factor. In this calculation it is set to 0.1. This means that the effective maximum keel depth is increased by 10% (2.4 m). Narrower structures would use a smaller factor because the rubble will be cleared sideways more easily. The other inputs relate to the parallel calculation of the crushing force associated with the consolidated layer (using a *Cr* value for sub-arctic) as well as the limit force load.

The graph in Figure 5.12 shows how the potential loads for the two modes of keel deformation vary with penetration. Initially at small penetration, the Dolgopolov load is small because a triangular shaped keel has small thickness at small penetration; whereas the load to create a plug failure is at its greatest because the plug failure planes have their greatest area.

With the above approach and inputs, the keel load is about 95 MN and adding the crushing load for the 3.5 m thick consolidated layer, the total ridge load is predicted to be as high as 334 MN. If the ridge is of trapezoidal shape with a significant bottom width, the keel load will trend to the Dolgopolov load at about 130 MN for a total ridge load of 369 MN.

The limit force load can be estimated assuming a ridge length of 1000 m (this is an estimate and any serious design load calculations would need to verify ridge lengths) and a level ice thickness of 1.4 m.

Using equation 5.2.8 for ridging force, the limit force load (shown on the spreadsheet in Figure 5.12) is about 188 MN. The range of possible limit force loads has been already been discussed in section 5.3

First Year Ridge Keel Loads

Developed by K.R. Croasdale and Associates Ltd.

- This version 1998 (with variable cohesion - assumed linear from bottom to top)
- To be used for research purposes only. Not valid for design without consultation with K.R. Croasdale
- To use the table, change the values in the Initial Data section of the worksheet. (all input numbers are blue)
- Top width calculated assuming 35 degree keel angle
- Dolgopolov Load modified to match ISO

Initial Data:

Structure width (m)	75
Keel depth (m)	24.3
Keel top width (m)	69.255
Keel bottom width (m)	0
Ice weight density (kNm^3)	8.90
Water weight density (kNm^3)	10.25
Rubble porosity	0.30
Rubble friction angle (deg)	0
Omega (deg)	0
Cohesion of keel (top) (kPa)	24
Cohesion of keel (bottom) (kPa)	12
Penetration increment (m)	1.731375
Max. penetration (m)	34.6275
Surcharge factor on width	0.1
Sheet ice thickness (m)	1.4
Refrozen layer thickness (m)	3.5
Cr	2.4
Strength reduction factor	0.9
Ice ridge length (m)	1000

Calculated Data:

Effective Buoyancy (kNm^3)	0.95
Pass. press. coeff.	1.00
Crushing Load (MN) (ISO) (SA)	238.45
Driving force (MN) (KRC A)	219.78
Driving Force (MN) (KRC B))	188.05

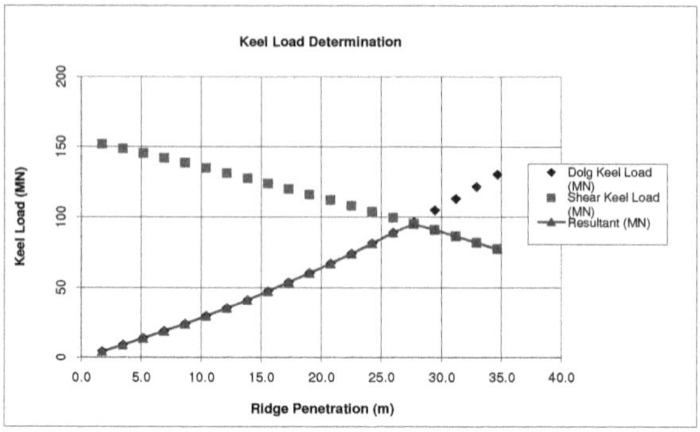

Max keel load **95.39 MN**
Total ridge load **333.85 MN**

Figure 5.12 Spreadsheet results for typical extreme first-year ridge off NE Sakhalin

5.5.3 *First-year Ridge Interaction on Upward Sloping Structures*

For keel interaction with an upward sloping structure, as a first approximation bounding calculation and unlikely to match reality, we can assume the configuration shown in Figure 5.13. It simply assumes that a dead wedge of ice rubble forms in such a way as to create an effective vertical face for the oncoming keel. The effective width of this vertical face is the average of the widths of the structure at the base and top of the keel. With this approach, the calculation method for the keel load just described for vertical structures can be applied.

The calculation method for level ice on sloping structures is used for failure of the refrozen layer, and then the two components are added. Using this method the total estimated load may or may not be lower than on a vertical face depending on the angle of the slope (and thence the average width exposed to the keel). Furthermore, after checking this initial load using the above method, the subsequent ice interaction process needs to be thought through. Where does the keel rubble go? In the case of an upward cone or slope it is undoubtedly driven up the slope to create a rubble pile. That should be analyzed using the methods described earlier for level ice failure in the presence of rubble build up. The difference is that the rubble volume on the slope will be greater because the keel is being scooped up. This is shown conceptually in Figure 5.14.

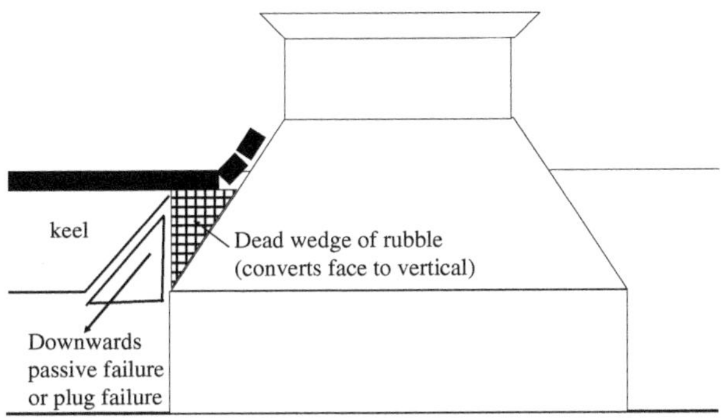

Figure 5.13 Idealisation assuming a dead wedge of rubble which allows methods for keel loads on a vertical structure to be applied to sloping structures.

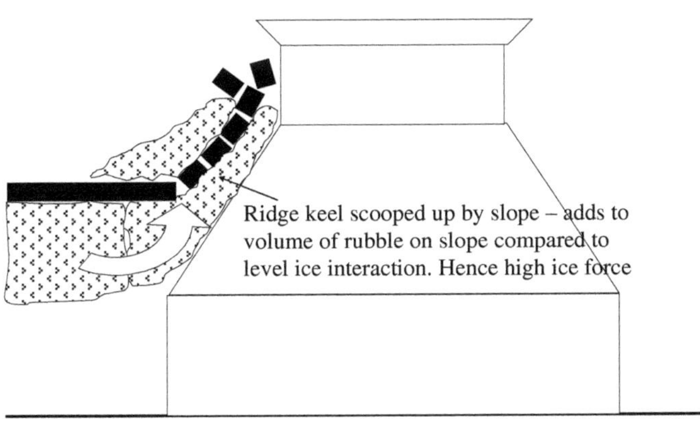

Figure 5.14 Ridge loads on sloping structures also need to be checked for the loads resulting from "scooping up" the keel creating a large rubble volume on the slope

One of the authors has used this approach for first-year ridge loads on sloping barriers in the Caspian Sea. As an example, on a 100 m wide barrier, the maximum flexural load was estimated at about 30 MN from level ice after a mature rubble volume had been established. The ridge load was calculated at 60 MN, for a higher rubble volume specified by pile height and a shallower rubble angle of repose. This approach also gets around the issue of how to apply the passive failure model for ridges in shallow water when there is no clearance between the keel bottom and the sea floor, as will be discussed in Section 5.6.2.

Based on these considerations, if first-year ridges are the design ice feature in an area and water level changes are not extreme, the structural configuration shown in Figure 5.15 may be superior to a structure with a slope extending to the base of the keel.

For this structure, the consolidated layer load is calculated using the methods for level ice (or a linear ridge) and the keel load is calculated against the vertical shaft (as previously described). The Sakhalin example is repeated for this configuration. The vertical shaft is kept at 75 m diameter, but a conical collar is added with a slope of 55 degrees. The collar is designed to fail the consolidated layer in bending and allow the keel to fail on the vertical shaft. Positioning the bottom of the collar at -

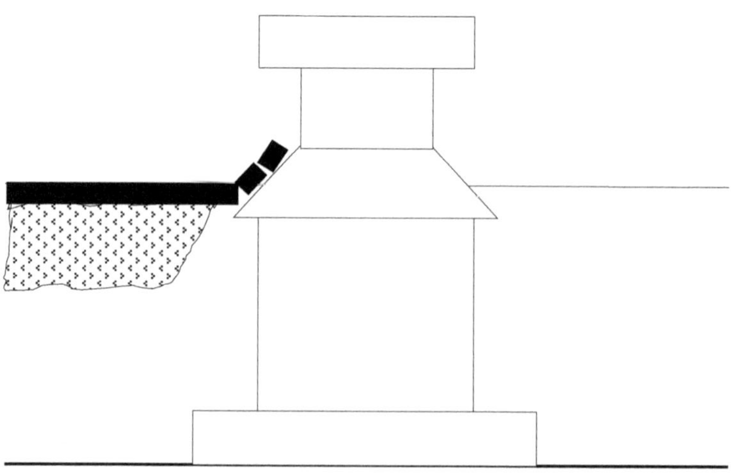

Figure 5.15 Platform with sloping section to fail the consolidated layer of a ridge with most of the keel failing against the vertical section.

4m and assuming a 50 m shaft on top gives the configuration shown with an ice line diameter of 71 m. The loads are shown from the KRCA spreadsheet in table 5.11 for two ride-up heights 15 and 20m. The higher ride-up gives a load of 138 MN to be compared with the 239 MN crushing load for the vertical structure. In this case the keel loads are the same at 95 MN or 130 MN depending on width for a total upper value load of 268 MN. It still appears that the limit force load will control design at 197 MN. In reality many ridges will get pushed by the structure at lower loads depending on keel depths, ridge widths and ride up heights achieved.

Table 5.11 Loads to fail the consolidated layer in bending: typical Sakhalin ridge

Initial Data:

	A	B
Flexural strength of ice (kPa)	**350**	**350**
Specific weight of ice (kN/m^3)	**8.98**	**8.98**
Specific weight of water (kN/m^3)	**10.20**	**10.20**
Young's modulus (kPa)	**5.00E+06**	**5.00E+06**
Poisson's ratio	**0.3**	**0.3**
Cone Angle (deg)	**55**	**55**
Rubble angle of repose (deg)	**45**	**50**
Rubble friction angle (deg)	**40**	**40**
Rubble height (m)	**15**	**20**
Waterline diameter (m)	**71**	**71**
Ice-cone friction	**0.1**	**0.1**
Ice-ice friction	**0.05**	**0.05**
Ice thickness (m)	**3.5**	**3.5**
Rubble porosity	**0.25**	**0.2**
Cohesion of rubble (kPa)	**0**	**0**

Results Summary

Horizontal Load (MN)	**119.80**	**137.65**
Vetical Load (MN)	63.13	73.14
HB	20.00	21.12
HP	0.24	0.12
HR	74.88	99.23
HT	8.82	8.82
HL	15.85	8.36
Total	119.80	137.65

Again this method may still be conservative, because the lifting of the consolidated layer by the slope may well disturb the keel enough to either weaken the cohesive bonds (which may be brittle) or to create a free surface which reduces the length of the failure planes.

These latter processes are apparently happening at the Confederation Bridge (described in Section 4.5) which has a "conical collar" geometry atop a near-vertical shaft. Evidence from marine growth and load panels on the vertical shaft indicate that the keel rubble is not contacting the shaft with any significant pressure (Lemee et al [36]). This is also confirmed by the global load measurements which show no correlation with keel depths (El Seify et al [37]). The interaction of first-year ridges against such geometries deserves more research. When extensive model testing of the Confederation Bridge piers was conducted, the keels were definitely loading the lower shaft with loads similar to those calculated using the passive failure and plug models, yet as mentioned above in the real world, the lower shafts apparently saw negligible ice pressures. This vignette also points out the potential futility of conducting model tests with model ice rubble without some full scale verification.

The method that follows for downward sloping structures can also be applied to upward slopes, especially for narrow structures where upward "scooping" of the keel material is limited because there will be side clearing.

5.5.4 *First Year Ridge Interaction on Downward Sloping Structures*

Figure 5.16 shows a downward-sloping structure that might be used in deeper water, and also might represent a spar floater through the ice line.

The downward failure of ridges ought to give lower loads, but how to calculate them? The use of the methods for vertical structures is clearly incorrect, although they can be used to estimate a first-approximation conservative load. The passive failure model can be adjusted to recognize the shorter failure planes for downwards failure. However, this approach should be used with caution as the presence of the consolidated layer may play a much bigger role in affecting how the rubble fails. A new, but very simple bounding approach has been developed [38]. It is to treat the first-year ridge as a composite beam.

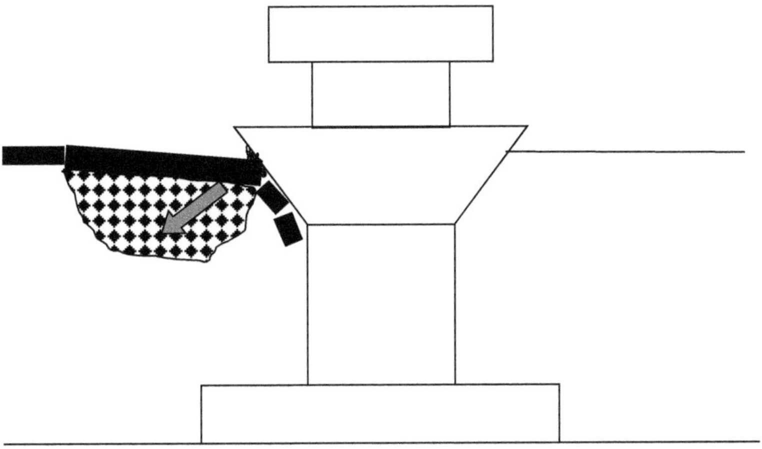

Figure 5.16 Down-breaking sloping platform showing a beam-like failure of a first-year ridge.

The composite beam is assumed to be made up of two layers, the refrozen layer and the keel rubble layer. Each has a thickness, modulus and strength which are the inputs required to solve the problem. The beam is shown as uniform in cross-section in Figure 5.17, but the method described here can also be used for a non-rectangular keel (the cross-section area and second moment of area will be different). Many structural engineering textbooks describe how composite beams should be treated in bending because this is a common problem in structural engineering. The method used here is based on [39].

The general approach is shown in Figure 5.17

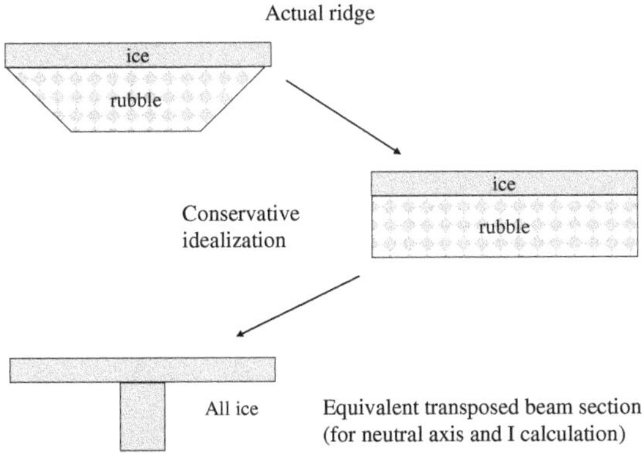

Figure 5.17 General approach treating a first-year ridge as a composite beam.

The method is quite straightforward. For the cross-section of the simplified first-year ridge as shown in Figure 5.18 it is necessary to specify the thickness and widths of the two materials as well as their elastic moduli (*E*). These are defined in the figure. Using these values, a transformed section is created for a single modulus of one of the materials. In this case the section is based on the modulus of solid ice.

The width of the solid ice equivalent of the keel is given by:

$$b = \frac{WE_k}{E_{cl}} \tag{5.5.5}$$

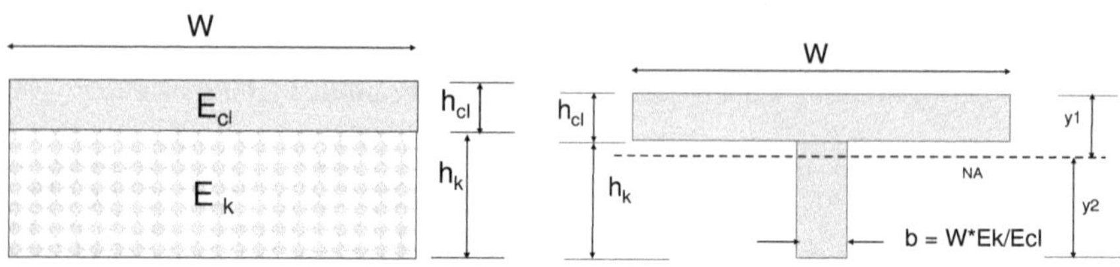

Figure 5.18: The idealized ridge with two materials and the transposed solid ice equivalent

For transposed section a neutral axis position and its second moment of area are calculated. Then the theory of bending is used to obtain maximum stresses in the composite beam. Then the stresses are corrected in the transposed material. For a given bending moment M, the stress in the unmodified material (in the case the consolidated layer, σ_{cl}) is

$$\sigma_{cl} = \frac{y_1 M}{I_T} \tag{5.5.6}$$

where I_T is the second moment of area of the transformed section

$$I_T = \frac{W(h_{cl})^3}{12} + Wh_{cl}\left(y_1 - \frac{h_{cl}}{2}\right)^2 + \frac{b(h_{cl})^3}{12} + bh_k\left(\frac{h_k}{2} + h_{cl} - y_1\right)^2 \tag{5.5.7}$$

and y_1 (the distance to the top surface from the neutral axis) is given by:

$$y_1 = \frac{W(h_{cl})^2}{2} + \frac{bh_k(h_{cl} + 0.5h_k)}{(Wh_{cl} + bh_k)} \tag{5.5.8}$$

The stress in the keel material (σ_k) is

$$\sigma_k = y_2\left(\frac{E_k M}{E_{cl} I_T}\right) \tag{5.5.9}$$

To assess the critical bending moment values, the previous approach for a solid beam or floating multi-year ridge will be used, treating it as a beam on an elastic foundation (Section 5.4).

The maximum bending moment M_1 for the infinite beam creates the centre crack, and is:

$$M_1 = \frac{V_1 L_c}{4} \tag{5.5.10}$$

where L_C is the characteristic length defined by

$$L_c = \left(\frac{4E_{cl} I_T}{W\rho_w g}\right)^{0.25} \tag{5.5.11}$$

(which assumes that the lift at which the beam breaks is less than the thickness of the consolidated layer)

Combining Equations 5.5.6 and 5.5.11 gives, for the consolidated layer,

$$V_1 = \frac{4\sigma_{cl} I_T}{y_1 L_c} \tag{5.5.12}$$

For the keel, we have

$$V_1 = \frac{4\sigma_k I_T E_{cl}}{y_2 L_c E_k} \tag{5.5.13}$$

The ridge will fail when one of the layers reaches its failure stress at its outer surface, at whichever value of the two values of V_1 is the lowest.

The same logic can be applied to the loads required to form the hinge cracks. If they form simultaneously, then the equations become, for the consolidated layer,

$$V_2 = \frac{6.17\sigma_{cl} I_T}{y_1 L_c} \tag{5.5.14}$$

and for the keel,

$$V_2 = \frac{6.17\sigma_k I_T E_k}{y_2 L_c E_{cl}} \tag{5.5.15}$$

Again, the controlling value for V_2 will be the smaller of the values given by Equations 5.5.14 and 5.5.15.

To apply this method to a first-year ridge, we need to know the modulus and flexural strength (tensile or compressive failure) of the rubble mass. Information on these properties in the context of this problem is rare. We can start with knowledge of rubble strength in tension. Some investigators who believe rubble to be strictly Mohr-Coloumb would say it is zero. The in-situ tests conducted by one of the authors already referred to in Section 3.6 indicate this is not the case. The pull up tests showed that to lift a block of ice cut through the refrozen layer required the bonds between the rubble blocks to be broken in tension. As reported in Section 3.6, the maximum tensile bond strength measured in all tests was 17.9 kPa. The average was 9.6 kPa with a standard deviation of 4.7 kPa. Using the average plus three standard deviations gives a value of 23.7 (rounded up to 25) kPa for the extreme tensile strength of ice rubble.

In this trial example we will use the above value. We can also assume a pseudo-compressive strength based on the measured rubble shear strengths. For this we will assume the compressive strength it is twice the shear strength. These values are those to assign to the extreme fibres of the beam (according to whether the bottom is in tension or compression). Now the question is what modulus to assign? This is very much a matter of judgment. It is proposed to bias it to the ratio of tensile strength because this should be a reflection of the actual area of a "solid" load path through the material (i.e. the area across any slice in which the blocks are bonded). Based on the above reasoning, table 5.12 shows some possibilities for the modulus ratio (ice to rubble). Using the values shown it is in the range of about 12 to 23, but in this example we will use 15.

Table 5.12 Ice properties assumed for the composite beam representing a first-year ridge

	Rubble	Rubble	Rubble	Solid ice	
	Average	ST. Dev	Extreme		Ratio
Tension (kPa)	9.60	4.70	23.70	350.00	14.77
Tension (kPa)	9.60	4.70	23.70	500.00	21.10
Shear (kPa)	10.79	4.29	23.65		
Compression 2xShear (kPa)			47.31	1500.00	31.71
BASED ON THE ABOVE WE WILL USE A RATIO OF 15 FOR E					
E (ratioed from solid ice)			333,333	5,000,000	

Using this approach and the ice properties given above gives some interesting results which are shown in table 5.13. The controlling vertical load using the new approach is calculated to be about 28 MN, and this is actually conservative because it assumes the hinge cracks occur simultaneously. If the downward slope is 45 degrees and the friction is 0.1, then the resolved horizontal load is about 35 MN. This compares to the conventional approach of about 104 MN, estimating the keel load using Dolgopolov as suggested in ISO 19906 and adding the load to fail the consolidated layer in bending).

One final observation on the composite beam approach can be made. It is theoretically correct to treat a beam composed of two materials in this way. It is well known that if a composite is treated as two separate beams and their capacities are added, the sum will be lower than if it is treated as a single composite beam. However, because keels are so much less stiff and weaker than solid ice, it is of interest to look at the potential error in treating the problem as two beams each of whose load capacity at failure can be added. This has been done and the results given in table 5.14. It shows that over a range of keel geometries, the average error in using the simplified approach of two beams is less than 5%. It is a useful check to calculate by both methods and if an answer is needed quickly, the two-beam approach will give a result which is well within the uncertainties of the input parameters.

Table 5.13 Keel load on sloping structure using composite beam method compared to the conventional approach

INPUTS		
Ridge width (W)	m	70
Consolidated layer thickness (h_{cl})	m	3.5
Keel thickness (h_k)	m	24.3
Modulus of consolidated layerv (E_{cl})	kPa	5 x 106
Modulus of keel ($E_k = E_{cl}/15$)	kPa	3.33 x 105
Width of transformed beam (b)	m	4.67
Flexural strength of consolidated layer	kPa	500
Flexural strength of keel based on tensile failure of rubble	kPa	25
Flexural strength of keel based on compression failure of rubble	kPa	50
RESULTS		
Centre crack load based on flex failure of CL ($V1_{cl}$)	MN	43.56
Centre crack load based on flex failure of keel ($V1_k$)	MN	9.38
Total hinge crack load based on flex failure of CL ($V2_{cl}$)	MN	67.2
Total hinge crack load based on flex failure of keel ($V2_k$)	MN	28.6 (controls)
Horizontal load (45 degree slope; $\mu = 0.1$)	MN	34.8
Dolgopolov keel load (assuming effective structure width of 75m) (From Figure 4.34)	MN	95
Flex failure of consolidated layer (to be added)	MN	9
Total ridge load by conventional methods (horizontal)	MN	104
Percentage of load using the new method of composite beams to conventional method	%	34

This observation also means that an areal type feature (rubble field) can be analyzed for flexural loads using the approach as for level ice (Section 4.5.2). One component of load is estimated from failure of the consolidated layer and the other from "flexural failure" of the keel; the two are then added. Note that for an areal feature (rather than the linear beam of a distinct ridge), the ice clearing components may be more significant in magnitude than the breaking terms.

Table 5.14 Composite-beam approach compared with the two-beam approach

	Ridge 1: W=70 m: hcl = 3.5 m: hk = 24.3 m	Ridge 1: W=60 m: hcl = 2.5 m: hk = 18 m	Ridge 1: W=50 m: hcl = 2 m: hk = 15 m	Ridge 1: W=40 m: hcl = 1.5 m: hk = 12 m
Composite beam approach (V2) (MN)	28.6	16.77	11.07	6.64
Separate beam approach				
V2 for CL (MN)	8.57	4.82	3.04	1.7
V2 for keel (MN)	19.01	11.2	7.43	4.5
Total	27.58	16.02	10.47	6.19
Difference (MN)	-0.62	-0.75	-0.6	-0.45
Difference %	-2.2	-4.47	-5.4	-6.8

5.6 Structures in Shallow Water

5.6.1 *Effects of Ice Rubble on Ice Loads*

One would expect that platforms in shallow water would be simpler to design and operate, and less costly than those in deeper water. This generally true, but there are complex issues relating to ice interaction which require careful consideration. The main problem is that ice rubble which will form around wide structures will not clear very easily and can become grounded. This rubble affects ice structure interaction and operational access. An example of ice rubble formed from a single ice movement event against an island in the North Caspian Sea is shown in Figure 5.19. It can be seen that access by vessel is now obstructed on two sides of the island. Evacuation is similarly hampered.

This phenomenon was first observed in the Canadian Beaufort Sea in the 1970s. Kry wrote a paper examining how such rubble can affect the overall stability of an island [41]. The general conclusion was that once fully established, grounded rubble is beneficial because it dissipates the ice load (now acting on the outside of the ice rubble) into the sea floor.

Figure 5.19: Ice rubble build-up around a shallow water platform (from [40]: photo by Derek Mayne)

The problem has been revisited for ice barriers and platforms in the Caspian Sea. Figure 5.20 shows ice rubble build up against a sloping barrier.

It has been established that care is required when using algorithms for wide sloping structures in shallow water. For a sloping structure the design engineer might assume that applying the ISO equations for a sloping structure will give the correct design loads. Croasdale's review [42] showed that this would be incorrect.

Figure 5.21 shows the stages of ice interaction for this case. In Stage 1, the ice fails in bending on a bare sloping face and rides up to the top of the slope. For 1 m thick ice acting on a 100 m wide barrier, the ice breaking load is about 2.5 MN (for a 45 degree slope, friction of 0.15, and a flexural strength of 500 kPa). As the ice rides up to the top of the slope at say 7 m above the ice line, the load increases to 16.6 MN. On a wide structure, an ice deflector may be needed to avoid overtopping, and in that case the ice falls back onto the oncoming ice and rubble builds up on the slope. In Stage 2, it is assumed that the ice keeps pushing through the ice rubble to continue to fail in bending on the slope. The sloping structure algorithm can still be used for this situation, but with a higher ride up height and rubble on top of the advancing ice. Using the ISO algorithm with a rubble height of 12 m and a rubble angle of repose of 35 degrees gives a total global load of about 30 MN.

Figure 5.20 Rubble in front of a wide barrier (from [42]: photo by Rune Nilsen)

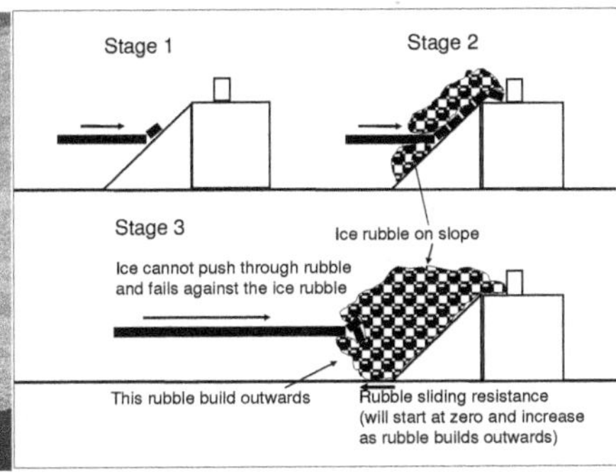

Figure 5.21 Stages in interaction

If the ice stopped moving at this point, the flexural failure ice load would be the design load. However, most ice movements will cause the rubble to continue to build to the point where the ice no longer pushes through the ice rubble to fail on the steel slope. It starts to fail on the outside of the rubble pile in the "rubbling" failure mode. This is shown as Stage 3 in Figure 5.21. For a short period of time when the growing rubble field still has minimal sliding resistance, the ice load is that due to the rubbling process. If this is greater than the Stage 2 ice load, then the rubbling load will control.

Rubbling loads are reviewed in detail in Section 5.3. We are proposing a refinement to the ISO equation for rubbling; for rubble that can start to ground in shallow water it is.

$$w = 4800 D^{-0.5} h^{1.1} \text{ repeat of} \qquad (5.2.7)$$

where w is the rubbling line load in kN/m, h is the ice thickness in m, and D is the width of interaction in m. The corresponding equivalent ice pressure is

$$p = 4.8 D^{-0.5} h^{0.1} \qquad (5.6.1)$$

where p is in MPa

For the scenario being considered (1 m thick ice; 100 m wide structure), the rubbling load from the above is 48 MN. This compares to the extreme flexural load of 30 MN, so the rubbling load controls, not the load calculated using ISO for the sloping barrier.

An idealized load trace with time for this scenario is shown in Figure 5.22. It also shows that the net load on the platform will likely reduce after the initial rubbling peak because the grounded rubble area will increase with ice motion and its sliding resistance will lower the net load on the platform or barrier. This process might be accounted for in a more sophisticated treatment, especially noting that the thickest ice might be acting against a grounded rubble field established earlier in the winter (with thinner ice). However, caution is recommended because the earlier rubble may be removed by operations.

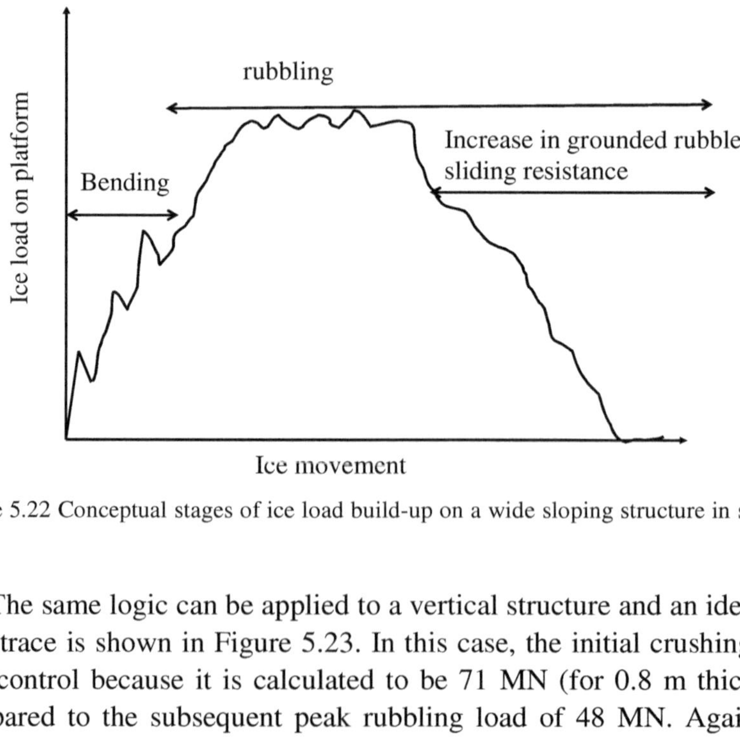

Figure 5.22 Conceptual stages of ice load build-up on a wide sloping structure in shallow water

The same logic can be applied to a vertical structure and an idealized load trace is shown in Figure 5.23. In this case, the initial crushing load will control because it is calculated to be 71 MN (for 0.8 m thick ice) compared to the subsequent peak rubbling load of 48 MN. Again this assumes equal thickness during both interactions, which may be conservative if initial rubbling occurs with thinner ice and is not removed by operations.

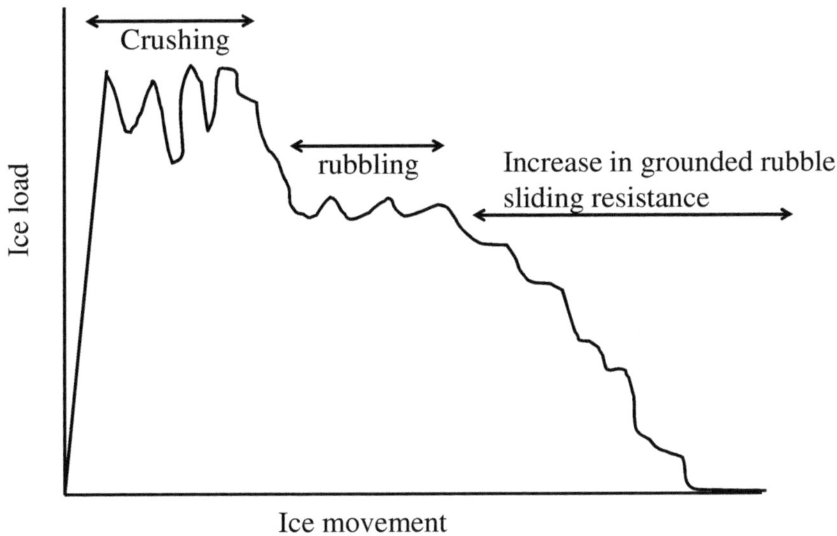

Figure 5.23 Conceptual stages of ice load build up on a wide vertical structure in shallow water

5.6.2 *First-year Ridge Loads in Shallow Water*

Ridge keels in shallow water will be limited by the water depth, so ridges in very shallow water are not likely to be a controlling ice feature for ice loads. There is a high probability that prior to ridge interaction, level ice will have reacted with the platform and formed a grounded rubble field which will protect the platform from the ridge.

Nevertheless, there may be some water depths and situations where there is no protective rubble and the keel is deep enough to require the keel loads to be considered. Considering the nature of the keel failure modes already discussed in Section 5.5, the critical interaction is when the keel is just clearing the sea floor. In this case the passive failure mode is restricted because the rubble cannot get pushed down. Initially it can on the flank of the ridge, but not when penetration has reached the maximum keel thickness. In reality the downward failure will be inhibited prior to a penetration corresponding to maximum keel depth because the failed rubble from the flank of the ridge will fill in the space between the flank of the ridge and the sea floor. If downward failure is inhibited by the sea floor then the load will be controlled by plug failure.

Depending on the width of the ridge this may increase the potential load significantly.

Two examples are reviewed. In the first, the ridge is triangular shaped and narrow. The scenario location is the Caspian Sea in 10 m water depth (accounting for surge) which limits the keel to 8.8 m (assuming a consolidated layer thickness of 1.2 m). Other input parameters are as given in the spreadsheet reproduced as Figure 5.24 where results are included. As indicated, as penetration proceeds, the effective ridge depth is initially less than the water depth due to the triangular shape of the keel. At about 5m penetration the effective keel depth is equal to the water depth and the Dolgopolov load is assumed to grow very rapidly because downwards failure of the keel in stopped by the seafloor. The actual load is now the plug failure load, and as shown in Figure 5.24 has a value of about 50 MN at this penetration. The total limit stress ridge load is then about 150 MN when the crushing load of the refrozen layer is added. However, the limit force for a 0.8 m thick ice sheet rubbling behind a 300 m long ridge gives a lower load of 65 MN, and will therefore control for the ridge scenario. A rafted ice scenario in the Caspian should also be checked, and the same scenario put into a probabilistic model gave about 75 MN (Jordaan et al [43]).

In the second example, the ridge is wider with a 20 m bottom width (Figure 5.25); now the plug load is even higher at about 105 MN for a total limit stress load of about 205 MN. However, limit force load is not increased assuming the ridge length is the same, so again this condition will control (or the rafted ice case).

This short review of ridges in shallow water has attempted to show that the normal method of Dolgopolov (as described in ISO [13]) will likely be non-conservative for the limit stress load assessment. But depending on ridge lengths and whether failure behind the ridge in the thinner ice is likely, the limit force load appears to control. Clearly, the above treatment is still simplistic and improvements are encouraged. One concern is whether the plug failure will be increased by friction of the plug on the sea floor. This would further increase the limit stress load. Conversely, perhaps the pushing down of keel rubble into the sea floor would initiate a heavily grounded situation which would shield the platform from the driving forces.

First Year Ridge Keel Loads

Developed by K.R. Croasdale and Associates Ltd.

- This version 1998 (with variable cohesion - assumed linear from bottom to top)
- To be used for research purposes only. Not valid for design without consultation with K.R. Croasdale
- To use the table, change the values in the Initial Data section of the worksheet. (all input numbers are blue)
- Top width calculated assuming 35 degree keel angle
- Dolgopolov Load modified to match ISO
- Keel just clearing sea floor: Dolgopolov load inhibited when pentration depth matches keel depth

Initial Data:

Structure width (m)	100
Keel depth (m)	8.8
Keel top width (m)	25.08
Keel bottom width (m)	0
Ice weight density (kNm^3)	8.90
Water weight density (kNm^3)	10.25
Rubble porosity	0.30
Rubble friction angle (deg)	0
Omega (deg)	0
Cohesion of keel (top) (kPa)	24
Cohesion of keel (bottom) (kPa)	12
Penetration increment (m)	0.627
Max. penetration (m)	12.54
Surcharge factor on width	0.1
Sheet ice thickness (m)	0.8
Refrozen layer thickness (m)	1.2
Cr	1.8
Strength reduction factor	1
Ice ridge length (m)	300

Calculated Data:

Effective Buoyancy (kNm^3)	0.95
Pass. press. coeff.	1.00
Crushing Load (MN) (ISO) (SA)	100.78
Driving force (MN) (KRC A)	65.04
Driving Force (MN) (KRC B)	45.90

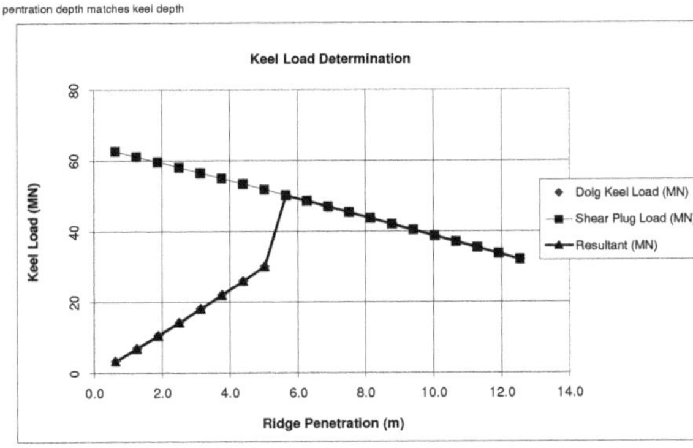

Max keel load 50.22 MN
Total ridge load 151.00 MN
(Limit Stress)

Figure 5.24 Triangular ridge in shallow water. At about 5 m penetration the Dolgopolov failure mode is inhibited by the sea floor and plug failure controls.

First Year Ridge Keel Loads

Developed by K.R. Croasdale and Associates Ltd.

- This version 1998 (with variable cohesion - assumed linear from bottom to top)
- To be used for research purposes only. Not valid for design without consultation with K.R. Croasdale
- To use the table, change the values in the Initial Data section of the worksheet. (all input numbers are blue)
- Top width calculated assuming 35 degree keel angle
- Dolgopolov Load modified to match ISO
- Keel just clearing sea floor: Dolgopolov load inhibited when pentration depth matches keel depth

Initial Data:

Structure width (m)	100
Keel depth (m)	8.8
Keel top width (m)	45.08
Keel bottom width (m)	20
Ice weight density (kNm^3)	8.90
Water weight density (kNm^3)	10.25
Rubble porosity	0.30
Rubble friction angle (deg)	0
Omega (deg)	0
Cohesion of keel (top) (kPa)	24
Cohesion of keel (bottom) (kPa)	12
Penetration increment (m)	0.627
Max. penetration (m)	12.54
Surcharge factor on width	0.1
Sheet ice thickness (m)	0.8
Refrozen layer thickness (m)	1.2
Cr	1.8
Strength reduction factor	1
Ice ridge length (m)	300

Calculated Data:

Effective Buoyancy (kNm^3)	0.95
Pass. press. coeff.	1.00
Crushing Load (MN) (ISO) (SA)	100.78
Driving force (MN) (KRC B)	65.04
Driving Force (MN) (ISO max)	104.31

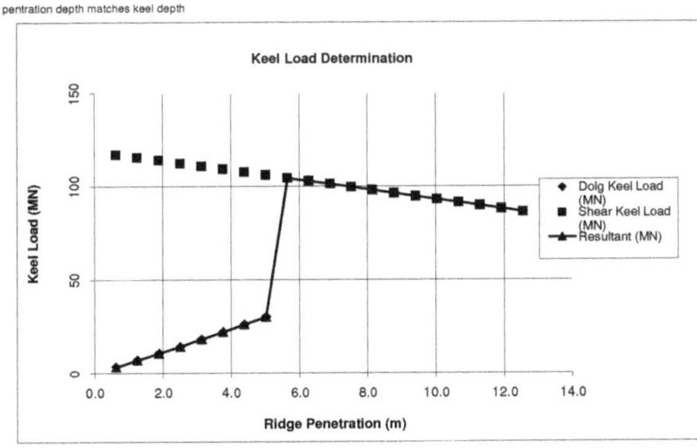

Max keel load 104.56 MN
Total ridge load 205.33 MN

Figure 5.25 Wide ridge (20 m bottom width) in shallow water

5.7 Multi-leg and Multi-hulled Platforms

5.7.1 *Multi-leg (with Vertical Legs)*

A multi-leg supported platform could be the upper part of a jacket or of a gravity base structure or a jack up. The potential advantages of a multi-leg through the water line (or ice line) are to optimize wave and ice loads and to give better support to the deck against seismic events. The disadvantage is the uncertainty surrounding ice interaction, which relates to how ice ridges interact and whether the rubble of the keels creates a plug of ice which blocks further ice motion between the legs. This is termed ice blocking, and creates an effective width which may be greater than the sum of the width of the legs.

Figure 5.26 shows that the effects of having a multi-leg structure can be complex. The global load will be less than the load on an individual leg multiplied by the number of legs. A leg factor less than 1 accounts for that effect, so that

$$\text{global load} = \text{individual leg load} \times \text{number of legs} \times \text{leg factor}$$

The leg factor will depend on the ice movement direction and leg spacing. Clearly, for a four legged structure, there is one direction when the leg factor can be as low as 0.5. For widely spaced legs, the leg factor can approach 1. In a study reported by Wang et al [44], based on model tests, the leg factor had a mean value of 0.71 and a standard deviation of 0.08. In this example a value of 0.7 will be used for four legs, 0.9 for three legs and 1 for two legs. In a probabilistic load calculation the leg factor can be given a distribution.

This approach relates to the overall global load on the structure. Each individual leg still needs to be designed for the full individual leg global load. A potential worst case would occur if the interior of the legs gets plugged with ice rubble. This is less likely to occur for level ice, but is more likely for a ridge given the large mass of keel rubble that has to be pushed through. There is a danger that if the ice plug persisted and became consolidated that the effective width of the structure increases to that of the ice plug plus the legs (see Figure 5.26). Unless we can be certain that plugging will not occur, the global loads should be checked for both the plugged and unplugged situation. In the plugged case, it may be more realistic to use the line load for rubbling (equation 5.2.8) rather than crushing over the sections of the width where the ice is failing against blocking ice.

The methods recommended in ISO [13] mirror the above points and give a similar leg factor range for four legs (about 0.67 to 0.79).

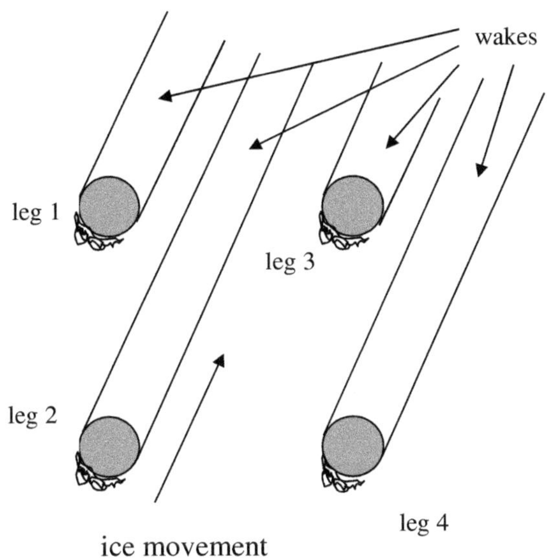

Figure 5.26 Typical normal ice action on a multi-leg structure

In the example above, legs 2 and 4 are subject to the full ice load, whereas legs 1 and 3 experience lower loads because of the closeness of the free edges. However, the effective global load reduction obviously depends on the ice movement direction as well as leg diameter and spacing.

The individual leg loads for a range of leg diameters are given in table 5.15 for a hypothetical (but severe) first-year ridge environment. These loads are based on the methods for crushing and keel loads already reviewed.

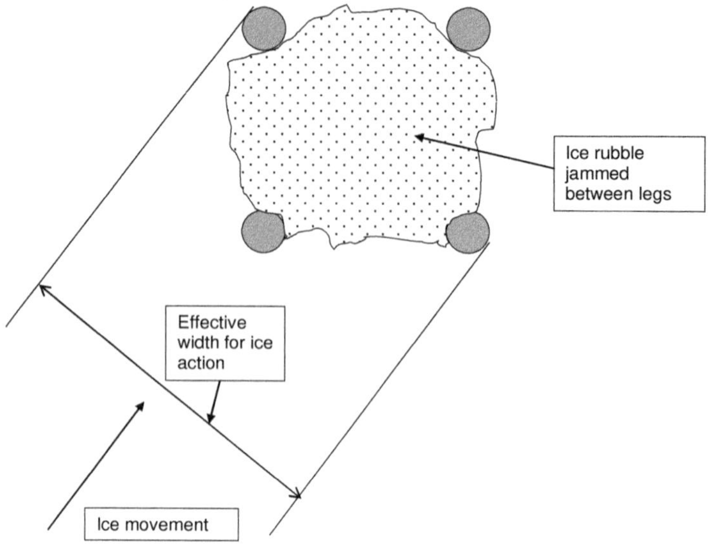

Figure 5.27 Increase in effective width due to ice blocking on a multi-leg structure

The global loads for a range of leg diameters, spacings and number of legs are given in table 5.16. Note that the angle of attack varies between cases examined in order to maximize the blocking width. The complexity of the table is an indication of the various contributions that have to be calculated as well as the range of possible outcomes that have to be considered.

Table 5.15 Individual leg loads; multi-leg platform – typical severe first-year ridge interaction

	Multi leg - Vertical: Individual leg loads					
	strength reduction factor =			0.75	Cr =	2.4
Leg Dia	RF Layer Thickness	Total Keel Depth	Rubble Depth	Leg Crushing Load	Leg Keel Load	Leg Total Load
(m)	(m)	(m)	(m)	(MN)	(MN)	(MN)
10	4.50	30.00	25.50	45.40	27.40	72.80
15	4.25	29.00	24.75	60.76	33.84	94.60
20	4.00	28.50	24.50	73.44	39.31	112.75
25	3.50	28.00	24.50	78.97	44.16	123.13
25	3.50	28.00	24.50	78.97	44.16	123.13

Table 5.16 Multi-leg global loads

Leg Dia	Leg CL Spacing	No of Legs	Leg Factor	Blocked Width (BW)	Keel Load on BW	RF Layer Thicknes	Crushing Global load on BW	Rubbling Global load on BW	Unblocked Global Load	Extreme Total Blocked Global Load	Rubbling Total Blocked Global Load	Limit Force Range
(m)	(m)			(m)	(MN)	(m)	(MN)	(MN)	(MN)	(MN)	(MN)	(MN)
10	40	4	0.7	66.4	73.44	3.50	179.39	86.03	200.56	252.83	159.47	188 - 245
15	50	4	0.7	85.5	83.78	3.50	221.83	101.65	253.90	305.61	185.43	188 - 245
20	60	3	0.9	80	80.93	3.50	209.78	97.29	279.21	290.71	178.22	188 - 245
25	75	2	1	100	90.90	3.50	253.03	112.72	248.83	343.93	203.63	188 - 245
25	80	4	0.7	137.8	107.42	3.00	290.11	117.56	328.53	397.53	224.99	188 - 245

5.7.2 *Multi-leg Structure with Conical Collars on the Legs*

The same general approach can be taken. It is likely that using the leg factors derived for vertical structures will be conservative for sloping collars. This is because the ice failure zone for a single leg will be wider when the ice fails in flexure compared to crushing. Again however, it will be difficult to dismiss the blocking situation, especially for a ridge, and this may therefore still control the global load for the whole platform.

5.7.3 *Multi-caisson Systems and Ice Barriers*

In recent years, the use of ice barriers to protect platforms in shallow water has been developed. As well, designs have been proposed with twin caissons supporting a common deck. These arrangements are not simple to analyse and guidance in ISO 19906 [13] is limited. The following is based on the paper published by Croasdale at POAC 2011 [42].

A typical deployment with two barriers on each side of a central platform is shown in Figure 5.28. In the example shown, the central platform is obviously completely protected from East-West ice movements (assumed across the page). North-South ice movements act on the ends of the platform, but these are of reduced dimension, so the design ice loads for the central platform can be lower than for an unprotected platform.

One concern is that oblique loads will act on a greater exposed width than the ends of the platform. On the other hand, the ice tongue shown in Figure 5.28 that acts on the side of the platform has a free edge, and is likely to be cracked because of prior interaction with the upstream ice barrier. It may also fail globally rather than in crushing against the side

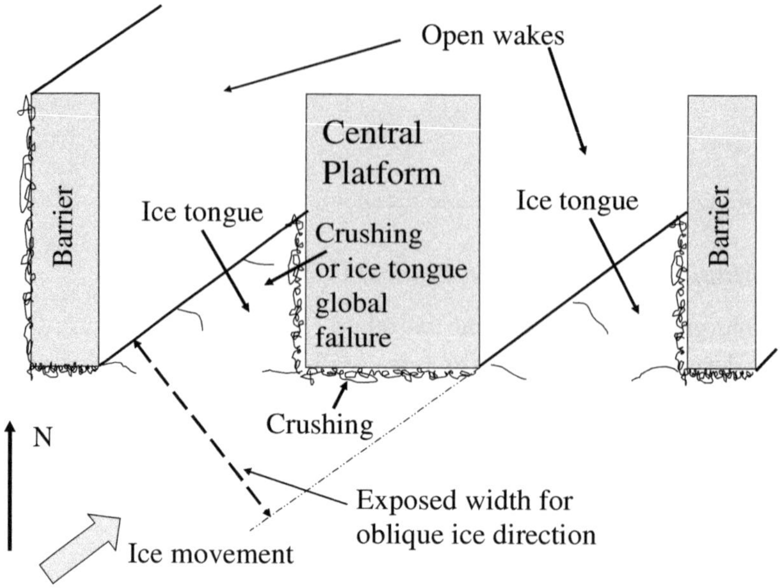

Figure 5.28 Oblique movement with ice barriers

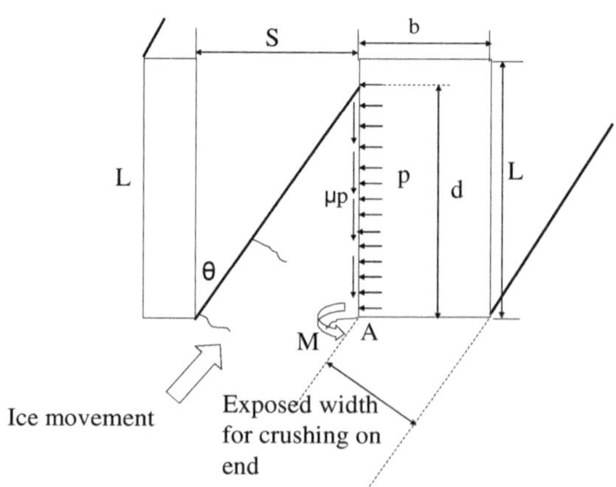

Figure 5.29 Geometric parameters

of the platform (by in-plane bending). The pressure p along the side of the platform (assuming vertical sides) will then be the lower of either crushing (possibly reduced from ISO values because of the free edge and cracking) or the pressure to fail the ice tongue in in-plane bending. ISO does not cover in-plane bending. The limit to p from in-plane bending can be derived by taking moments about the corner A in Figure 5.29 and is

$$M_A = 0.5pd^2h \tag{5.7.1}$$

where M_A is the bending moment at A, p is the ice pressure, d is the length of ice contact and h is the ice thickness.

From elastic bending theory, the flexural stress σ_f can be related to bending moment and the cross-section of the beam as follows,

$$\sigma_f = \frac{6M_A}{hS^2} \tag{5.7.2}$$

where S is the gap between barrier and platform. Combining the above equations

$$p = 0.33\sigma_f\left(\frac{S}{d}\right)^2 \tag{5.7.3}$$

and

$$d = \frac{S}{\tan\theta} \tag{5.7.4}$$

where θ is the angle of ice direction from the long axis; d cannot exceed the length L of the platform. So, by substitution,

$$p = 0.33\sigma_f \tan^2\theta \tag{5.7.5}$$

except that for angles of θ that would lead to d greater than L, p will be independent of θ and given by.

$$p = 0.33\sigma_f\left(\frac{S}{L}\right)^2 \tag{5.7.6}$$

There is also a frictional shear associated with p. The ice load F_s on the side of the structure resolved in the direction of motion is therefore

$$F_S = pdh\sin\theta + \mu pdh\cos\theta \tag{5.7.7}$$

Note that p is the lower of either that derived for the ice tongue failing in in-plane bending or the value based on crushing. Crushing will control for small values of d (high values of θ) because the moment arm to fail the ice tongue in bending is much reduced.

The ice load F_E due to crushing on the end of the structure in the direction of motion is estimated using the projected width normal to the ice motion (see Figure 5.29).

$$F_E = \sigma_c b\cos\theta \tag{5.7.8}$$

where σ_c is the ice crushing pressure as derived using ISO and b is the end width of the platform (see Figure 5.29).

The total global load is given by $F_S + F_E$. Figure 5.28 shows the results of using the above theory for the following inputs; platform length, 100 m; platform width, 50 m; barrier length, 100 m; separation distance, 80 m; ice thickness, 1 m; ice flexural strength, 350 kPa; side friction coefficient, 0.1. A strength reduction of 0.8 is applied for crushing on the platform side because of the free edge; this is adjustable based on judgment.

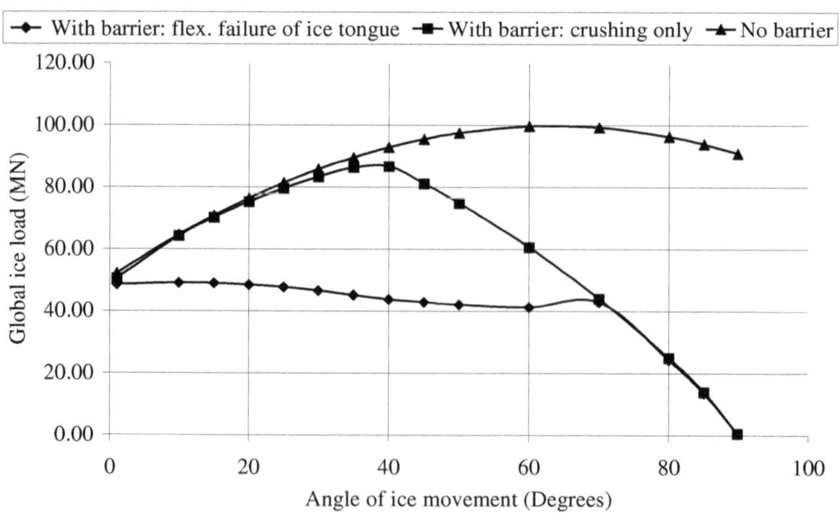

Figure 5.30 Loads on a platform with and without barriers as function of ice direction and analysis methods

The results in Figure 5.30 show that using the approach using the global failure of the ice tongue gives a maximum ice load of about 49 MN for 100 m barriers placed 80 m from the platform long side. This compares with a load of about 100 MN for the same structure with no barriers. Prior to this new theory, the predicted load with barriers would have been about 86 MN.

The approach can be applied to investigate how the loads vary with barrier separation; this is shown in Figure 5.31. The results show that the separation can be increased from 80 to 100 m and even to 120 m without any significant increase in maximum load.

The same principles of analysis can also be applied to more complex arrangements, for example using more barriers, various barrier lengths and platforms with spaced substructures supporting a common deck.

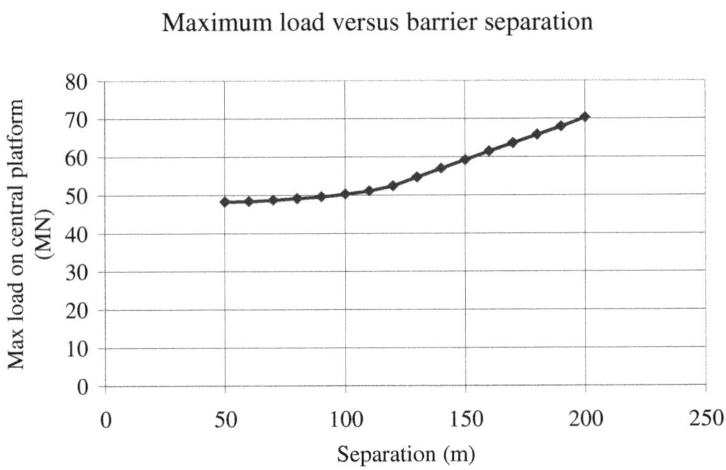

Figure 5.31 Example of how the load on a central platform can vary with separation of protective barriers from the platform

5.8 Limit momentum (limit energy) Ice Loads

5.8.1 *Principles and Application to a Vertical Structure*

In the original discussion of ice load limits in Croasdale [5], the term limit momentum was coined for the interaction of an isolated ice mass such as an iceberg or a thick floe with a platform; recognizing that the resulting impact load on the platform would be governed by the initial momentum of the ice feature and how it is dissipated. Practically, it is easier to conduct the calculation by considering the dissipation of kinetic energy, so the term limit energy is now preferred (ISO 2010 19906 [13]).

Overall, it is very simple mechanics: the loss of kinetic energy of the ice as it is brought to rest is equated to the work done by the ice force. The work done by the ice force is the integration of the ice force multiplied by relative penetration between ice and structure. A simple example would be a circular structure and a rectangular floe as depicted in Figure 5.32. Shown in the figure is a thinned edge (because most multi-year floes will not have a vertical edge equal to full floe thickness). The ridging force of thinner ice "*w*" on the back of the floe is also shown and has to be accounted for in the calculation, if the thick feature is surrounded by pack ice.

For head-on loading, the most conservative scenario, the general equation is,

$$\frac{MgV^2}{2} = \int p_x B_x h_x dx - \int w_x L dx \tag{5.8.1}$$

where M is the floe mass, g is the acceleration due to gravity, V is the floe velocity prior to impact, p_x is the ice pressure across width B_x at penetration x, h_x is the thickness across B_x, w_x is the ridge building force behind the floe at penetration x and L is the floe width and length dimension for an equivalent square floe.

The equation is solved for B_{max} which occurs when the floe has stopped. If B_{max} is less than D (the structure width) then the peak impact load will be less than the limit stress load based on full relative penetration. After the impact, if pack ice is present, the load when the floe is stopped will be that given by the pack ice driving force.

For a square floe and circular structure, the width B_x at relative penetration x is given by,

$$B_x = (4x(d-x))^{0.5} \tag{5.8.2}$$

If there is no driving force (an isolated floe), the solution of equations 5.8.1 and 5.8.2 is achievable by simple integration, assuming too that the relationships for h_x and p_x are also simple functions of x). The build-up of driving force as a function of x is not simple. It might be assumed that the driving force builds up to a value given by equation 5.2.8 as the volume of ice rubble generated behind the floe reaches a value equal to

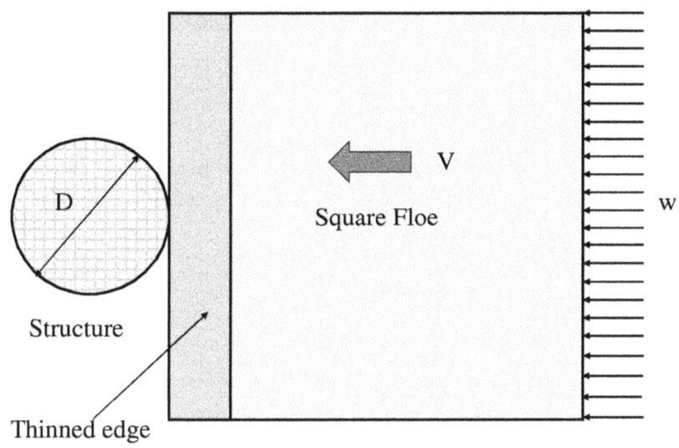

Figure 5.32 Idealized impact case: Circular structure and square floe

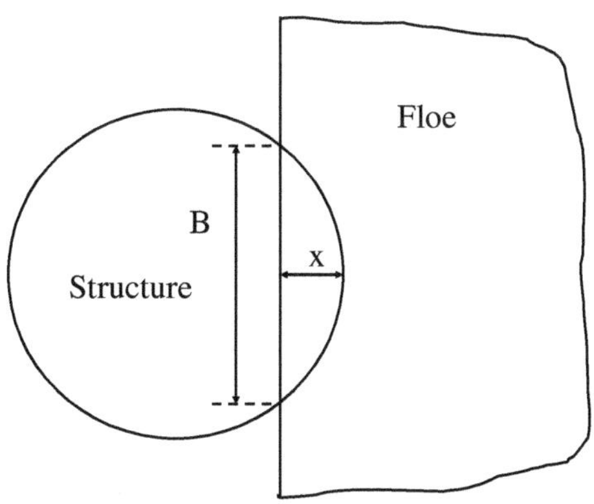

Figure 5.33 Geometric parameters: simple penetration of circular structure into a floe

the volume of ice rubble compatible with the assumptions in deriving that equation. It is possible to develop such a relationship, but not so easy to apply as a general closed form solution. The authors therefore decided to set up the calculation numerically by stepping through small increments of x and calculating the force on the structure for each value of x until the floe stops. In this calculation, the driving force is assumed to build up to its maximum steady state value after a specified number of increments of relative penetration (between 20 and 50).

Table 5.17 Spreadsheet inputs and outputs for impact loads of a thick ice feature on a circular structure; surrounded by pack ice 1.6 m thick.

Limit Momentum Calculation with Driving Force

For R&D Only: Not to be used for design or contract work without consulting with Ken Croasdale

Formulated by K. R. Croasdale & Associates Ltd.

Spreadsheet: moment_d3 B

V 1 Initial Spreadsheet without driving force- October 1995
V 2 Revised - December 1996
V3 Formatting and Graphs - January 1997
V4 Driving Force builds up over 20 penetration increments - June 1997
V5 CSA Ice Crushing formula added 2006
V6 ISO crushing formula added 2011 — p = Cr *(t^-0.14)*(D^-0.16)
V7 Driving force - KRCA new formula B 2011 — w = 1360*(d^-0.34)*(t^1.1)

Input Data:

Structure Dia. (m)	80
Floe Thickness (m)	12
Floe Dia (m)	1,000
Ice Density (kg/m^3)	900
Floe Speed (m/s)	0.25
Added mass factor	1.1
Cr	2.8
Pack Ice Thickness (m)	1.6
Pack Ice Force (kN/m)	218
Initial 'KE (MN-m)	371.25
Pen. Incr. (m)	0.2
Edge Thick. (m)	1.5
Edge dist.(m)	5
Limit Force Load (MN)	217.81
Mass + added mass (mln. kg.)	11,880

Results Summary

Limit Momentum Load	**(MN)**	**420**
Ice pressure - max penetration	**(MPa)**	**1.140**
Limit Stress Load	**(MN)**	**942**
Limit Force Load	**(MN)**	**218**

Results by penetration

x	B	A	t	p	Crushing Load	Driving Force	Δ 'WD	Total WD	Vx
0	0.000	0.000	1.500	0.000	0.000	0.000	0	0	0.25
0.2	7.990	15.341	1.920	1.833	28.115	10.890	1.722	1.722	0.249
0.4	11.285	26.408	2.340	1.687	44.545	21.781	3.999	5.721	0.248
0.6	13.804	38.100	2.760	1.596	60.807	32.671	5.090	10.811	0.246
0.8	15.920	50.625	3.180	1.529	77.424	43.561	6.200	17.011	0.244
1	17.776	63.995	3.600	1.477	94.503	54.452	7.391	24.403	0.242
1.2	19.448	78.183	4.020	1.433	112.061	65.342	8.677	33.080	0.239
1.4	20.980	93.151	4.440	1.396	130.084	76.232	10.057	43.137	0.235
1.6	22.400	108.864	4.860	1.365	148.550	87.123	11.528	54.665	0.231
1.8	23.728	125.286	5.280	1.336	167.436	98.013	13.085	67.750	0.226
2	24.980	142.386	5.700	1.311	186.719	108.903	14.724	82.474	0.220
2.2	26.166	160.134	6.120	1.289	206.376	119.794	16.440	98.913	0.214
2.4	27.294	178.502	6.540	1.268	226.387	130.684	18.228	117.142	0.207
2.6	28.372	197.468	6.960	1.249	246.733	141.574	20.086	137.228	0.198
2.8	29.405	217.007	7.380	1.232	267.397	152.465	22.009	159.237	0.189
3	30.397	237.099	7.800	1.216	288.364	163.355	23.994	183.231	0.178
3.2	31.353	257.726	8.220	1.201	309.619	174.245	26.038	209.270	0.165
3.4	32.276	278.867	8.640	1.187	331.148	185.136	28.139	237.408	0.150
3.6	33.169	300.508	9.060	1.174	352.939	196.026	30.293	267.701	0.132
3.8	34.033	322.632	9.480	1.162	374.980	206.916	32.498	300.199	0.109
4	34.871	345.225	9.900	1.151	397.261	217.807	34.752	334.951	0.078
4.2	35.685	368.272	10.320	1.140	419.772	217.807	38.142	373.093	0.000
4.2	35.685	368.272	10.320	1.140	217.807	217.807	0.000	373.093	0.000
4.2	35.685	368.272	10.320	1.140	217.807	217.807	0.000	373.093	0.000
4.2	35.685	368.272	10.320	1.140	217.807	217.807	0.000	373.093	0.000
4.2	35.685	368.272	10.320	1.140	217.807	217.807	0.000	373.093	0.000
4.2	35.685	368.272	10.320	1.140	217.807	217.807	0.000	373.093	0.000
4.2	35.685	368.272	10.320	1.140	217.807	217.807	0.000	373.093	0.000
4.2	35.685	368.272	10.320	1.140	217.807	217.807	0.000	373.093	0.000
4.2	35.685	368.272	10.320	1.140	217.807	217.807	0.000	373.093	0.000
4.2	35.685	368.272	10.320	1.140	217.807	217.807	0.000	373.093	0.000
4.2	35.685	368.272	10.320	1.140	217.807	217.807	0.000	373.093	0.000
4.2	35.685	368.272	10.320	1.140	217.807	217.807	0.000	373.093	0.000
4.2	35.685	368.272	10.320	1.140	217.807	217.807	0.000	373.093	0.000
4.2	35.685	368.272	10.320	1.140	217.807	217.807	0.000	373.093	0.000
4.2	35.685	368.272	10.320	1.140	217.807	217.807	0.000	373.093	0.000
4.2	35.685	368.272	10.320	1.140	217.807	217.807	0.000	373.093	0.000
4.2	35.685	368.272	10.320	1.140	217.807	217.807	0.000	373.093	0.000
4.2	35.685	368.272	10.320	1.140	217.807	217.807	0.000	373.093	0.000
		Maximum impact load =			**419.772**	**MN**			

Table 5.17 is a spreadsheet with typical inputs and outputs for this scenario. The results provide some interesting insights. First, for this size of floe and thickness (1000 m by 12 m), and floe speed (0.25 m/s), surrounded by pack ice, 1.6 m thick, the full limit stress load of 942 MN is never reached. The load is controlled by either the limit momentum load 420 MN or limit force load 218 MN, whichever is the larger. In this case, the limit momentum load of 420 MN will control.

The floe speed of 0.25 m/s is about 22 km/day which is fairly fast for the average drift of winter pack ice in the Arctic (ten – tenths coverage). If we drop the impact speed to 0.15 m/s, the various loads are as shown in table 5.18. In this case, the value for limit momentum is 337 MN and the limit force load is again 218 MN.

For extreme design, it would be hard to justify such a low drift speed, even in 10/10 ice. The floe size of 1km may be low. A range of sensitivities is shown in Table 5.19 with impact speeds up to 0.5 m/s and floes to 5 km. Also shown are loads for isolated floes as well as for floes within the pack ice.

Table 5.18 Loads for a lower impact speed of 0.15 m/s

Limit Momentum Calculation with Driving Force				
	For R&D Only: Not to be used for design or contract work without consulting with Ken Croasdale			
	Formulated by K. R. Croasdale & Associates Ltd.			
	Spreadsheet: moment_d3 B		V 1 Initial Spreadsheet without driving force- October 1995	
			V 2 Revised - December 1996	
Input Data:			V3 Formatting and Graphs - January 1997	
			V4 Driving Force builds up over 20 penetration increments - June 1997	
Structure Dia. (m)		80	V5 CSA Ice Crushing formula added 2006	
Floe Thickness (m)		12	V6 ISO crushing formula added 2011	p = Cr *(t^-0.14)*(D^-0.16)
Floe Dia (m)		1,000	V7 Driving force - KRCA new formula B 2011	w = 1360*(d^-0.34)*(t^1.1)
Ice Density (kg/m^3)		900		
Floe Speed (m/s)		0.15		
Added mass factor		1.1		
Cr		2.8		
Pack Ice Thickness (m)		1.6		
Pack Ice Force (kN/m)		218		
Initial 'KE (MN-m)		133.65		
Pen. Incr. (m)		0.15		
Edge Thick. (m)		1.5		
Edge dist.(m)		5		
Limit Force Load (MN)		217.81		
Mass + added mass (mln. kg.)		11,880		
Results Summary				
Limit Momentum Load	**(MN)**	**337**		
Ice pressure - max penetration	**(MPa)**	**1.184**		
Limit Stress Load	**(MN)**	**942**		
Limit Force Load	**(MN)**	**218**		

Table 5.19 Sensitivity analysis

Case 1: Isolated floes					
Floe dia.	Impact load (0.25ms-1)	Impact load (0.5ms-1)	Limit stress load (12m)	Ridge load (25m)	
m	MN	MN	MN	MN	
1000	267	535	942	1770	
3000	632	861	942	1770	
5000	795	942	942	1770	
Case 2: Floes in 1.6 m pack ice					
Floe dia.	Impact load (0.25ms-1)	Impact load (0.5ms-1)	Limit stress load (12m)	Ridge load (25m)	Static limit force
m	MN	MN	MN	MN	MN
1000	420	589	942	1770	218
3000	773	933	942	1770	450
5000	935	942	942	1770	630

The results in this table 5.19 also give us some good insights into the complex situation of multi- year floe interactions with a vertical structure. The results indicate that in an extreme situation with floes greater than about 3 km moving within pack ice at 0.5 m/s or greater, the simply calculated limit stress load of 942 MN will be the deterministic design ice load. The lower limit force load of 493 MN cannot be invoked because before this load occurs, the initial kinetic energy of the thick floe has to be dissipated as it is brought to rest. The associated limit momentum load is as high as the limit stress load.

However, this analysis does tell us that the extreme ridge load of 1770 MN shown in the table may never occur unless the ridge is within about 50 m of the floe edge. This is because the floe will be stopped within that edge distance. This insight may not help in deterministic design, but when considering design loads at certain probability levels, such a consideration becomes important. This review also helps us realize that energy absorbing devices such as underwater berms and stepped structures may reduce the limit momentum load, potentially allowing the limit force load to control.

The model presented so far is quite simple. More complex versions have been developed which account for eccentric hits and more complex shapes for both ice and structure. That can be important when looking at iceberg impacts; sensitivities to eccentric hits are given in table 5.20 for the 3 km floe case at two speeds. It can be seen that at maximum eccentricity for a 3 km floe, the load is reduced by about 16%; so as a first approximation the complexity of accounting for eccentricities may not be worthwhile.

Table 5.20 Effects of eccentricity on Limit momentum (energy) loads.

floe speed	impact load (MN)		
	eccentricity 0	eccentricity 1000 m	eccentricity 1500 m
0.25 m/s	632	570	528
0.5 m/s	861	777	710

5.8.2 *Sloping Structures*

The same overall approach can also be applied to sloping platforms. In fact, more energy will likely be absorbed prior to peak load than with a vertical structure. This is because the peak load on a sloping structure is not the initial breaking term but is associated with mature ride up and rubble formation which requires more relative movement between ice and structure and hence more work is done by the ice force as it builds up more gradually to its potential peak value.

An approximate approach for a sloping platform is to assess how much ice motion is required for the mature ride-up, and thence for the peak load to be reached. The peak ice load is derived using the methods for sloping structures described in section 4.5. As an approximation, we can assume the peak ice load builds up linearly, so that at any value of penetration the work done by the ice force is half the product of the ice load and the penetration. That approach is used to modify the spreadsheet previously developed for a vertical platform. In this case the peak ice load is entered together with the estimated penetration. As an example, we use the scenario examined in section 4.5.5, where a 12 m thick floe acts on an 80 m wide sloping structure. In that example it was proposed that the peak load would be associated with a ride up of 25 m. We can approximate this as a relative horizontal penetration of about 35m.

Results are shown in table 5.21 for a 2,500 m floe moving at 0.25 m/s surrounded by pack ice 1.6 m thick. As the floe slows down on impact, the ridge building force starts to build up. It is assumed that about 100 increments of penetration are required to build up to the maximum limit force as estimated in section 5.3. In this example, the

Table 5.21 Spreadsheet inputs and outputs for impact loads of a 12 m thick ice feature on a typical Arctic sloping structure; floe surrounded by pack ice 1.6 m thick.

Limit Momentum Calculation Sloping Structure with Driving Force

For R&D Only: Not to be used for design or contract work without consulting with Ken Croasdale

Formulated by K. R. Croasdale & Associates Ltd.

Spreadsheet: moment_d3 B — Sloping structure

V7 Driving force - KRCA new formula B 2011

V8Modified for sloping structure Aug 2011

V9 Driving force builds up in 100 increments

Input Data:

Structure Dia. (m)	80	
Floe Thickness (m)	12	
Floe Dia (m)	2,000	
Ice Density (kg/m^3)	900	
Floe Speed (m/s)	0.25	
Added mass factor	1.1	
Peak Flexural load (MN)	400	
Ice penetration to peak load (m)	35	
Pack Ice Thickness (m)	1.6	
Pack Ice Force (kN/m)	172	
Initial 'KE (MN-m)	1,485.00	
Pen. Incr. (m)	0.6	
Limit Force Load (MN)	344.15	calculated
Mass + added mass (mln. kg.)	47,520	

Results Summary

Limit Momentum Load	**(MN)**	**267**
Limit Stress Load	**(MN)**	**400**
Limit Force Load	**(MN)**	**344**

Results by penetration

x	Impact Load	Driving Force	Δ 'WD	Total WD	Vx	Time
0	0.000	0.000	0	0	0.25	0
0.6	6.857	3.442	1.025	1.025	0.250	2.400
1.2	13.714	6.883	3.074	4.099	0.250	4.802
1.8	20.571	10.325	5.123	9.222	0.249	7.208
2.4	27.429	13.766	7.173	16.395	0.249	9.618
3	34.286	17.208	9.222	25.617	0.248	12.035
3.6	41.143	20.649	11.272	36.889	0.247	14.461
4.2	48.000	24.091	13.321	50.210	0.246	16.897
4.8	54.857	27.532	15.370	65.580	0.244	19.345
5.4	61.714	30.974	17.420	83.000	0.243	21.808
6	68.571	34.415	19.469	102.469	0.241	24.286
6.6	75.429	37.857	21.518	123.987	0.239	26.783
7.2	82.286	41.298	23.568	147.555	0.237	29.301
7.8	89.143	44.740	25.617	173.172	0.235	31.843
8.4	96.000	48.181	27.667	200.838	0.232	34.410
9	102.857	51.623	29.716	230.554	0.230	37.006
9.6	109.714	55.064	31.765	262.320	0.227	39.634
10.2	116.571	58.506	33.815	296.134	0.224	42.297
10.8	123.429	61.947	35.864	331.998	0.220	45.000
11.4	130.286	65.389	37.913	369.912	0.217	47.746
12	137.143	68.830	39.963	409.874	0.213	50.541
12.6	144.000	72.272	42.012	451.886	0.209	53.390
13.2	150.857	75.714	44.061	495.948	0.204	56.299
13.8	157.714	79.155	46.111	542.059	0.199	59.275
14.4	164.571	82.597	48.160	590.219	0.194	62.326
15	171.429	86.038	50.210	640.429	0.189	65.462
15.6	178.286	89.480	52.259	692.688	0.183	68.696
16.2	185.143	92.921	54.308	746.996	0.176	72.040
16.8	192.000	96.363	56.358	803.354	0.169	75.512
17.4	198.857	99.804	58.407	861.761	0.162	79.133
18	205.714	103.246	60.456	922.217	0.154	82.933
18.6	212.571	106.687	62.506	984.723	0.145	86.946
19.2	219.429	110.129	64.555	1049.278	0.135	91.224
19.8	226.286	113.570	66.605	1115.883	0.125	95.838
20.4	233.143	117.012	68.654	1184.537	0.112	100.899
21	240.000	120.453	70.703	1255.240	0.098	106.592
21.6	246.857	123.895	72.753	1327.993	0.081	113.273
22.2	253.714	127.336	74.802	1402.795	0.059	121.837
22.8	260.571	130.778	76.851	1479.646	0.015	138.091
23.4	267.429	134.219	78.901	1558.547	0.000	138.091
23.4	0.000	137.661	0.000	1558.547	0.000	138.091
23.4	0.000	141.102	0.000	1558.547	0.000	138.091
23.4	0.000	144.544	0.000	1558.547	0.000	138.091
23.4	0.000	147.985	0.000	1558.547	0.000	138.091
23.4	0.000	151.427	0.000	1558.547	0.000	138.091
23.4	0.000	154.869	0.000	1558.547	0.000	138.091
23.4	0.000	158.310	0.000	1558.547	0.000	138.091
23.4	0.000	161.752	0.000	1558.547	0.000	138.091
23.4	0.000	165.193	0.000	1558.547	0.000	138.091
23.4	0.000	168.635	0.000	1558.547	0.000	138.091
23.4	0.000	172.076	0.000	1558.547	0.000	138.091
Maximum impact load =	267.429	MN				

impact load at 267 MN does not control, but the limit force load at about 344 MN will control. With larger floes and higher speeds the controlling loads may change order, but the load cannot exceed the limit stress load, in this case 400 MN.

5.8.3 *Iceberg Impact Loads*

The method described above can also be applied to icebergs, with and without surrounding pack ice. The interaction shape of an iceberg is potentially more irregular and random than an ice floe. However, the method in the spreadsheet that incorporates a tapered edge can be used to bound a range of interaction areas and penetration for icebergs. A version has been developed for spherical ice shapes, which will generally be a reasonable approximation for well-ablated ice bergs. The worst case will be a tabular berg which can in the same way as an ice floe, only thicker.

There is a good discussion in ISO 19906 [13] on iceberg impacts, as well as in the papers by Nevel [45] and Fuglem et al [46]. The general principles are the same as reviewed here.

References

1 Parmerter, R. R. and Coon, M. D. (1973) On the mechanics of ice ridge formation.*Proceedings, Offshore Technology Conference,* Houston. OTC Paper 1810.

2 Hibler, W.D. III (1980) Modelling a variable thickness sea ice cover. *Monthly Weather Review.* Vol 108, No 12, Dec 1980.

3 Rothrock, D.A. (1975) The energetics of plastic deformation of pack ice by ridging. *J.Geophys. Res.* 80, 4514 – 4519, 1975.

4 Croasdale K. R. and Marcellus R. W. (1981) Ice forces on large marine structures. *Proceedings, IAHR Ice Symposium,* Quebec City, Canada.

5 Croasdale, K. R. (1984) The limiting driving force approach to ice loads.*Proceedings, Offshore Technology Conference,* Houston. OTC Paper 4716.

6 Metge, M., Strilchuk, A. and Trofimenkoff, P. N. (1975) On recording stresses in ice. *Proceedings, IAHR Ice Symposium,* Hanover, N.H.

7 Croasdale, K.R., Comfort, G., Graham, B.W and Marcellus, R.W. (1986) Evaluation of ice pressure sensors and strain meters in a large test basin. *Proceedings, IAHR Ice Symposium,* Iowa City, IA.

8 Croasdale, K.R., Comfort, G., Frederking, R., Graham, B.W. and Lewis, E. L. (1987) A pilot experiment to measure Arctic pack-ice driving forces. *Proceedings, Ninth International Conference on Port and Ocean Engineering under Arctic Conditions,* Fairbanks, AK.

9 Croasdale, K. R. (2011) Ice rubbling and ice interaction with offshore facilities. *Cold Regions Science and Technology*; 76–77 (2012) 37–43.

10 Croasdale, K. R., Frederking, R., Wright, B. and Comfort, G. (1992) Size effects on pack ice driving forces. *Proceedings, IAHR Ice Symposium*, Banff.

11 Croasdale K. R. (2009) Limit force Ice loads – An update. *Proceedings, International Conference on Port and Ocean Engineering under Arctic Conditions*, Luleå, Sweden.

12 Comfort, G., Singh, S, and Dinovitzer, A. (1998) Limit force ice loads and their significance to offshore structures in the Beaufort Sea. *ISOPE Journal* 8.(1) (ISSN 1053-5381)..

13 International Standards Organisation (2010) *Petroleum and natural gas industries — Arctic offshore structures*. ISO 19906:2010.

14 G.W. Timco, K. Croasdale and B. Wright. (2000) *An overview of first-year sea ice ridges*.Technical Report HYD-TR-047. PERD/CHC Report 5-112.

15 McKenna, R., et al Ice encroachment in the North Caspian Sea. (2011) *Proceedings, 21st International Conference on Port and Ocean Engineering under Arctic Conditions*, Montréal.

16 Verlaan, P., Croasdale, K.R., (2011) Ice issues relating to the Kashagan Phase 2 Development; North Caspian Sea. *Proceedings, 21st International Conference on Port and Ocean Engineering under Arctic Conditions*, Montréal.

17 Sudom, D., Timco, G., Sand, B. and Fransson L. (2011) Analysis of first - year and old ice ridge characteristics *Proceedings, 21st International Conference on Port and Ocean Engineering under Arctic Conditions*, Montréal.

18 Johnston, M., Masterson, D., Wright, B. (2009) Multi-year ice thickness: Knowns and unknowns. *Proceedings, 20th International Conference on Port and Ocean Engineering under Arctic Conditions*, Luleå, Sweden.

19 Hetenyi, M. (1946) *Beams on elastic foundation.* Scientific series, The University of Michigan Press, University of Michigan Studies, Ann Arbor.

20 Croasdale, K. R. (1975). Ice forces on marine structures. *Proceedings, Third Intl. Symposium on ice problems, IAHR*, Hanover NH, USA.

21 Ralston, T.D. (1980) Plastic limit analysis of sheet ice loads on conical structures. *Proceedings, IUTAM (International Union of Theoretical and Applied Mechanics) Symposium on physics and mechanics of ice,* Copenhagen, pp. 289-308

22 Wadhams, P. (2000) *Ice in the ocean*. Gordon & Breach, London.

23 Lewis. J.W. and Croasdale, K.R. (1978) Modelling the interaction between pressure ridges and conical shaped structures. *Proceedings of IAHR Ice Symposium*, Lulea, Sweden. Part I, pp. 165 - 196.

24 Metge, M. and Tucker, J.R. (1990) *Multifaceted cone tests: Year two: 1989 -1990.* Esso Resources Report ERCL.RS.90.11. Part of Joint Industry/Government Project.

25 Wang, Z. (1997) Ice forces on a multifaceted conical structure. PhD Thesis. Memorial University. St John's, Nfld.,Canada.

26 Croasdale, K.R. Muggeridge, D.B.(1993) A collaborative research program investigate ice loads on multifaceted conical structures. *Proceedings, Twelfth International Conference on Port and Ocean Engineering under Arctic Conditions*, Hamburg.

27 Wang, Y.S. (1984) Analysis and model tests of pressure ridges failing against conical structures. *Proceedings, IAHR Ice Symposium,* Hamburg.

28 Smirnov,V.,Sheikin, I.B., Shushlebin, A., Kharitonov, V., Croasdale, K. R., Metge, M., Ritch, R., Polomoshnov, A., Surkov, G., Wang, A., Beketsky, S., and Weaver, J.S. (1999) Large scale strength measurements of ice ridges: Sakhalin, 1998. *Proceedings of 6th Intl. Conf. on Ships and Marine Structures in Cold Regions. ICETECH'2000.* St Petersburg Russia, 2000.

29 Croasdale, K. R, Allyn, N.F.B. and Marcellus, R. W. (1990) Thermal response of ice rubble – predictions and observations. *Proceedings, IAHR Ice Symposium,* Espoo, Finland.

30 Brown, T.G., Croasdale, K. R. and Wright, B. (1996) Ice loads on Northumberland Strait Bridge piers - an approach. *Proceedings, Sixth Intl. Offshore and Polar Engineering Conf.,* Los Angeles, May 1996.

31 Dolgopolov, Y.V., Afanasiev, V. P., Koenkov, V.A., and Panilov, D. F. (1975) Effect of hummocked ice on the piers of hydraulic structures. *Proceedings, Third IAHR Ice Symposium,* CRREL, Hanover, NH.

32 Croasdale, K. R. (1980). Ice forces on fixed rigid structures. *In Working Group on Ice Forces on Structures, IAHR. A State of the Art Report, CRREL Special report 80-26.* Hanover, NH, USA

33 Cammaert, A.B., Jordaan, I., Bruneau, S., Crocker, G., McKenna, R., and Williams, A. (1993) *Analysis of ice loads on main span piers for the Northumberland Strait crossing.* Contract report for J. Muller International – Stanley Joint Venture Inc. CODA Contract number 5022.01.

34 Bruneau, S. (1997) *Development of a first-year ridge keel load model.* PhD thesis, Memorial University of Newfoundland, St. John's, Canada.

35 K.R.Croasdale & Associates Ltd. (1998) *In-situ ridge strength measurements - 1998.* A study sponsored by NRC (PERD) and Exxon Production Research Co.

36 Lemée, L. and Brown, T.G.. (2005) Review of ridge failure against the Confederation Bridge. *Cold Regions Science and Technology,* 42 (1) pp.1-15.

37 El-Seify, M., and Brown, T.G. (2006) Ice ridges interaction with Confederation Bridge piers – results of detailed video and sonar data analysis. *Proceedings, IAHR Ice Symposium,* Sapporo, Japan, 1, pp.159-166.

38 Croasdale, K.R. (2012) A simple model for first-year ridge loads on sloping structures. *Proceedings, IceTech,* Banff, 2012.

39 Beer, F.P., Johnston, E.R. and DeWolf, J.T. (2006) *Mechanics of Materials.* McGraw Hill, New York.

40 Croasdale, K., Jordaan, I. and Verlaan, P. (2011) Offshore platforms and deterministic ice actions: Kashagan Phase 2 Development: North Caspian Sea, *Proceedings, 21st International Conference on Port and Ocean Engineering under Arctic Conditions,* Montréal.

41 Kry, P. R. (1977) Ice rubble fields in the vicinity of artificial islands. *Proceedings, Ninth International Conference on Port and Ocean Engineering under Arctic Conditions,* St. John's, Canada.

42 Croasdale, K. R. (2011) Platform shape and ice interaction: a review. *Proceedings, 21st International Conference on Port and Ocean Engineering under Arctic Conditions,* Montréal.

43 Jordaan, I., Stuckey, P., Bruce, J., Croasdale, K. and Verlaan, P. (2011) Probabilistic modelling of the ice environment in the NE Caspian sea and associated structural loads. *Proceedings, 21st International Conference on Port and Ocean Engineering under Arctic Conditions,* Montréal.

44 Wang, A., Poplin, J. and Heuer, C. E. (1993) Hydrocarbon production concepts for dynamic annual ice regions. *Russian Arctic Offshore Conference*, St. Petersburg.
45 Nevel, D., (1986) Iceberg impact forces. *Proceedings, IAHR Ice Symposium*, Iowa City, IA 3, pp.345-369.
46 Fuglem, M., Muggeridge, K. and Jordaan, I.J. (1994) Design load calculations for iceberg impacts, *International Journal of Offshore & Polar Eng.*, 9 (4) (1999).

Chapter Six

Ice Forces On Floating Platforms

6.1 Introduction

Ice forces on a floating moored platform system are more complex than on a fixed platform, but the same general principles of ice mechanics apply. In designing a floating system, in addition to the global load applied from the ice to the vessel, the mooring line loads are also required. These are not necessarily the same, because of the inertia and compliance of the system. If the ice engineer can supply the ice loading ramp describing how the ice load builds up with relative distance moved by the ice, then the marine engineer can calculate the response of the system and the mooring loads. In many floating systems there can be a dynamic magnification of the global ice load as it is transferred to the moorings, but this is usually less than about 1.5.

In reality, floating systems will not be designed for interaction of the worst ice features in a region because mooring lines have limited load capacity. It is usual to employ ice management (either breaking or towing an ice feature) and also the ultimate procedure of disconnection, in order to avoid overloading the mooring lines and the vessel itself. Ice management involves a complex combination of detection, monitoring, ice breaking and ice towing within the context of alert zones which require certain procedures to safeguard the well (and the crews), as the hazard level increases. That topic is beyond the scope of this book. Readers are referred to ISO 19906 [1] for more information and guidance.

This chapter gives some background on sea ice management for the operation of floating systems in ice, and goes on to develop ice loads on floaters for a range of scenarios. Parts of the chapter are adapted from a POAC paper by Croasdale *et al.*[2].

6.2 Background to Use of Floaters in Sea Ice

The use of drilling vessels in ice-covered waters began off Canada's East Coast and West Greenland in the early 1970s. Those operations generally took place without sea ice but in the presence of icebergs. The ice management methods involved detecting, tracking and ultimately towing the icebergs away from the vessels. If significant sea ice occurred, the vessels would usually disconnect, depending on the ice severity. At that time no systematic attempts were made to manage the sea ice.

Beginning in the Beaufort Sea in about 1976, Dome Petroleum brought in three drill ships to operate in the summer. In 1977, the Canadian Coast Guard icebreaker *John A MacDonald* was chartered for the winter months. Using this vessel, as well as smaller icebreaking supply boats, the drilling was extended by several weeks into the beginning of the ice season. That is now a well-established technique. A large icebreaker upstream of the moored vessel breaks the ice into large fragments, and two or more smaller icebreakers break those large fragments into smaller ones that flow around the moored vessel. The larger, more powerful vessel would also focus on the thick ice features and leave the thinner ice to the smaller vessels.

In the Canadian Beaufort Sea, a newly designed drillship was introduced and drilled 12 wells in the period 1983 to 1993. The Kulluk has a round hull with a waterline diameter of 70 m and a water line slope of 30 degrees. Typically the Kulluk was supported with two or four CAC 2 icebreakers. The Kulluk generally started drilling in late May and could continue into late November to mid December. Figure 6.1 shows the Kulluk in managed ice. As described by Wright [3], the Kulluk had an ultimate mooring capacity of 1000 tonnes (10 MN), although in the drilling mode its maximum load tolerance would have been about 750 tonnes. As will be described later, the mooring lines were instrumented and this has provided valuable data on ice loads in managed ice.

Figure 6.1 Kulluk in managed ice in the Canadian Beaufort Sea (Photo courtesy Brian Wright)

After the decline in Beaufort Sea operations in the early 1990s, it was to be seasonal oil production off Sakhalin that provided the next significant experience in managing ice around stationary vessels. These operations have been described by Keinonen et al [4]. Starting in 1999 oil was exported by tanker from the Molikpaq (fixed platform) during the summer months, but with seasonal extension during ice break-up and freeze-up. The tanker was loaded with oil from an FPSO moored to a SALM buoy. The FPSO accepted production from the Molikpaq through an on-bottom flexible pipeline laid each year.

Another noteworthy milestone in operational sea ice management was the coring of the Arctic seafloor (Lomonosov ridge) in 2004. This has also been described by Keinonen [5].

In recent years, several organizations have been studying the year-round use of floating production systems in ice covered regions. These are generally being considered where water depths are greater than about 100 to 150 m. This is the water depth range at which fixed platforms become very costly. Floating systems also have advantages in seismically active regions such as the deeper waters off Northern Sakhalin. In these studies, both symmetric (Kulluk type) and ship-shaped vessels are being considered.

6.3 Loads on Floaters in Unmanaged Ice

In any assessment of ice loads on floating systems it is useful to first estimate ice loads from unmanaged ice of various severities in order to have some appreciation of the severity of the ice conditions beyond which the use of ice management becomes necessary.

For symmetric (circular) downward-breaking hulls, the methods for sloping structures already developed can be applied. For downward breaking, the equations in 4.5.2 are applied with the unit weight of ice (9 kN/m3) replaced by the buoyant unit weight (about 1.2 kN/m^3). As examples, consider both a SPAR and a Kulluk-like vessel which we will refer to as a round ship. The SPAR has a water line diameter of 40m with a slope of 45 degrees. The round ship has a 70 m diameter at the ice line and a 30 degree slope. Ice rides down to -15 m for the SPAR, and to -20 m for the round ship, with some underwater rubble build up in both cases. The flexural strength is taken as 500 kPa and friction as 0.1.

The KRCA spreadsheet with the sloping structure equations is applied to these examples for a range of ice thickness from 0.5 to 3m. The results are given in Table 6.1. For the SPAR, it can be seen that maximum mooring loads range from about 3.66 MN (366 tonnes) for 0.5 m ice to 15.17 MN (1517 tonnes) for 3 m ice. For the round ship the corresponding values are 10.3 MN and 24.33 MN, primarily because the ice line diameter is larger.

As mentioned earlier, the main advantage of these platforms is that they offer the same general shape and width to the ice regardless of its direction. This has a potential advantage in eliminating the use of a rotating turret which is required on a ship-shape. Nevertheless, ship-shapes are often favoured, possibly because of the familiarity that operators have with such vessels, which are used frequently in open water for drilling and production. They can also usually transit under their own power, which is an advantage if they have to disconnect.

In terms of ice interaction, considerable work has been done on the transit resistance of ice breaking vessels and this might be applied to the case of a stationary ship-shape with ice moving past it. There are differences however; one is that the momentum of a moving ship can smooth out ice load peaks. Whether the compliance of a mooring system can do the same will depend on mass and stiffness. Another difference is that relative speeds are greater for transiting ships than for a moored vessel. Changes in ice movement direction on a stationary vessel and the time it takes to vane into the ice movement direction, together with the associated ice loads, are critical issues.

Table 6.1 Unmanaged ice loads for level ice on SPARs and round ships

Level Ice Loads On Downward Sloping Structures

Level Ice Loads On Sloping Structures
K R Croasdale & Associates Ltd.

Initial Data

	Spar	Spar	Spar	Round ship	Round ship	Round ship	Round ship
Flexural strength of ice (kPa)	500	500	500	500	500	500	500
Specific weight of ice (kN/m^3)	8.89	8.89	8.89	8.89	8.89	8.89	8.89
Specific weight of water (kN/m^3)	10.10	10.10	10.10	10.10	10.10	10.10	10.10
Buoyant Weight (kN/m^3)	1.21	1.21	1.21	1.21	1.21	1.21	1.21
Young's modulus (kPa)	5.00E+06	5.00E+06	5.00E+06	5.00E+06	5.00E+06	5.00E+06	5.00E+06
Poisson's ratio	0.3	0.3	0.3	0.3	0.3	0.3	0.3
Cone Angle (deg)	45	45	45	30	30	30	30
Rubble angle of repose (deg)	35	35	35	20	20	20	20
Rubble friction angle (deg)	45	45	45	45	45	45	45
Rubble depth (m)	15	15	15	20	20	20	20
Waterline diameter (m)	40	40	40	70	70	70	70
Ice-cone friction	0.1	0.1	0.1	0.1	0.1	0.1	0.1
Ice-ice friction	0.05	0.05	0.05	0.05	0.05	0.05	0.05
Ice thickness (m)	0.5	1.5	3	0.5	1.5	2	3
Rubble porosity	0.2	0.2	0.2	0.2	0.2	0.2	0.2
Cohesion of rubble (kPa)	5	5	5	5	5	5	5
Results Summary							
Horizontal Load (MN)	3.66	7.13	15.17	10.30	14.88	17.69	24.33
Vertical Load (MN)	2.96	5.68	11.87	13.89	19.57	22.88	30.39
Clearing load (MN)	0.90	2.31	4.74	2.94	5.91	7.59	11.36

Methods for downward breaking sloping structures are modified for ship-shapes as follows.

The bow is assumed to act as a downward breaking sloping face.

Friction of ice and rubble along the ship side is based on a friction coefficient and the level of lateral ice pressure present.

For ridges, the method just described for downward breaking of a composite beam is proposed.

Figure 6.2 depicts the calculation scheme for a ship-shape in level ice.

The KRCA spreadsheet for downward breaking has been adapted for this configuration and applied to a vessel with a 50 m beam and 200 m long. Typical inputs and results are shown in Table 4.24. The results show that maximum mooring loads range from about 4.21 MN (421 tonnes) for 0.5 m ice with small ice pressure to 17.43 MN (1743 tonnes) for 3 m ice with ice pressure. This range is very similar to loads for SPARs and circular ships. However, ship shapes are sensitive to pressured ice because of the long sides. The issue of pressured ice will be reviewed later when managed ice loads are considered.

First-year ridge loads can be estimated using the new method given in Section 5.5.4, treating the ridge as a composite beam or plate, failing in bending as it is pushed down by the vessel. Ridge loads are estimated for a range of ridges up to 28 m thick and with consolidated layers up to 3.5m and shown in Table 6.3.

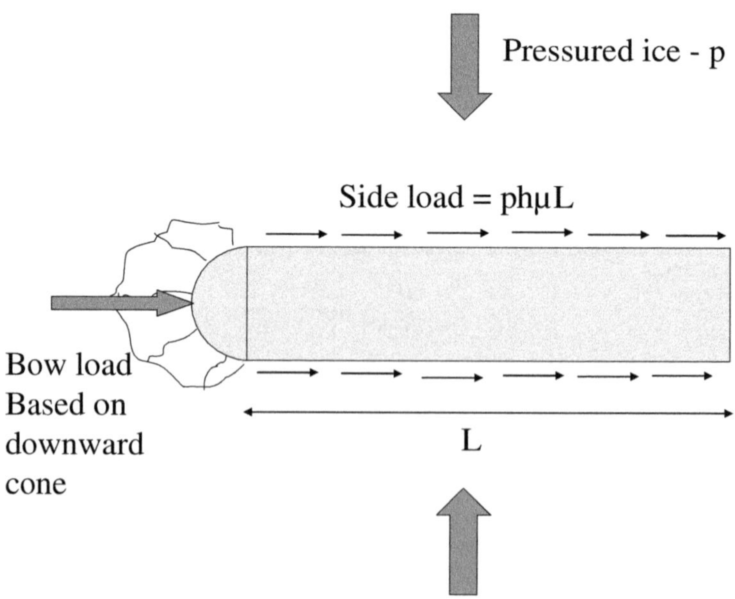

Figure 6.2 Scheme for calculating head-on loads on a ship shape

Table 6.2

Level Ice Loads On Ship-Shape Floaters

Level Ice Loads On Sloping Structures
K R Croasdale & Associates Ltd.

Initial Data:

Flexural strength of ice (kPa)	500	500	500	500	500	500	500
Specific weight of ice (kN/m^3)	8.89	8.89	8.89	8.89	8.89	8.89	8.89
Specific weight of water (kN/m^3)	10.10	10.10	10.10	10.10	10.10	10.10	10.10
Buoyant Weight (kN/m^3)	1.21	1.21	1.21	1.21	1.21	1.21	1.21
Young's modulus (kPa)	5.00E+06	5.00E+06	5.00E+06	5.00E+06	5.00E+06	5.00E+06	5.00E+06
Poisson's ratio	0.3	0.3	0.3	0.3	0.3	0.3	0.3
Bow Angle (deg)	30	30	30	30	30	30	30
Rubble angle of repose (deg)	20	20	20	20	20	20	20
Rubble friction angle (deg)	45	45	45	45	45	45	45
Rubble Depth (m)	15	15	15	15	15	15	15
Beam at Bow (m)	50	50	50	50	50	50	50
Ice -ship friction	0.1	0.1	0.1	0.1	0.1	0.1	0.1
Ice-ice friction	0.05	0.05	0.05	0.05	0.05	0.05	0.05
Ice thickness (m)	0.5	0.5	0.5	1.5	1.5	3	3
Rubble porosity	0.2	0.2	0.2	0.2	0.2	0.2	0.2
Cohesion of rubble (kPa)	2	2	2	2	2	2	2
Waterline length (m)	200	200	200	200	200	200	200
Pressured Ice (kPa)	5	15	30	5	30	5	30

Results summary

Bow load (MN)	4.11	4.11	4.11	7.11	7.11	13.83	13.83
Side load (MN)	0.10	0.30	0.60	0.30	1.80	0.60	3.60
Total Load (MN)	4.21	4.41	4.71	7.41	8.91	14.43	17.43
Bow clearing load	1.26	1.26	1.26	2.95	2.95	6.19	6.19

A major concern with a ship-shape floating production unit (FPU) is its response to changing ice direction. If this is gradual, then positioning the turret between the bow and centre of the ship will allow the vessel to "ice vane" so it always faces the oncoming ice. As well, accompanying ice breakers can assist in this requirement. A problem can arise if the ice stops for a while and then starts up in an unexpected and adverse direction (e.g. broadside). If this occurs, the vessel will be subject to broadside loads. They may be transient, because ice vaning should still occur, but even so they will need to be estimated so that the response of the moorings can be assessed for such a situation.

A calculation scheme for this situation is shown in Figure 6.3. Most FPUs designed for ice will have a small downward slope at the ice line along their sides as shown. This should lead to initial bending failure of the ice sheet in the broadside loading situation. Loads can be calculated for this failure mode using the methods already described for flexural failure. A concern would be if the vessel did not rotate fast enough under this loading and the failure mode changed to rubbling at higher loads.

Table 6.3 Ridge loads for various ridge sizes and bow (slope) angles (μ= 0.1)

	Ridge 1: W=70m: hcl = 3.5m: hk = 24.3m	Ridge 2: W=60m: hcl = 2.5m: hk = 18m	Ridge 3: W=50m: hcl = 2m: hk = 15m	Ridge 4: W=40m: hcl = 1.5 m: hk = 12m
Vertical load to fail ridge (V2) (MN)	28.6	16.77	11.07	6.64
Horizontal force for bow angle of 25 degrees (MN)	16.9	9.9	6.53	3.9
Horizontal force for bow angle of 35 degrees (MN)	24.6	14.42	9.52	5.71
Horizontal force for slope angle of 45 degrees (MN)	34.9	20.46	13.51	8.1

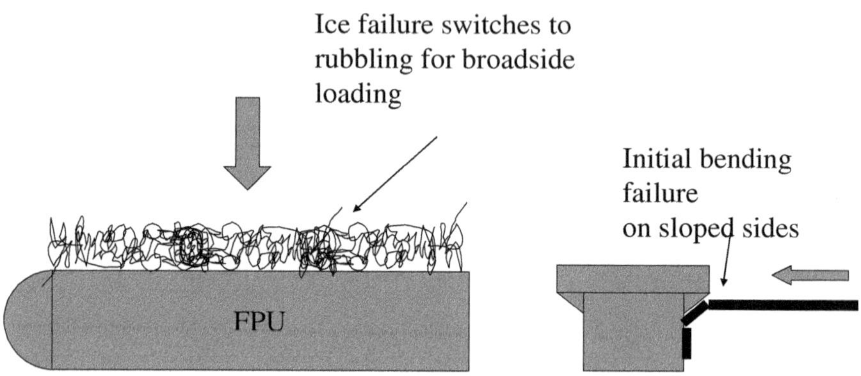

Figure 6.3 Scheme for broadside ice loading on a ship-shape floating production unit

Sample loads for both bending and rubbling are given in Table 6.4 for a 200 m long vessel for a range of ice thickness. In this scenario, the rubbling loads are based on the average maximum keel depths for the relevant ice thickness not the extremes. This is because it is assumed that the vessel will have ice-vaned to head into the ice before such extremes could occur. In a real design or operating situation this could be checked by investigating the response time for this and the associated ice movement (hence time) associated with each stage of ice loading.

A ridge load in the broadside position would be very rare, but could be estimated by assuming a limit-force scenario using the rubbling equation (5.2.8) against the length of the ridge. If the ridge is longer than the vessel length then this would give a higher load than those shown in Table 6.4, but can be calculated using the same approach with rubbling on the back of the ridge.

Table 6.4 Broadside loads in unmanaged ice

ice thickness	initial bending load	bending with rubble to -10m	rubbling load
m	MN	MN	MN
0.5	1.75	7.6	12.2
1.0	4.5	13.1	26.0
1.5	7.76	19.5	40.6
2.0	11.65	26.6	55.7

6.4 Loads on Floaters in Managed Ice

Although it is useful to estimate loads on floaters from unmanaged ice (to help define limits), it is not very likely that a floater would be deployed without some form of ice management, and therefore managed ice loads need to be considered.

Methods have been developed using bounding approaches based on simplified scenarios. These methods are described in detail in Croasdale *et al.* [2], and the two most promising are described here. The recommended method for general use has been calibrated against Kulluk measurements. Again, methods are categorized according to whether the floating platforms are symmetric or ship-shaped. Examples of symmetric floating platforms are vessels similar to the Kulluk, SPARs, shallow draft buoys, and some multi-leg Semis. The main feature of these platforms is that they offer the same general shape and width to the ice regardless of its direction. As mentioned earlier, this has a potential advantage in eliminating the use of a rotating turret which is required on ship-shape structures.

Ice regimes of interest will generally be mobile ice (pack ice) in which thicker ice features are embedded. In first-year ice regimes, the thicker ice features will be first-year pressure ridges and rubble fields. In multi-year ice regimes, there will also be second-year and multi-year floes and ridges.

For the exercise of developing ice load models in managed ice, it is not necessary to go into more detail than to say that a managed ice regime will consist of thick ice features broken into pieces of defined sizes which will be surrounded by ice rubble and/or ambient level ice of a given thickness, which also may be broken into smaller floes. In an aggressively managed ice field, even the thick fragments may be reduced to ice rubble.

Managed ice loads are mostly due to the clearing of broken ice. They are thus much more sensitive to pressured ice fields than loads calculated in the traditional way for unbroken ice features. Therefore, the general approach to estimate ice loads in managed ice needs to account for pressured ice. In this method the scenario to be analyzed is depicted in Figure 6.4. The stationary "false prow" is often seen when some pressure exists in the ice (Wright [6]). The calculation scheme is shown in Figure 6.5. The formula developed is similar to that for a mature limit-force load as developed by Croasdale and Marcellus [7].

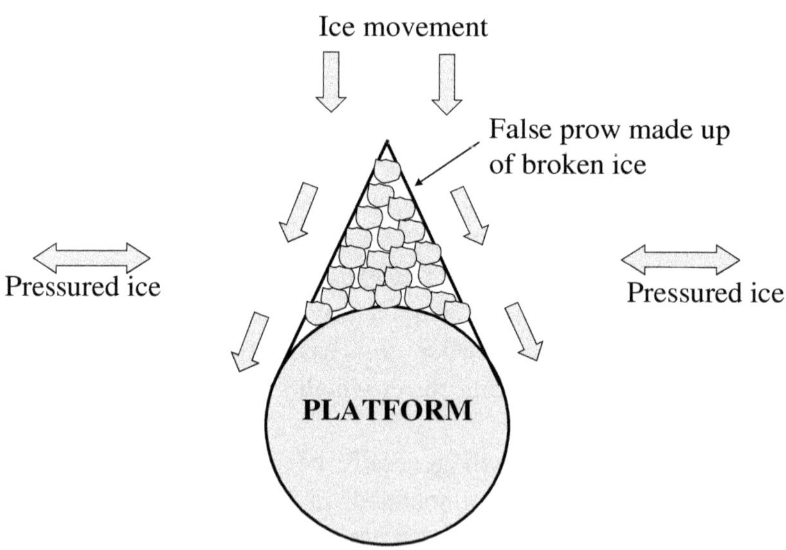

Figure 6.4 Scenario showing how managed ice can act on and clear around a platform in the presence of some confinement (pressured ice)

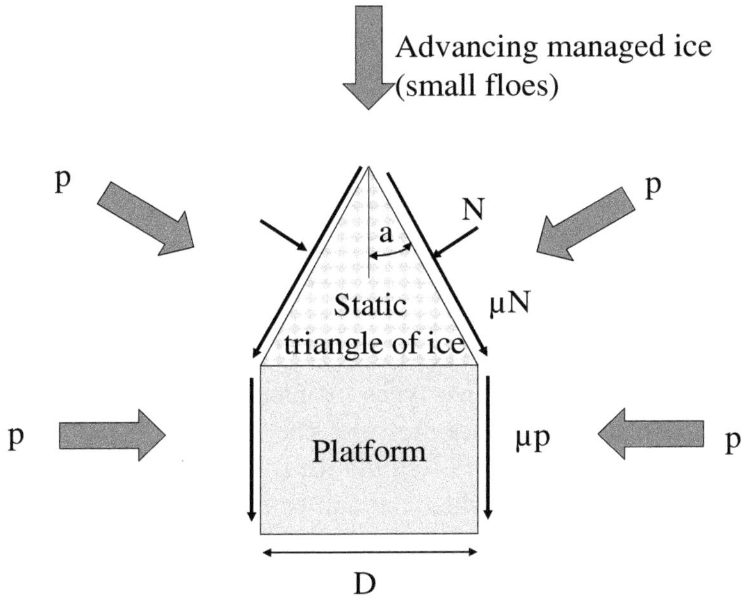

Figure 6.5 Calculation scheme for stationary wedge (false prow) and small piece sizes (10/10 ice)

The ice load on the stationary ice wedge is approximated as:

$$F_{Sw} = pDh\left(1+\frac{\mu}{\tan a}\right) \tag{6.1}$$

In this scenario, the friction on the side of the structure may also be important. This can be included to give:

$$F_{S3total} = pDh\left(1+\frac{\mu}{\tan a}\right)+2pDh\mu \tag{6.2}$$

A modification is to add a background "cohesion" along the sliding surfaces. This way, it may be possible to better match the Kulluk data in low pressure, as is discussed below.

The equation for total loads becomes

$$F_{S3total} = pDh\left(1+\frac{\mu}{\tan a}\right)+\frac{cDh}{\tan a}+2Lh(p\mu+c) \qquad (6.3)$$

6.5 Calibration Against the Kulluk Data

In order to assess the validity of the above equation and to help in choosing input values, it is useful to look at the data from the Kulluk which had instrumented moorings and good information on the ice parameters when loads were recorded.

Wright [6] categorizes two types of interaction. The first is in ice with good clearance and no ice pressure. The second is in ice with some pressure. For non-pressured ice (but full ice coverage), the Kulluk data is shown in Figure 6.6. The dashed line will be taken as a reasonable fit to the data (at the higher ice thickness). It gives,

$$F_{knp} = 116.7h \qquad (6.4)$$

in tonnes force, where h is ice thickness in m. A line through zero thickness and zero load makes more physical sense than a best fit line with a remnant load at zero ice thickness)

For pressured ice, the Kulluk data is shown in Figure 6.7. Again the dashed line is a reasonable fit to the data at the higher ice thicknesses. It gives,

$$F_{kp} = 200h \qquad (6.5)$$

again in tonnes force, where vh is ice thickness in m. After several trials with various combinations of parameters, results from equation 6.3 are compared with the measured correlations in Figures 6.6 and 6.7. This suggests that equation 6.3 can be used for "non-pressured" ice if a low value of pressure of about 4 kPa is combined with a remnant cohesion of 2 kPa. The measured correlation for "pressured ice" can be matched by equation 6.3 if a pressure between 10 and 15 kPa is used.

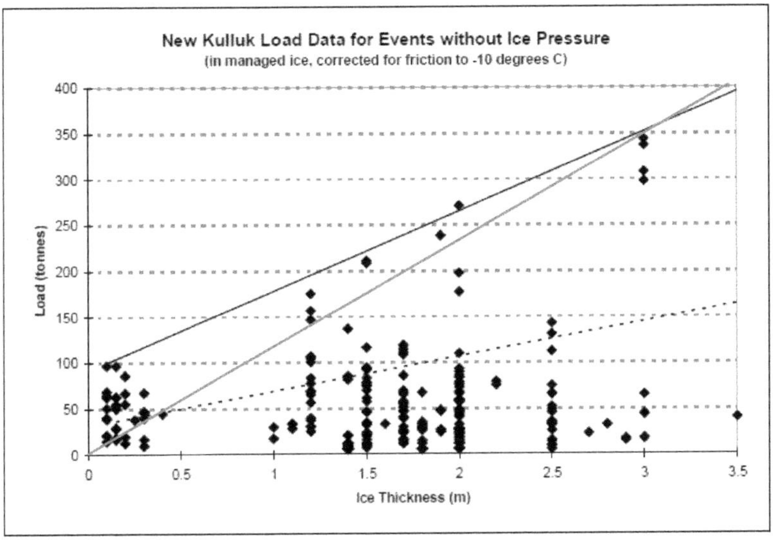

Figure 6.6 Kulluk data without ice pressure (from Wright [6])

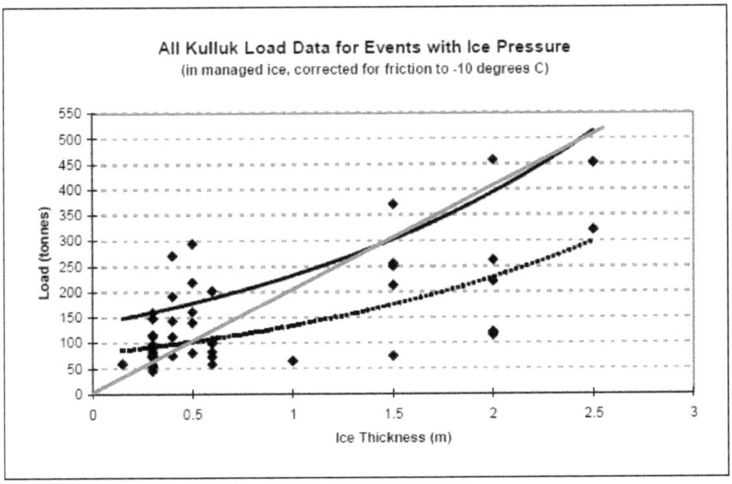

Figure 6.7 Kulluk data for pressured ice (From Wright [6])

Based on this comparison it is proposed that equation 6.3 be used for managed ice loads in 10/10 ice. The comparison with Kulluk data is shown in Figure 6.8. The best matches with Kulluk are obtained if the following values are used for the following input parameters: *c* 2 kPa; *a* 15°; μ 0.1

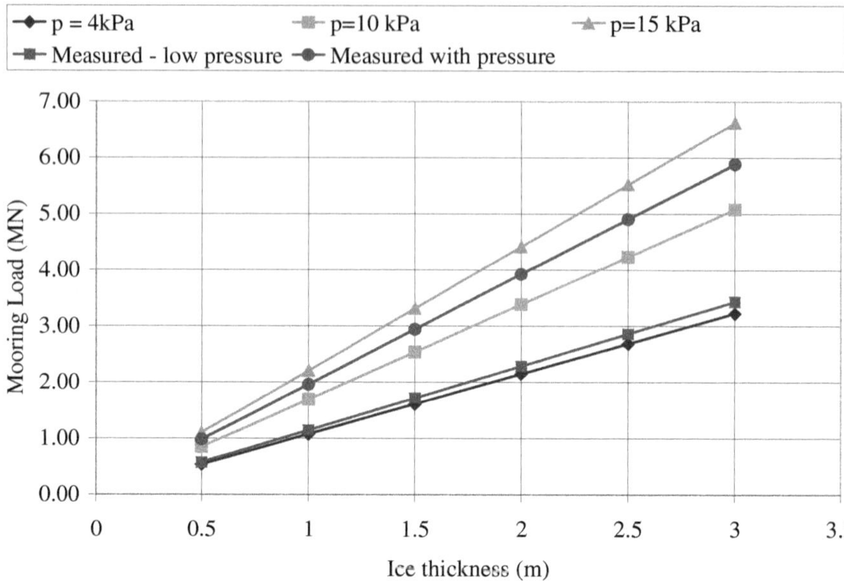

Figure 6.8 Comparison of Kulluk data with the KRCA model (equation 6.1.3)

6.6 Influencing Parameters

Before applying this method, the effects of other influencing parameters should be reviewed, for example ice concentration, ice speed, thickness, ice pressure and floe size.

Concentration

In Wright [6] it is stated that with ice concentrations less than about 8/10th, mooring loads are small. Comfort and others [8], are quoted as saying "All of the test data in *managed* ice show that the pack ice concentration is the most important factor. The loads rise rapidly at ice concentrations greater than about 8/10. The loads increase slightly with ice thickness, and they are not very dependent on speed". This observation is from model tests, and it is expected that the confining effect of the model tank walls may lead to build up of ice pressure more quickly than in the real world. Wright makes the same point for well-managed ice. If there are large thick ice features which have not been managed, their momentum may need to be considered.

The higher the drift speed, the more ice breaking resources are required to manage the same oncoming ice. The effects of drift speed are not accounted for in the KRCA model.

Ice strength and thickness

The ice is broken into pieces by the icebreakers and ice strength may have some influence on icebreaking requirements. However, the model proposed here is developed for processes which move already broken ice around the platform, and so the underlying ice strength is not an input. On the other hand, friction and cohesive forces of the broken ice and ice rubble are important.

The managed ice load equations have ice thickness as an input. This thickness should be taken as the average thickness of the managed ice floes or the resulting ice rubble.

Ice rubble thickness may be nominally thicker than ambient level ice because the thicker ridges have been broken into ice rubble which floats to the surface. When limit force loads are calculated on the basis of ice ridging behind a blocking feature, a range of actual ice thickness should be used.

Pressured ice

The model has a pressured ice term and the mooring loads are very sensitive to this parameter. Wright too noted that mooring loads increase rapidly with ice pressure. The appropriate values to use for managed ice are uncertain. It has been shown that matching the predicted loads and measured loads for the Kulluk in "tight" ice is possible with values between 10 and 15 kPa.

We have already reviewed pack ice pressures in the context of limit-force, and the recognise that internal pack ice pressure is limited by ridging in the thinnest ice in the region, but given a sustained converging ice field, the thin ice can get used up and the ridging forces could progressively increase. Table 6.5 shows pack ice pressures calculated from wind stress, at the same time recognising limits due to ridge building in a given ice thickness.

The wind stress (T) acting over the ice is given by:

$$T = \rho_a C_{10} V_{10}^{\ 2} \tag{6.6}$$

where ρ_a is the air density (1.3 kg/m^3), C_{10} is the drag coefficient based on the air speed at +10 m (approximately 0.002) and V_{10} is the wind speed at +10 m. T will be in Pa.

The internal stress or pressure in the ice (p_w) due to the wind is:

$$p_w = \frac{Ts}{h} \tag{6.7}$$

where s is the fetch length and h is the ice thickness

The ridge building pressure is calculated using the approach for limit force ice loads as reviewed in section 5.2 In this scenario it is proposed to use the ridging line loads associated with average maximum sail heights and keel depths for a given ice thickness and for a width value of 5000m. The equation is:

$$p = \frac{(1360)(5000)^{-0.34} h_r^{\ 1.1}}{h_v} \tag{6.8}$$

where p is the effective pressure ice acting on the vessel kPa; h_r is the ice thickness in m at the location where ridging is limiting the pressure in the region, and h_v is the local ice thickness at the vessel.

Table 6.5 Pressured ice values due to wind and limits due to ridge building in thin ice

Wind speed	Lateral wind fetch	Ice thickness at vessel	Potential ice pressure due to wind	Thinnest ice in region	Pressure limit due to ridge building
m/s	km	m	kPa	m	kPa
5	5	1	3	0.3	20
5	10	1	6	0.3	20
5	15	1	10	0.3	20
5	30	1	19	0.3	20
10	5	2	6	0.5	18
10	10	2	13	0.5	18
10	15	2	19	0.5	18
10	30	2	38	0.5	18
25	250	2	1,993	0.5	18
10	20	2	26	0.75	27
10	30	2	38	0.75	27
10	30	2	38	1	38
10	30	2	38	1	38

The values in Table 6.5 suggest that with strong winds and large fetches, potential internal pressures in the ice could reach 2 MPa. However, the values in the table also suggest that thin ice (for example in refrozen leads) can limit internal pressures in the ice to less than 40 kPa.

In the experimental work in the Beaufort Sea in 1986 (Croasdale et al [9]), average internal pressures in the ice were negligible as an instrumented floe drifted with the pack ice parallel with the coast line. Only when onshore winds occurred and the pack ice was pushed into the landfast ice did the instrumented floe show the presence of an internal stress. The internal pressure was in the range 19 to 29 kPa, in April when the Beaufort pack ice had reached a maximum thickness of about 1.7 m, but there may also have been thinner ice in refrozen leads.

The issue of how pressured ice can develop in the open ocean is complex and is a book topic in itself. Ice pressure on a moored FPU is different from a vessel trying to move through pressured ice. A free vessel's preferred and only safe route may be normal to the wind (and the resulting ice motion and/or pressure). With an FPU, on the other hand, it is lateral ice pressure, normal to the ice movement direction, that will cause difficulties. It should be less than the maxima generated by wind stress and ridge building in the direction of ice motion.

Nevertheless, it is clear that pressured ice, if it occurs, will significantly influence the managed ice loads. It will be prudent for operators to review its potential occurrence and possible magnitudes. The use of ice management vessels to reduce ice pressure should also be investigated and be available if proven effective.

Effect of floe size

For the Kulluk, Wright recommended that target piece sizes for managed ice should be in the range 10 to 50 m when the ice concentration is high. At low concentrations floe size is less important because ice can divert and or rotate around the platform. The model used here assumes floe sizes which are less than the width of the platform.

6.7 Typical Managed Ice Loads

Typical managed ice loads will now be calculated using the developed equations and then discussed. The examples chosen will be those already examined for unmanaged ice. Note that these are "quasi-static" loads not accounting for the response of the mass-spring system ande up of the platform and the moorings.

First is the SPAR with a 40 m water-line diameter. The managed ice loads, using equation 6.3 are given in Table 6.6.

Table 6.6 Managed ice loads for a 40 m diameter spar as a function of thicknesses of managed ice and ice pressure

Managed ice thickness	Mooring loads for various pressures				Clearing	Unmanaged
	5 kPa	10 kPa	15 kPa	30 kPa	load limit	load
m	MN	MN	MN	MN	MN	MN
0.5	0.34	0.48	0.63	1.07	0.9	3.66
1.5	1.01	1.45	1.89	3.22	2.31	7.13
3	2.02	2.9	3.79	6.44	4.74	15.17

For comparison, the unmanaged ice loads are also shown and range from 3.66 MN for 0.5 m ice to 15.17 MN for 3 m ice, and so up to about 15 kPa pressured ice, the managed ice loads are limited to about 25% of the unmanaged ice loads. Table 6.6 also are the clearing loads from the unmanaged ice calculations. The reason for referencing them is because it is proposed that the managed ice loads cannot exceed these values. The logic is that the static triangular wedge, on which most of the load is acting, is itself acting against the slope of the vessel. As the load on the static wedge increases with lateral ice pressure, there will be some value at which it gets pushed downwards under the hull and clears. For small managed ice pieces, this load is approximately the clearing load shown in the calculation for unmanaged ice (the HR, HP and HT components in the bending load calculations reviewed in section 4.5.2).

Managed ice loads for a 70 m diameter round ship are shown in Table 6.7. They can be compared to the unmanaged ice loads in the same table. The loads associated with the clearing load limit suggest that if the stress in the pressured ice becomes higher than about 30 kPa, the loads will not get much higher and will be controlled by the clearing load limit.

Table 6.7 Managed ice loads for a 70 m round ship

Ice	Mooring loads for various pressures				Clearing	Unmanaged
thickness	5 kPa	10 kPa	15 kPa	30 kPa	load limit	load
m	MN	MN	MN	MN	MN	MN
0.5	0.59	0.85	0.98	1.7	2.94	10.3
1.5	1.77	2.54	2.95	5.11	5.91	14.9
3	3.53	5.08	5.89	10.22	11.36	24.3

For a ship-shape the situation is more complex. Table 6.8 shows the calculated loads for 0.5 m of ice acting on a vessel with a 50 m beam and 200 m long. Two bow loads are shown, one based on the static wedge (using Equation 6.3, the first term only), the second based on ride-down and clearing of the already broken (managed) ice pieces that make up the static wedge. The managed ice load is then taken as the smaller of these bow loads plus the load on the sides. In this case, the unmanaged ice loads are also a function of ice pressure. For an ice pressure just greater than about 30 kPa, the static wedge gets pushed under the bow and the bow clearing load controls.

Table 6.8 Managed ice loads for ship-shape; head-on in 0.5 m ice; beam 50 m; length 200m

Ice	Ice	Bow wedge	Bow clearing	Load on	Total	Total
Pressure	Thickness	Load	load	sides	Managed	Unmanaged
kPa	m	MN	MN	MN	MN	MN
5	0.5	0.36	1.26	0.50	0.86	4.2
10	0.5	0.53	1.26	0.60	1.13	4.3
15	0.5	0.70	1.26	0.70	1.40	4.4
30	0.5	1.22	1.26	1.00	2.22	4.7
40	0.5	1.56	1.26	1.20	2.46	4.9

Tables 6.9 and 6.10 show results using the same approach for ice thicknesses of 1.5 m and 3 m.

Table 6.9 Managed ice loads for a ship-shape; head-on in 1.5 m ice; beam 50 m; length 200m

Ice	Ice	Bow wedge	Bow clearing	Load on	Total	Total
Pressure	Thickness	Load	load	sides	Managed	Unmanaged
kPa	m	MN	MN	MN	MN	MN
5	1.5	1.07	2.95	1.50	2.57	7.4
10	1.5	1.59	2.95	1.80	3.39	7.7
15	1.5	2.10	2.95	2.10	4.20	8.0
30	1.5	3.65	2.95	3.00	5.95	8.9
40	1.5	4.68	2.95	3.60	6.55	9.5

Table 6.10 Managed ice loads for ship-shape; head-on in 3 m ice; beam 50 m; length 200m

Ice	Ice	Bow wedge	Bow clearing	Load on	Total	Total
Pressure	Thickness	Load	load	sides	Managed	Unmanaged
kPa	m	MN	MN	MN	MN	MN
5	3	2.15	6.19	3.00	5.15	14.43
10	3	3.18	6.19	3.60	6.78	15.03
15	3	4.21	6.19	4.20	8.41	15.63
30	3	7.30	6.19	6.00	12.19	17.43
40	3	9.36	6.19	7.20	13.39	18.63

It is also of interest to estimate the broadside loads on a ship-shape when the ice has been managed into smaller floes. In fact, an operational tactic when an FPU is in a situation when the ice has stopped moving will be to break up the ice within a few ship's lengths into small floes. In the worst case situation, the ice starts to move broadside and a static wedge of ice will now start to form on the upstream side. The broadside load can be calculated in this situation using the same method, but now the width of interaction is the ship's length rather than its beam.

Table 6.11 shows managed ice loads calculated using this approach. The unmanaged ice loads are also shown for comparison.

Table 6.11 Worst-case broadside loads in managed ice: Ship-shape 200m long

Ice	Mooring loads for various pressures				Unmanaged
thickness	5 kPa	10 kPa	15 kPa	30 kPa	loads
m	MN	MN	MN	MN	MN
0.5	1.56	2.27	2.98	5.12	12.20
1.5	4.67	6.81	8.94	15.35	26.00
3	9.35	13.62	17.89	30.70	87.10

These loads are probably worst-case because they that assume the vessel stays broadside as the static wedge builds up. In reality as the load starts to build up, the vessel can be expected to ice vane into the direction of ice motion and reduce its exposed width. Another bounding approach, not dependent on the build up of the static wedge is to assume a rotational failure of the managed ice similar to a footing failure in soil mechanics. This approach is described in Croasdale et al 2009 [2], and the approximate equation is:

$$F_b = 2\pi Lqh \tag{6.9}$$

where F_b is the broadside load, L is the ship length, q is the shear strength or cohesion between the managed ice floes or rubble, and h is the managed ice thickness.

In Croasdale [2] this approach was calibrated against Kulluk data and it was found that the best match was with values for q in the 1 kPa to 1.5 kPa range. The results from using this approach are shown in Table 6.12. In these cases, values of q of 1.5 and 3 kPa were used to err on the conservative side. The loads are similar to those derived using the first method.

Table 6.12 Broadside loads on a ship shape in managed ice using the "rotational" failure model.

Ice	Broadside loads (MN)	
thickness	q = 1.5 kPa	q = 3 kPa
1.5	6.4	12.7
3	12.7	25.4

In any scheme to look at mooring loads in managed ice, it is also advisable to consider extreme and abnormal situations which could arise. The ice alert procedures which could lead to a disconnect need to identify and monitor such situations. It is relatively easy to do this by using the relatively simple models just described.

Clearly, the managed ice loads are significantly affected by ice pressure. It is desirable for operators to understand how and when ice pressure could build up at a particular location. One approach is to study satellite imagery in which converging ice might be identified. Another is to create an ice movement model for the region, which also if validated could then be used to predict pressured ice and its potential magnitudes. Naturally, ice management techniques and vessel-active devices that can locally reduce ice pressure on a vessel should also be considered.

The other concern is whether a thick floe can be broken into sufficiently small pieces to reduce managed ice loads to the levels predicted in the foregoing. In general, if a manage floe size is larger than the vessel size (in the direction of ice motion), then ice loads will be higher than predicted here. Larger floes may be acceptable if ice concentration is not high and the floes can be diverted around the FPU.

In tight ice and with an unplanned large thick floe, the abnormal load can be estimated using the same methods as described, but substituting the floe width for the vessel width. In this way, knowing the capacity of a given mooring, the ice alert procedures can establish what size of thick floe will exceed that capacity.

To close this section, we can note that the methods have been developed using bounding approaches based on simplified scenarios and mechanics. The methods apply to extreme situations of high ice concentration. They will enable approximate required mooring capacities to be estimated for various platforms and ice conditions. The loads are quasi-static, and actual mooring loads may be magnified by the dynamic response of the system. The methods can be further developed and improved by using numerical techniques, focused model tests and full scale trials (e.g as reported by Bonnemaire et al. [10]).

References

1 International Standards Organisation. (2010) *Petroleum and natural gas industries — Arctic offshore structures*. ISO 19906:2010.

2 Croasdale K.R , Bruce, J. and Liferov, P. (2009) Sea ice loads due to managed ice. *Proceedings, 20th International Conference on Port and Ocean Engineering under Arctic Conditions*, Luleå, Sweden.

3 Wright, B., (2001) Ice loads on the Kulluk in managed ice conditions. *Proceedings, 16th International Conference on Port and Ocean Engineering under Arctic Conditions*, Ottawa.

4 Keinonen, A. and Martin, E. (2008) Ice management experience in the Sakhalin offshore with azimuth icebreakers, *Proceedings, Icetech 2008*, Banff, Canada.

5 Keinonen, A., Evan, M., Neville, M. and Gudmestad, O. T., (2007) Operability of an Arctic drill ship in ice with and without ice management, *Proceedings, 19th Annual Deep Offshore Technology International Conference & Exhibition (DOT)*, Stavanger.

6 B Wright and Associates, (2000) *Full Scale Experience with Kulluk: Stationkeeping Operations in Pack Ice*. PERD CHC Report pp.25-44.

7 Croasdale K. R. and Marcellus R. W. (1981) Ice forces on large marine structures. *Proceedings, IAHR Ice Symposium*, Quebec City, Canada.

8 Comfort, G. (2001) Moored vessel station-keeping in ice-infested waters: an assessment of model test data for various structures and ship shapes. *Proceedings, 16th International Conference on Port and Ocean Engineering under Arctic Conditions*, Ottawa.

9 Croasdale, K.R., Comfort, G., Frederking, R., Graham, B.W. and Lewis, E. L. (1987) A pilot experiment to measure Arctic pack-ice driving forces. *Proceedings, Ninth International Conference on Port and Ocean Engineering under Arctic Conditions*, Fairbanks, AK.

10 Bonnemaire, B., Lundamo,T., Serre, N. and Frederiksen. (2011) Combining numerical simulations and ice basin test to assess the response of a moored vessel in variable ice drift. *Proceedings, 21st International Conference on Port and Ocean Engineering under Arctic Conditions*, Montréal.

Chapter Seven

Arctic Marine Pipelines and Export Systems

7.1 Introduction

An Arctic offshore production system has to have a means of exporting the oil and gas that it delivers. It often also needs flowlines and intrafield systems for water injection and gas transfer. In many Arctic locations, seabed pipelines offer the simplest, cheapest and most reliable way to move fluids. In marginal and light ice conditions, on the other hand, it may be possible to move fluids by export tankers, and that option is examined in section 7.5.

Another book [1] discusses the general questions raised by marine pipeline design and construction away from the Arctic. A pipeline has to be large enough to transport the fluid, it has to be made of a material that will not corrode internally, it has to be protected against external corrosion, it has to be strong enough to withstand the various pressures and forces that will be applied to it, it has to be heavy enough to be stable on the seabed, and so on. Above all, it must be constructible at an acceptable cost, and must be safe against damage and the risk of environmental pollution. There is of course an established technology of marine pipelines, and most of the issues are reasonably well understood, though there is much scope for technological development and improvement [2], and current practice in the central topic of stability is hard to defend [3,4]. Many of the influences that apply to pipelines apply equally to umbilicals and cables.

Several special factors come into play in the Arctic [5], and further complicate the designers' and constructors' tasks. The principal additional factors are seabed ice gouging (examined in section 7.2), strudel scour (7.3) and construction (7.4). It would however be wrong to think that every aspect of design and construction is necessarily more difficult in the Arctic. Some problems become smaller: if the sea is always ice-covered, lateral instability induced by wave action will not occur, and ships will not try to anchor or fish.

7.2 Seabed Ice Gouging

7.2.1 *Introduction*

Gouging happens when ice runs aground and cuts into the seabed. The mechanism is sketched in Figure 7.1. Figure 7.2 is an acoustic plan-view picture made by side-scan sonar, together with a vertically-exaggerated cross-section of part of the same area. In Canada, and sometimes elsewhere, the phenomenon is called 'ice scour' rather than 'ice gouging', but we prefer the term 'gouging', because 'scour' suggests a potentially confusing link with sediment transport scour features, well known in rivers and shallow seabeds [6,7] but produced by quite different mechanisms.

Gouging has been observed in many parts of the contemporary Arctic. Elsewhere there are relic gouges formed when the climate and the sea level were different from today, and there are thought even to be relic ice gouges on Mars [8]. Seabed pipelines in the Arctic were thought about in the 1960s and 1970s, and from the first there was concern about gouging. If ice could cut into the seabed itself, then it could reach a seabed pipeline, drag it across the bottom, and cut into it, even if it were buried. Much research was carried out at that time: see, for example [9,10]. Weeks [11] thoughtfully summarises the present state of knowledge, but with an emphasis on the oceanographic and geological aspects of the problem rather than engineering.

The depth to the bottom of level ice is rarely more than 2 or 3 m, and often much less. The mechanisms described in chapter 2 push ice plates towards each other, and the ice fractures in compression at the intersections. The broken ice fragments pile together and form pressure ridges. Later the fragments freeze together and consolidate, creating a much deeper and structurally stronger ridge, which may survive over the

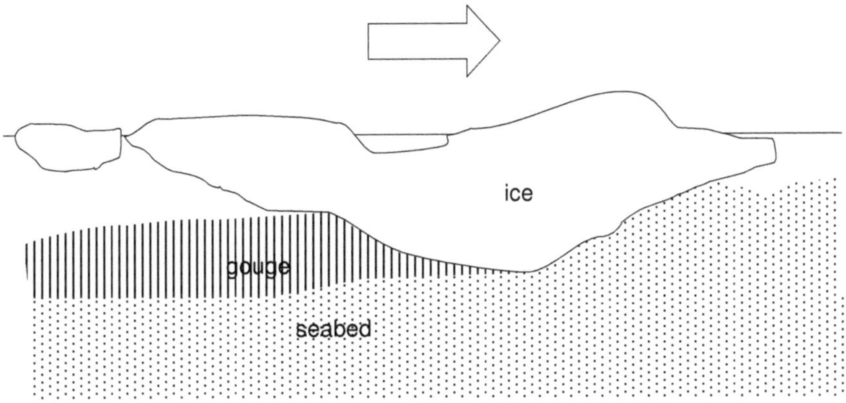

Figure 7.1 Ice gouging: schematic

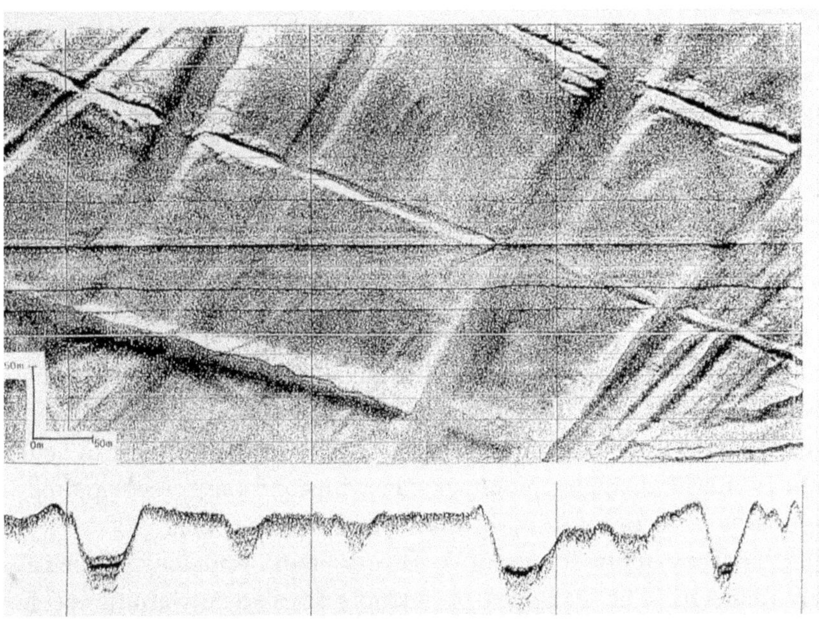

Figure 7.2 Ice gouging: sidescan sonar picture of gouged seabed (upper picture) and seabed cross-section (lower picture)

following summer to move around as a multi-year ridge. Ridges of that kind account for the gouges observed out to water depths of 40 m or more. The most intense level of deep gouges is generally seen at water depths between 10 and 25 m, but that is not a reliable guide to gouge intensity, because gouges in shallower water are rapidly infilled by wave and current action: that important factor is discussed below in section 7.2.3.

Some of the dangers to seabed pipelines were not at first taken seriously. A certain amount of wishful thinking went on. It was argued that the observed gouges were relic gouges formed by conditions in the past rather than the present, and that the ice could not be strong enough to damage a pipeline, because the base of the ice would break off before it could apply a large enough force. Those hopes turned out to be illusory, and in reality gouging later turned out to be a still more acute problem than had been suspected. Lewis [9] carried out repetitive gouge mapping, in which he made a side-scan survey after one winter, returned after the following winter, made a second side-scan, and checked that he was in the same place by comparing the gouge images. The results showed new gouges, and confirmed unequivocally that deep gouges are cut by processes that occur today. Other repetitive gouge mapping studies reached the same conclusion [12-15]. In parallel, it was straightforward to calculate an estimate of how much force is needed to cut a gouge, using geotechnical theory developed for cutting operations such as ploughing. A rough estimate [10] used a two-dimensional theory to demonstrate that to cut a gouge 4 m deep and 40 m wide in a soil with a shear strength of 50 kPa (typical of a medium clay) requires a force of 32 MN (3200 tonnes). That force is one or two orders of magnitude larger than the forces that are known to cause severe damage to seabed pipelines. Large-diameter pipelines can withstand the forces applied by trawlboards and beam trawls (typically no more than 100 kN), but marine anchors apply forces of 1 MN or more, and they often cause severe damage. The example of anchors is also a warning of how expensive it might be to protect a pipeline. If the anchor risk is judged to be extreme, as sometimes happens in port approaches, it is possible to safeguard a seabed pipeline by trenching it deeply and backfilling the trench with rock. This has occasionally been done, in port approaches in

Australia and Singapore for example, but only at a high cost, typically several M$/km.

There was a modest element of truth in the arguments that the gouging problem had been exaggerated. Some of the gouges that are seen in deeper water may indeed be relic, and may have been cut in the distant past when the climate was different. The extensive gouge fields found today off the coast of north-east Greenland [16] are definitely relic, and were cut by icebergs at the end of the last glaciation: no icebergs reach there today. The forces needed to cut deep gouges are indeed large enough to break fragments away from a weak fragment assemblage at the base of the gouge, where the fragments are almost at the same temperature as the sea, and are only loosely held together by their buoyancy and by some freezing at contact points. The effect will be to tear off jagged fragments at the ridge base, and to smooth off the bottom surface until more solid ice is reached. Atkins [17] discusses breakable cutting implements, which are important in other contexts such as drilling. In marginal cases, a pipeline might be strong enough to bring to a stop ice that drifts against it, though that possibility might enhance the risk that ice might accumulate on the upstream side, creating the limit-force mechanism described in chapter 4. Finally, and if the seabed soil is extremely soft, the ice might push the pipeline downwards and ride safely over it, though it would be risky to rely on this mechanism to save a pipeline from damage.

7.2.2 *Ice Gouging: The First Model*

The first analyses of what would need to be done to safeguard a seabed pipeline were based on a simple idealisation. It was thought that a pipeline would be at risk if it were touched by an ice mass strong enough to cut a gouge, but that it would be safe if the ice crossed above it. If that model is correct, the design problem resolves itself into the task of determining the deepest gouge that might intersect the pipeline alignment during the design life of the pipeline. A pipeline trenched below that depth would be safe. Much effort was given to surveys of gouged areas, to assembling gouge depth statistics, and to extrapolations to estimate the extreme gouge depth.

Gouge depth statistics sometimes appear to fit an exponential distribution, though that is purely an empirical observation: there is no physical reason why the exponential distribution should be expected to fit. The fit to an exponential distribution is often better for shallow gouges than it is for deep gouges, but it is of course the deep gouges that are important to an engineer.

Figure 7.3 is an example of a distribution of gouge depths, redrawn from a distribution in [11] from measurements of 16620 gouges. The horizontal axis is the gouge depth z. The vertical axis is the number of gouges less deep than z, on a logarithmic scale. If a survey crosses 16620 gouges, the deepest gouge is 2.9 m and the second deepest 2.7 m (as in this instance), a probability that can be associated with a 2.9 m gouge is $1/16620 = 0.6\times10^{-4}$, the probability associated with a 2.7 m gouge is $2/16620 = 1.2\times10^{-4}$, and so on. That approach plainly contains an arbitrary assumption, and it can be argued that the probabilities ought instead to be (1/2)/16620 and (3/2)/16620, or something else. Similar questions always arise when offshore engineers attempt to estimate extreme waves such as the '100-year wave' from wave heights measured over a shorter period of time, and then have to extrapolate from low estimated probabilities to much lower probabilities. Sarpkaya and Isaacson [18], discuss the issues involved. An estimate of extreme gouge depth found by straight-line extrapolation from a plot like Figure 7.3 can be no more than a best estimate, and the calculation ought to be supplemented by an estimate of confidence limits. Instead of saying that the 100-year gouge depth is 4.3 m, it would be much better to say that the probability is 95% that the 100-year gouge is less deep than 5.5 m, and that the probability is 5% that the 100-year gouge is less deep than 2.5 m. St. Denis [19] and Borgman [20] describe how to estimate confidence limits, in the context of extreme waves. Extreme gouges are at least as important as extreme waves, and engineers ought always to calculate confidence limits, but rarely do so, maybe because the results are so discouraging. There are also physical limits: the force required to cut a very deep gouge is huge, especially in sand, and so there may not be enough driving force available (another instance of the limit force case discussed in chapter 4). Moreover, the ice may not be strong enough to cut the gouge, particularly if it is first-year and only loosely frozen together.

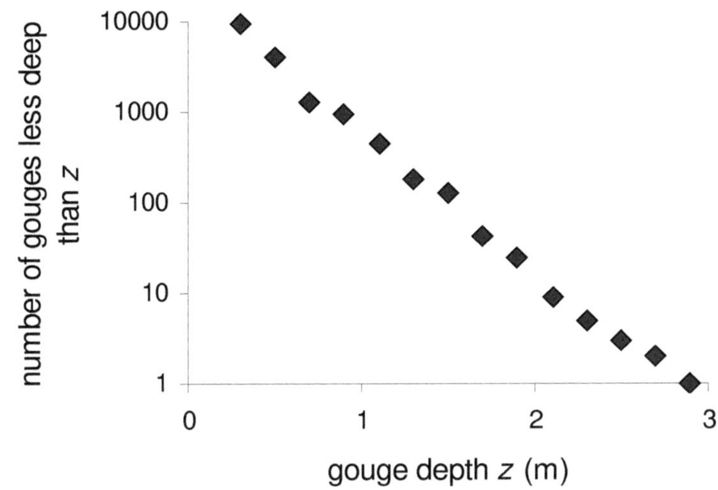

Figure 7.3 Gouge depth distribution, redrawn from Figure 13.6 in Weeks [11]

It is often found that the distribution of gouge depths below the mudline is negative exponential, as it is here and in three other instances cited in [11], so that if there are N gouges the number of individual gouges deeper than z is $N\exp(-\lambda z)$; in Figure 7.3 λ is 3.2 /m. That distribution may not apply to the deepest gouges, because sometimes the frequency of those gouges is significantly greater than an exponential distribution based on the shallower gouges would lead one to expect: that is particularly unfortunate for engineering applications. The same distribution applies to pressure-ridge keel depths. It is not known if there is a physical reason for those distributions. Gouge infill by sediment transport must play a part, because then the distribution of observed gouge depths is systematically different from the original as-cut gouge depths; infill is described below in section 7.2.3.

The λ parameter characterises the depth distribution, but is not itself enough for an estimate of the probability of an extreme gouge depth during the life of a seabed pipeline. That estimate requires a measure of the frequency of gouging, in terms of the number of gouges of a given depth that cross a line in a given time, That kind of estimate is most easily obtained by repetitive gouge mapping.

A possible alternative way to estimate the maximum gouging depth is to start from data of the draft of free-floating ice masses. There are two serious difficulties with that strategy. The first is that the large ridges that cut most deeply are infrequent, and so it would be necessary to survey a very wide area, not in the open-water period but earlier in the year when the ridges are at their largest. The second difficulty is that when the ice contacts the bottom and begins to cut a gouge, the very large horizontal cutting forces that occur are accompanied by large vertical forces. Analysis indicates that the horizontal and vertical forces are of the same order of magnitude [10]. The vertical force can easily be shown to be large enough to lift the ice significantly, so that it does not cut as deeply as it would if it were free-floating. Some field observations suggest that when ice grounds on a gently shelving seabed and is driven further, the vertical force ensures that the base of the does not move horizontally, but instead moves roughly parallel to the seabed.

The alternative of estimating gouge depths from free-floating keel depths is therefore unlikely to be useful for ice gouging by pressure ridges, though it might be in the rather different context of gouging by icebergs.

7.2.3 *Ice Gouging: Gouge Infill by Seabed Sediment Transport*

Later it came to be realised that the simple model outlined above is defective in at least two important ways, and that it seriously underestimates the risk to marine pipelines. The sea, the ice, and the seabed together form a dynamic system that is continuously evolving.

The first factor is that the gouge record observed in a marine survey is not a reliable guide to the original gouge depth.

Gouging occurs in the winter and spring, when the sea ice is at its thickest and pressure ridges are moving about. Most surveys are carried out in a short open-water period in the Arctic summer, because it is much easier and cheaper to make long survey profiles from a ship than from the ice surface, where equipment has to be laboriously dragged across the ice or landed from helicopters. Two or three months usually go by between a gouging event and the survey that measures the gouge depth. Within the open-water period there will often be storms, a storm generates waves, and even a modest wave can start to move the seabed. Sediment moving across the bottom will be carried into a gouge and begin to fill it in, so that a later survey will measure a gouge depth significantly less than the original depth.

Think of the analogy of graffiti on a subway car. If the car is never washed, the graffiti last for ever, and a graffito only disappears when it is overwritten. If the car is washed occasionally and carelessly, some graffiti are erased and some remain. If the car is washed continuously, a graffito is erased at once, and an observer might mistakenly conclude that no one is writing on the car.

Imagine then a section of seabed in a depth of 10 m located 40 nautical miles (74 km) shoreward of the ice edge, and make some simple calculations by the Sverdrup-Munk-Bretschneider wave forecasting method [7] and linear wave theory [7,18]. A 40-knot (20 m/s) wind begins to blow from the ice edge towards the shore. The deep-water significant wave height is 2.1 m after 2 hours, 3.1 m after 4 hours, and 3.4 m after 4.8 hours, after which it is fetch-limited. At that time the significant wave period is 7.3 s, and the corresponding deep-water wavelength is 83 m. In a depth of 10 m, a 3.4 m wave will start the seabed moving if the seabed is finer than 25 mm particle diameter (very coarse gravel). Wave heights generally follow a Rayleigh distribution [18], and one wave in a thousand has a height greater than 1.8 times significant, in this instance 6.1 m, and will move cobbles.

That analysis applies the Kamphuis-Sleath method [6]. Seabed sediment transport is a notoriously controversial subject, and other methods would give somewhat different results, but the broad conclusion is the same. The same method has been used to show that seabed movement is an important factor in assessing the stability of an untrenched pipeline under waves, and that in most locations the seabed begins to move long before the design conditions for a pipeline are reached [4].

There are of course more sophisticated methods of calculating the wave heights, the wave spectrum, and the sediment transport response.

Palmer and Niedoroda [21] carried out more elaborate calculations of the rate at which a gouge would infill. They considered a V-trench in fine sand (D_{50} 0.125 mm) in water 20 m deep, under 1 m high 8 s waves acting simultaneously with a 0.1 m/s steady current superposed on a 0.2 m/s tidal current. The analysis applied the Channels Model, a time-dependent 2-D coupled process-based hydrodynamic, sediment transport and morphodynamic model based on a numerical solution to the RANS (Reynolds-averaged Navier-Stokes) equations with shallow-water assumptions. A gouge initially 2 m deep fills in so that after 15 days it is only 0.5 m deep, and after 60 days 0.3 m deep. The evolution of gouge depth was examined as a function of water depth in the range 13 to 28 m, and it was found that a gouge initially 2 m deep had after 15 days invariably filled in to less than 0.8 m deep.

A clear conclusion is that a storm does not need to be unusually severe for it to move the seabed and cause a gouge to infill. A second conclusion is that gouge statistics determined by sounding are likely severely to underestimate gouge depths, unless it is certain that there has been no open-water period or high current between the gouging events and the survey.

Early research on gouging came to the conclusion that gouging is most severe in water depths between 10 and 25 m. That conclusion was based on direct observation of gouges. It may have been distorted by wave-induced infill, which is depth-dependent because the seabed velocity induced by a given wave height decreases as the depth increases. It does remains true that very deep gouges are unlikely to occur in water shallower than 5 m, because there is not enough driving force available to push into shallow water the large ice fragments required to cut deep gouges. Deep gouges in water deeper than 25 m may be genuinely uncommon, because there they require a combination of a large ice fragment with a deep draft and a very large ice force able to drive the fragment forward.

It can tentatively be concluded that gouge depth statistics are of little value as an indicator of original gouge depth, unless there has been a correction for infill.

7.2.4 *Ice Gouging: Subgouge Deformation*

A further difficulty is that a pipeline that lies below the maximum depth the ice can reach is still not necessarily safe. A gouging ice feature drags some of the seabed soil along in the gouging direction. This phenomenon is called subgouge deformation. It is sketched in Figure 7.4, which shows two mechanisms, one for a low-angle ice keel at 15° to the horizontal, and the other for a steeper high-angle keel at 45°. The mechanisms are quite different. The 15° keel presses the seabed soil downward, and the soil rises up on either side and forms berms along the sides of the gouge. The 45° keel carries forward a dead wedge of soil, and soil flows first upwards in front of the dead wedge and then sideways again to form berms.

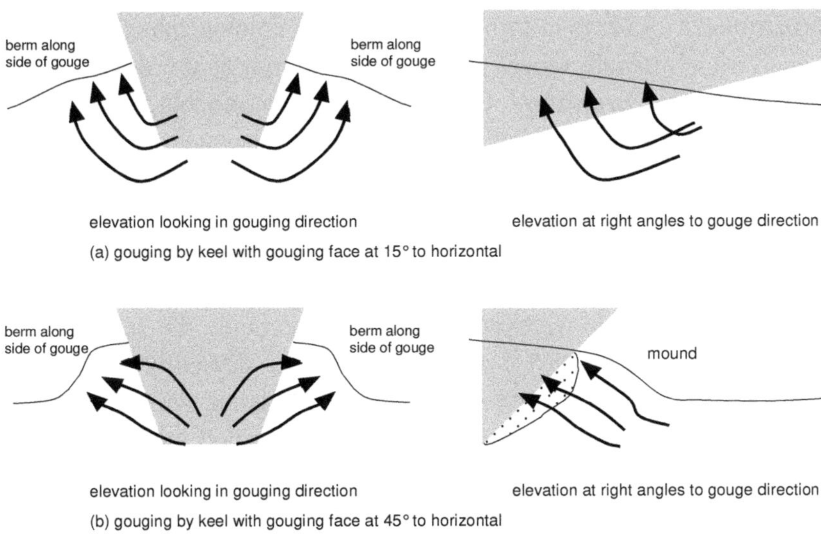

Figure 7.4 Ice gouging mechanisms for 15° and 45°keels

Subgouge deformation has been the target of much research since its importance was first recognised in the 1990s [22], but is still not fully understood. It is important to seabed pipelines because a pipeline might be carried with the soil under a gouging ice mass, and might be severely bent under the gouge, even though the pipeline is buried far enough down not to be contacted directly by the ice. It may then be necessary to trench and bury the pipeline more deeper than the maximum gouge depth.

Different methods have been applied to estimate the depth of subgouge deformations, among them small-scale centrifuge modelling, 1-gravity tests at a larger scale, numerical finite-element analysis, and observations of relic gouges onshore in Canada. A limitation is that conventional soil mechanics is seldom concerned with large deformations, except occasionally in the contexts of cutting processes such as ploughing [11, 24] and of uplift movement of pipelines [25]. Most of the soil constitutive models that are applied in finite-element analysis are not well suited to the deformations that occur under gouges. Sometimes they produce absurd results: if, for example, a Drucker-Prager soil model with an associated flow rule is applied to large deformations, it can lead to grossly exaggerated volume changes [26].

The first centrifuge model tests [8] used kaolin as the model seabed soil. The results were used to create tentative models of subgouge deformation. They were somewhat discouraging, because they appeared to indicate that large and potentially damaging displacements might extend to one gouge depth below the gouge base. If the extreme gouge were 1.5 m deep, and the pipeline 0.8 m in diameter, then the trench depth would then need to be $2 \times 1.5 + 0.8 = 3.8$ m. If on the other hand subgouge deformation were not taken into account, the trench depth would need to be only $1.5 + 0.8 = 2.3$ m. The difficulty and cost of trenching increase rapidly with increasing trench depth. A 3.8 m trench is going to be much more expensive than a 2.3 m trench, and it would be beyond the reach of many types of trenching equipment: this issue is discussed further in section 7.4. However, it is not certain that centrifuge tests can model full-scale behaviour correctly, particularly when deformations are important, because of the significance of shear zones and fracture (Palmer [25, 27]). The results of centrifuge tests need to be interpreted cautiously.

A later programme [28] was on a larger scale but at 1 gravity. That programme was carried out as part of research for the Kashagan project in the northern Caspian, where the mean gouge depth is only 0.32 m. It can be thought of either as a near full-scale simulation of gouging at Kashagan, or as a larger-scale model simulation of the much deeper gouges observed elsewhere. The soil was a sandy clay, a mixture of illitic E-Ton and Asser sand, with a water content of 22.3% and a shear strength between 35 and 38 kPa. Extruded blocks modelled the seabed soil, in a dredging flume in the Delft Hydraulics laboratory. Ice keels were simulated by rigid steel indenters. Since gouging by a symmetrical keel ought to be symmetric about a vertical plane through the gouge centreline, the test exploited that symmetry to simulate half a gouge, and a glass wall on the centreline made it possible visually to observe the subgouge movements.

Figure 7.5 plots the observed movements under low-angle (15°) and high-angle (30 ° and 45°) keels. The vertical axis is the depth below the gouge base divided by the gouge depth (measured from the original mudline). The horizontal axis is the horizontal movement of the soil below the centre of the gouge, divided by the gouge depth.

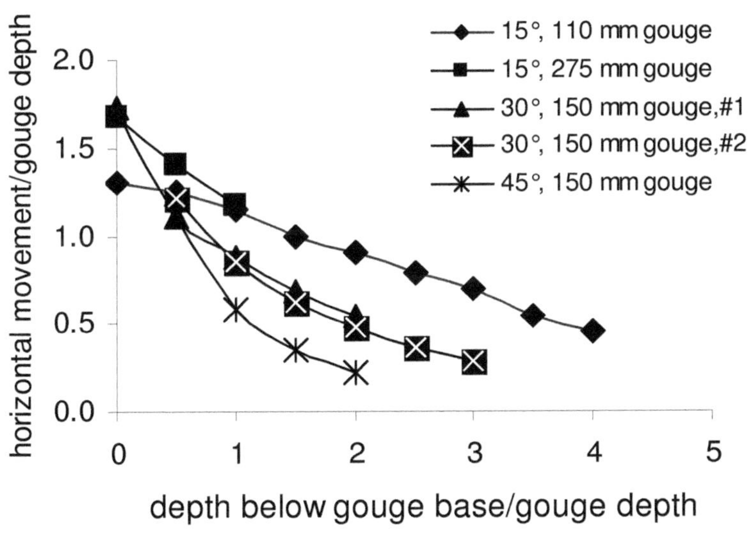

Figure 7.5 Distributions of horizontal movements under gouges with different keel angles

The comparison allows some tentative conclusions:

(1) the displacement at the gouge base is about 1.5 times the gouge depth;

(2) significant horizontal movements extend at least two gouge depths below the gouge base;
(3) the vertical distribution of horizontal movement is influenced by the keel face angle: low-angle keels induce subgouge movements that die off less rapidly with increasing depth;
(4) the results are roughly comparable with the earlier PRISE centrifuge model [8] study, which suggested that the horizontal movement at depth z is proportional to exp(-(2/3)z/D), where z is gouge depth, but the agreement is far from exact and the scatter is very marked;
(5) the formula in [8] for horizontal displacement at the gouge base is not to be relied on..

Recent large-scale tests [29] on behalf of the Kashagan project dragged a large steel and concrete indenter with a medium-angle 30 ° keel across compacted sand (USCS classification SP, dry unit weight 15 to 17.3 kN/m^3) and across clayey sand (SC, LL 35 to 40, PL 13 to 15, undrained shear strength 25 to 70 kPa). There were 17 tests in all. The tests measured the pulling forces, the pore pressure response in the soil, and the bending strain response in a buried 168.3 mm (nominal 6-inch) pipeline as the indenter was pulled over it. The quantitative results are proprietary and have not been published. The paper says that:

> "In general, empirical parameterisation of the subgouge displacements may not be applicable and subgouge displacement equations such as provided by Woodworth-Lynas et al. (1996)[8] and Been et al. (2008)[28] for clays should be interpreted with caution"

The writers agree.

Gouging remains incompletely understood, and much more needs to be done. The distance to which subgouge deformations extend clearly depends on the keel angle, and must also depend on the soil type and the extent to which the keep is able to lift in response to vertical forces. The subject progresses slowly, and it is sobering to reread the conclusions of early studies, for example [9, 10], that raised central questions that 20 years later are still waiting for answers.

7.2.5 *Ice Gouging: Alternative Routes to a Choice of Safe Gouge Depth*

Sections 7.2.3 and 7.2.4 describe one possible pathway towards a decision on the trench depth required to keep a pipeline safe against gouging. The analyst first has to determine the depth of the deepest gouge that may cross the pipeline during its design life. The input data are the statistics of gouge depth determined by seabed survey, but allowing for the interaction between gouge depth and water depth, taking account of infill between gouging events and marine survey, and ideally including confidence limits. The analyst them has to estimate how far gouge deformations extend below the bases of the extreme gouge event, taking account of the seabed soil, and carrying out a numerical analysis to determine the consequences for a pipeline.

This process is cumbersome, and almost every step is beset with uncertainty. It may be possible to shortcut much of the analysis by choosing a different starting point and relying on different data. Farmers know that if the same field is repeatedly ploughed, always to the same depth, then the soil within the ploughing depth is repeatedly remoulded by the plough. Below it the soil remains undisturbed, and is bounded above by a hard layer. Farmers call that 'plough pan', and take steps to break it up by ploughing deeper with chisel ploughs. In the same way, repeated gouging and subgouge deformation create a remoulded surface layer, but below it the soil is undisturbed. That difference can be expected to create a "break" in soil properties, indicated by an increase in soil strength, and that increase can be detected by a cone penetrometer or a vane shear test. This idea exploits the fact that the soil itself holds a record of the deformation it has been subject to. Palmer [30] suggested that this might be an alternative and less unreliable pathway to a decision about trench depth: the pipeline should simply be trenched into the stronger layer, to a depth at which the top of the pipeline is level with the top of the stronger layer. This appears to be a potentially attractive option, but it has not yet been confirmed and exploited.

7.2.6 *Methods for Minimising Required Trench Depth*

It is not practicable to eliminate ice gouging. In some cases, it might be practicable to protect a pipeline by creating a strong barrier, for instance by placing rock embankments on either side of the pipeline, or by driving parallel lines of sheet piles and placing the pipeline between them. That would inevitably be expensive, and it is unlikely to be economically attractive, except over short distances close to shore crossings. Another difficulty is that it would lead gouging ice masses to stop moving and pile up against the upflow side, so that the force on the pipeline would progressively increase as ice continued to accumulate.

It might be easier and more economical to protect a pipeline against subgouge deformation, by interposing a weak layer below the gouge base and above the pipeline. The weak layer would deform severely when a gouging ice mass crossed the pipeline, but it would be too weak to transmit much shear stress downward, and the pipeline and the soil around it would remain almost undisturbed. Palmer, Tjiawi, Chua and Chan [31 – 33] investigated this scheme in small-scale model tests, and found that it significantly reduces pipeline deformations and bending stresses. The weak layer might be a soft clay tremied into a trench above the pile, or it might be layers of geotextile lubricated so that one layer could easily side over the next: both possibilities have been investigated, and both worked well in the model tests. An usually deep gouging ice mass would heavily distort the weak layer when it crossed above the pipeline, but it would be extraordinarily unlikely that the design ice mass would cross twice in the same place.

7.3 Strudel Scour

Strudel scour is a phenomenon peculiar to highly localised areas in the Arctic. In the Arctic spring, the rivers thaw first while the sea is still frozen. River water floods out over the sea ice, and from time to time it encounters a holes or a crack in the sea ice. The river water flows downward though the hole, and forms a strong rotating vortex, called a strudel, the German for “whirlpool” (and the same strudel as in apple strudel, which is pastry rolled around a fruit or poppy seed filling). Reimnitz [34] described strudels off the north coast of Alaska, near the mouths of the Kuparuk, Colville and Sagavanirktok rivers, initially at and seaward of the 2 m depth contour, but later closer to shore as the bottom-fast ice lifted off the seabed. The deepest strudel scour hole he observed was circular and 12 m across, and reached 4.3 m below mudline in sandy silt and silty sand in a water depth of 3 m. A survey dive found that the sides of the scour were nearly vertical, that the bottom had been partially in-filled with mud (presumably by wave and current action, like the gouges described in section 7.2), and that the hole was bounded by a small ridge 0.6 m high formed during the scouting process.

It may at first sight seem unexpected that the water should flow downwards, but how that happens can be understood by imagining a simple thought-experiment (or indeed by carrying out a physical experiment). Think of a cylindrical drinking-glass filled with water, and in the glass a loosely-fitting circular piece of ice with a hole in it. Press the ice downwards in the glass until water floods over the upper surface, to a depth of say 50 mm above the top surface, and then release the ice. The ice floats upwards, and water flows downwards through the hole. The ice is less dense than either the river water or the seawater, and therefore floats in either. It is the density difference between the water and the ice that drives the flow, not the density difference between the fresh river water on top and the salt seawater underneath.

The pressure difference across the hole is the thickness of the ice multiplied by the difference in unit weight between ice and water. If the ice is 2 m thick, and the unit weights of ice and water are 900 kg/m^3 and 1015 kg/m^3 (half-way between salt and freshwater), the pressure difference is 2.8 kPa. If the hole is 0.5 m in diameter and 2 m long, and has a hydraulically smooth surface, the pressure difference induces a flow with a mean velocity of roughly 10 m/s. Below the ice the flow will created a downward-directed jet. The jet will spread out, but if the seabed is not far below the ice the velocityu will easily be enough to erode the seabed, as Reimnitz observed.

Figure 7.6 is an air photograph of a strudel. It would clearly be extremely hazardous to approach across the ice or in a boat, and that is one of the reasons why not much is known about strudels.

Figure 7.6 Strudel scour (photo by courtesy of Craig Leidersdorf, Coastal Frontiers)

One of the potential problems for a pipeline intersected and exposed by a strudel scour is vortex-induced oscillation [1] of a free span. A typical oil export pipeline in such an area might be 508 mm (20 inches) outside diameter, and have a wall thickness of 19.05 mm, a polypropylene anti-corrosion coating 3 mm thick, and a weight coating of 50 mm of 3050 kg/m^3 concrete. Its mass per unit length filled with 810 kg/m^3 oil is 646 kg/m, and allowing for the added mass effect [1] its effective mass per unit length is 950 kg/m. The flexural rigidity is 183 MN m^2. If that pipeline becomes spanned diametrally across the 12 m strudel scour described by Reimnitz, the natural frequency is 10.7 Hz, and if the water velocity is 10 m/s, the reduced velocity is 1.5. That is well below the level at which cross-flow oscillations could occur, even if the velocity were uniform across the 12 m diameter. In reality a rotating flow in the scour hole acts in one direction across one half of the pipeline span and in the opposite direction on the other half, which makes vortex-induced oscillation even less likely. On the other hand, the reduced velocity is inversely proportional to the square of the span length, and so if the scour hole were much larger than 12 m, and the water velocity about the same, the reduced velocity might be raised into the range beyond 3 where vortex-excited vibration is less improbable [1].

Another and rather more likely mechanism by which a strudel scour might lead to problems is interaction with lateral buckling [1,35]. The scour frees the pipeline from lateral constraint, and the pipeline might buckle laterally under the effect of axial force induced by pressure and temperature. A further possibility is that the pipe might be overstressed by its own weight and yield and bend downward, though thick-walled pipes are quite robust against bending buckling and that kind of bending is unlikely to lead to buckling or rupture.

The distance over which a pipeline might be vulnerable to strudel scour is quite short. It might be practicable to protect it, by trenching and backfill with rock too large to be eroded by the strudel jet, or by laying mattresses over it, or by a combination of backfill and a geotextile.

It has been suggested that the presence of an underwater pipeline might increase the likelihood of formation of a strudel. A pipeline will probably be warmer than the soil that surrounds it or the water above it, and the pipeline could therefore create a convection cell that would lift a plume of warmer water to the underside of the ice above. That warm water would melt and thin the ice, and create above the pipeline a line of thinned and weakened ice, or perhaps a line of open water. It is plainly important that this issue should be resolved and quantified, because it would much increase the likelihood of damaging interactions.

7.4 Construction

7.4.1 *Introduction*

Figure 7.7 illustrates the different ways of constructing marine pipelines.

The methods can be divided into two groups. One group, sketched on the left-hand side of the Figure, is based on a laybarge. Short lengths of pipe, typically 12 or 24 m long, are transported to the barge by specialised supply vessels, and one by one welded to the end of the pipeline. The barge moves forward, one length at a time. The pipeline

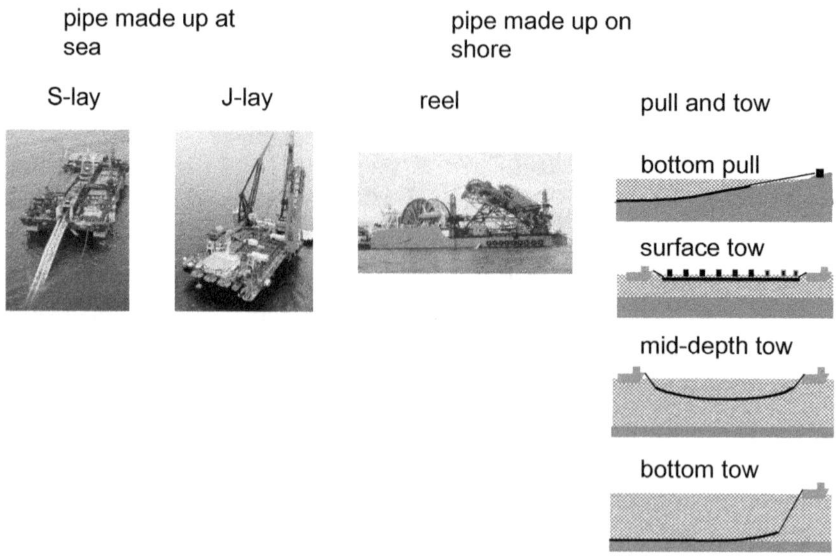

Figure 7.7 Alternative methods of marine pipeline construction

moves through a series of welding stations, the welds are X-rayed or tested ultrasonically, an anti-corrosion coating is applied over the weld, the gap in the concrete weight coating is filled, and the pipeline leaves the barge, either steeply in the J-lay version or nearly horizontally in the S-lay version, where it leaves the barge over a long bridge-like structure called a 'stinger'. The pipeline moves down through the water until it reaches the bottom, and on its way it is supported by tension applied by the barge. Some barges are dynamically positioned by thrusters, and others are positioned by mooring lines and anchors.

The other group of methods, sketched on the right-hand side of Figure 7.7, welds together on shore a much greater length of pipe, called a 'string' in pipeline jargon, typically several km long. They then transport that length to its required position, either by winding it onto a reel on a ship or by towing it between tugs or by dragging it along the seabed.

The details of these methods are described in another book [1]. Most pipelines are constructed by one of the laybarge methods, which have proved themselves versatile and adaptable. Laybarges come in different sizes, from very large barges able to lay large-diameter pipe in water 3000 m deep, down to much smaller barges that use the same basic methods but are able to work in shallow water in estuaries and shore crossings. The reel and pull methods are mostly used for relatively short pipelines.

Returning now to marine pipeline construction in the Arctic, all these methods can in principle be applied. The laybarge and reelship methods depend on floating vessels. The vessel has to be able to move along the pipeline route, and to hold its position and direction accurately. The laybarge layrate is generally between 2 and 5 km/day, and so most pipelines require weeks or months. It has to be possible to continue to supply the barge with pipe during that period. A reelship lays much more rapidly, typically at 2 knots (1 m/s, 3.6 km/h), but can only carry a limited length of pipe, depending on the pipe diameter but typically about 10 km. In either case, the vessel must not wander away from its prescribed path, because any sideways movement or yaw rotation rapidly increases the bending moment at the lift-off point where the pipeline leaves the stinger, and at the touchdown point where it reaches the seabed, and the pipeline can then easily buckle and kink.

In the Arctic context, laybarge and reelship construction methods are then most suitable when the pipeline route has an open-water period long enough to allow the pipeline to be completed in one summer. These conditions are met in parts of the eastern Canadian Arctic, in the Barents and Kara seas, and sometimes in the Chukchi and Beaufort Seas. A barge is an expensive piece of equipment, and it will plainly be costly and disruptive if the barge cannot complete a pipeline and has to wait on location until the following summer, or if it should inadvertently become trapped in the sea ice and compelled to overwinter.

A further problem is that most potential construction locations in the Arctic are a long way from the offshore centres where barges are based. The mobilisation of the barge has to be carefully planned, so that it does not arrive too early and cannot reach the site because of ice, or does not arrive so late that valuable laying time is lost. In some instances a potential mobilisation route might be temporarily blocked at an area on the way where the ice conditions are more severe. In the North American Arctic, for example, the Beaufort Sea off Point Barrow often remains ice-covered while the sea off the Mackenzie Delta is already free of ice, and in the Russian Arctic the sea off the Mys Chelyushkin promontory is blocked while navigation is open both to the east and to the west.

The possibility of continuing to lay pipe in a sea partially or completely covered by ice has been investigated, but the conclusions have been discouraging. The barge has to remain fixed in position, and it cannot manoeuvre or weathervane to avoid the ice. Ice can apply large forces, as we have seen in chapter 4. If the vessel is dynamically positioned, the system has to be unusually powerful to enable the vessel to stay in position. If the vessel is moored, it is difficult to make anchors and mooring lines strong enough to withstand the forces applied by moving ice. This has been investigated in model tests, but the results are not encouraging. Ice fragments can also accumulate around the mooring lines and the stinger, though that problem can be eliminated by locating the mooring line fairleads under water, and some barges have their stingers almost entirely submerged.

The natural alternatives are various pull, tow and lowering-in methods.

Surface tow and near-surface tow float a pipeline just below the water surface, supported by pontoons, and moves it into position by tugs. That method might be practicable, but it would be difficult to control the position of the pipeline if the sea were partially covered by ice, and the pipeline still has to be lowered to the seabed, by progressively removing the pontoons or flooding the pipeline with water. The method is not often used, and then only when the water is shallow and protected against wave action. It is unlikely that it would be more attractive in the Arctic.

Mid-depth tow (controlled-depth tow) hangs the pipeline in a long catenary between two tugs. The tugs carry the pipeline to its final location, and then lower it. The pipe has to be very light, and the sag at the mid-point of the catenary is significant, typically 50 m, unless the tension applied by the tugs is very high. This method has been widely applied to install pipeline bundles in the North Sea, and the maximum length installed in this way is 8500 m. It is unsuitable for shallow water because the suspended pipe must be held well clear of the seabed. It is unlikely to be appropriate for an Arctic pipeline project.

Bottom pull and bottom tow drag the pipeline along the bottom. The difference between them is that bottom pull applies the required force from a winch mounted on shore, or on a floating barge, or on the sea ice, whereas bottom tow uses a free-floating tow vessel. Pull is often applied to construct shore crossings and estuary crossings. Figure 7.8 is a photo of a pull in Singapore, across a 4 km wide channel between Jurong and Bukom islands: the photo shows the line going into the water at the Jurong end where the pipe was made up. The longest recorded pull was 32 km in Iran, between the mainland and Khargu Island. Tow is not a widely popular method, but has been applied in the North Sea, the Gulf of Mexico and Australia. It requires the seabed to be unobstructed, and the pipeline has to have a rugged external coating that can resist abrasion.

Figure 7.8 Bottom pull construction

Pull completes the pipeline relatively rapidly. The pipe can made up onshore and rolled onto a launchway, and those operations can be completed in almost any weather. The pull cable is then connected between the winch and the pipeline, and the pull can begin at once. In one example of a pulled shore crossing in eastern England, the pull force was applied by linear winches, through a two-part cable looped around a sheave at the pullhead at the leading end of the pipeline. The pipeline was pulled at 0.1 m/s (3.6 km/h).

Pull lends itself to Arctic marine pipeline construction because of its simplicity, speed, and relative insensitivity to ice and weather. It was applied to construct the Panarctic Drake F76 pipeline described in section 7.4.2 below.

One further option can be applied in the Arctic but is not available elsewhere. If the sea ice is strong and stable, the pipeline can be constructed on the sea ice and then lowered-in, through a slot in the ice and into a pre-cut trench in the seabed. This is very similar to land pipeline construction: a pipeline is welded together, supported just above the ground, and sidebooms lift it and then lower it sideways into a trench. This method was applied to construct the Northstar pipeline described in section 7.4.3 below.

7.4.2 *Panarctic Drake F-76 Pipeline*

This was the first pipeline constructed under the Arctic ice, as long ago as 1978. Panarctic Oils is a company that brought together most of the companies that had been engaged in exploration in the Canadian Arctic islands. Over a period of several years, it had found gas in the Hecla and Drake fields, to the west and east of the Sabine Peninsula of north-east Melville Island, as well as close to King Christian Island further to the north-east. Some of both the Hecla and Drake fields lie under the peninsula, but large areas lie under water, out to depths of 400 m. The sea is covered by ice almost the whole year round. The field has 1.6×10^{11} Nm^3 (5.7 Tcf) proven and probable reserves [36], roughly half offshore and half onshore.

There is clearly little point in finding gas if you cannot produce it, and the Panarctic participants wished to be reassured that it would be possible to produce the gas and to pipeline it to shore, without waiting for some future technology that had yet to be developed.

It was decided to build as a demonstration project a pipeline between Melville Island and a wellhead, Drake F-76, in 55 m of water 1100 m from shore in the Drake field, and to construct and connect the pipeline, applying technology that did not depend on the use of divers and could therefore be used in the deepest segments of the field. Divers can of course easily descend to 55 m, but they could not reach 400 m (and cannot do so now). Some of the technology had been proposed for construction of large-diameter pipelines between the Arctic Islands. 55 m was chosen instead of 400 m, because the pipeline would be much shorter, and the technological jump would be smaller. 55 m was chosen instead of a smaller depth because ice islands are occasionally encountered in the area; they originate from land ice on Axel Heiberg and Ellesmere islands much further north, and drift south through the channels of the archipelago. Ice islands were believed to have a maximum draft of 45 m, and so a 55 m water depth would leave several metres clearance, allowing for the height of the wellhead. Secondary factors in the choice of location were the wish to have a relatively steep shore crossing, a need for a reasonably level make-up area onshore in line with the offshore pipeline route, and level seabed close to the wellhead. The seabed is a plastic soft silty clay, with some less plastic clay and some silt. During the construction season from February to April the sea ice is landfast and some 2 m thick.

The project is described in detail in several technical papers [23,38-40] and a movie [41]. The pipeline bundle consists of two 168.3 mm (nominal 6-inch) flowlines, together with one 51 mm annulus access line, three 25 mm hydraulic control lines, and one 25 mm methanol/glycol line for hydrate suppression, all seven lines within a 606.4 mm (24-inch) carrier. Both the flowlines have thermal insulation, and both have heat tracing so that hydrates will not form and the lines could if necessary be warmed up from cold. Figure 7.9 is a cross-section. The design meets Canadian standard CSA Z-184 'Gas pipeline systems'. The seamless pipe specification for the flowlines was based on API 5LX, but with additional restrictions on joint length, carbon and manganese content, carbon equivalent and Charpy impact strength. The carrier is ERW longitudinal seam pipe, with a slightly less restrictive specification.

The bundle was placed in a 1.5 m deep trench for the first 300 m. Further out the water depth was more than 15 m. It was thought that the risk of seabed ice gouging in that deeper water was acceptably low, and that if a future gouge did reach to such a great depth the gouging ice force would be so large that it would be impracticable to protect the bundle.

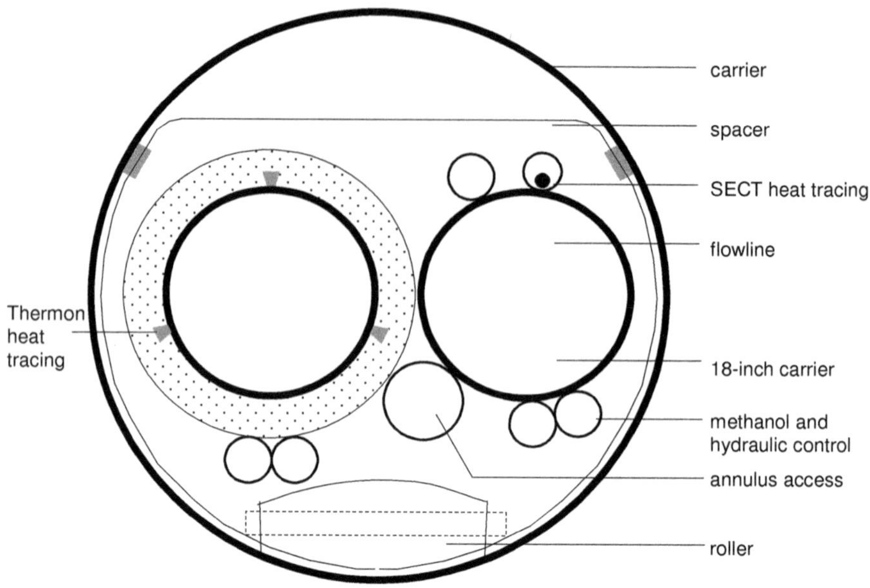

Figure 7.9 Panarctic Drake F-76 pipeline bundle: cross-section more than 300 m from shore

The bundle was put together in a heated tent near the shoreline, and then pulled landwards along a launchway of regularly-spaced rollers. The ice at the wellsite was thickened by pumping water onto the ice surface, leaving a moonpool through which the well could be drilled. The well was drilled and completed from a derrick above the moonpool, and the wellhead assembly was lowed and attached. A narrow slot was cut along the pipeline route, and a pull cable was lowered through the slot. The 300 m trench was cut by a plough, and a wider hole in the ice was cut near the shore, so that the plough could be lowered to the bottom. The plough was then pulled along the route to cut the trench. The pipeline followed a week later. A connection module was welded to the leading end of the bundle. The first 150 m behind the connection module were held off the bottom by pontoons, but held down by drag chains, so that the pipe floated 2 m above the bottom. The main pull was carried out first, along a path selected so that the line of the pipe was offset 50 m to one side of the wellhead. The buoyant flexible section could be deflected sideways, without moving the rest of the bundle, by relatively small forces applied by cables led through seabed sheaves from winches on the ice above. Once the connection module was aligned with the wellhead, cables pulled the module into the wellhead structure, and mechanical connectors joined the bundle piping to the wellhead pipework.

The pipeline had additional protection against ice. The pipeline was in a trench, and had a second outer carrier pipe. Refrigerated coolant was circulated though the annulus between the two casings, in order to freeze back the surrounding soil and form an outer annulus of frozen mud. This scheme is sketched in Figure 7.10. It was inspired by a dam on the Irelyakh river in Siberia [42] with an artificially frozen core, kept frozen by circulating winter air through the centre of the core.

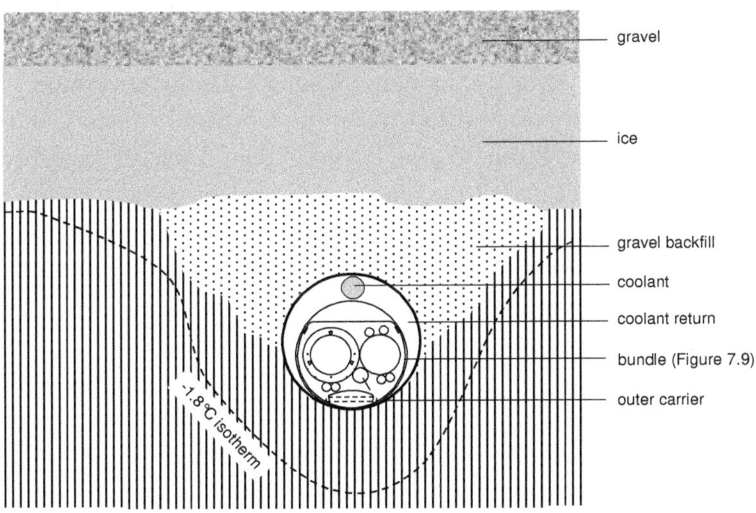

Figure 7.10 Panarctic Drake F-76 pipeline bundle: cross-section less than 150 m from shore

In the shallowest water, less than 150 m from shore, blocks of ice were piled above the pipeline to form a berm 150 m square, and the berm was insulated by a layer of gravel. The performance of the freezeback system turned out to be disappointing. The external refrigeration system was more complicated and expensive than the original concept, and in retrospect the scheme ought to have been reconsidered when it grew to be far more complex than the original concept. After a year, the artificial refrigeration was turned off, and the frozen mud reverted to the sea temperature.

At the time, it was expected that within a few years there would be a transportation system for Arctic Islands gas. Drake F-76 with its flowline system would have been one of the wells that supplied the gas to those projects. One candidate system was the Arctic Pilot Project liquefied natural gas project described in section 7.5 below. Its gas liquefaction plant at Bridport Inlet on the coast of Melville Island was to be connected to Drake Point by a 558.8 mm (22 inch) pipeline. The other candidate system was the Polar Gas Project, a 914.4 mm (36-inch) pipeline from Melville Island to markets in the south [37]. One route alternative for Polar Gas was to go south-westward across McClure Strait to Victoria Island or Banks Island, and thence across Amundsen Gulf or Dolphin and Union Strait to mainland Canada. The other option was to go eastward across Byam Martin Channel to Cornwallis Island, and from there across Barrow Strait to the Boothia Peninsula, or possibly still further east to Quebec across the entrance to Hudson's Bay.

In the event the gas price did not rise in the way that had been anticipated. Neither project was built, and at the time of writing (2012) it seems unlikely that either will be revived for many years, but economic and political change might prove that prediction wrong.

Panarctic decided to close and remove its base at Rea Point on Melville Island, and to plug the Drake F-76 well permanently. It did so in 1995-6. The abandonment operations are described in an interesting paper [43] that explains the systematic decision process that was followed, and the difficulties of working with a system that had been left unmaintained for 17 years. The flowlines were left empty.

That could have been an opportunity to survey the pipeline bundle and to discover what had happened to it in the nearly twenty years since it had been constructed. The industry would have learned valuable lessons that could be applied to Arctic marine pipelines in the future. In particular, light could have been thrown on the gouging damage issue.

The Drake F-76 design was not particularly conservative. If the survey had found that the pipeline was undamaged, or if it had suffered only light damage that would not have led to leaks, that would have led to a more soundly based confidence, at least for environmentally similar areas. Understandably but regrettably, the opportunity was not taken.

7.4.3 *Northstar Pipeline*

Northstar is an artificial island in 11 m of water off the coast of Alaska. Built in 1982 and originally called Seal Island, it was renamed for reasons of environmental tact. A 9.7 km pipeline bundle of twin 273 mm (10-inch) lines from Point Storkersen to the island was built in 2000. One line exports oil from the island, and is connected onward to Pump Station 1 of the TAPS (Trans-Alaska Pipeline System) described above. The second line carries gas to the island. Lanan [44] describes the project in detail.

The pipelines are API 5L grade X52 (yield stress 358 MPa) seamless, wall thickness 15.1 mm (0.594 inch), coated with 1 mm of dual-layer fusion-bonded epoxy and cathodically protected by aluminium anodes. The submerged weight is 355 N/m (36.2 kg/m) empty and 830 N/m (84.6 kg/m) filled with 42° API oil. The specific gravity empty is 1.60 referred to seawater: that high value is desirable to avoid possible flotation when the trench is backfilled. The large wall thickness means that the pipeline is lightly stressed, with a factor of 2.8 on the code requirement for internal pressure containment. A further advantage is that a pipe with a low D/t ratio of 18 has a high resistance to bending buckling.

The deepest ice gouge observed in 10 years of summer open-water surveys was 0.6 m. The maximum predicted 100-year gouge depth was estimated to be 1.1 m. The minimum depth of cover was chosen as 2.1 m in the area beyond the barrier island, and 1.8 m in the Gwydir Bay lagoon area. That is more than enough to stabilise the pipe against possible upheaval buckling [1] resulting from the difference between the installation pipe temperature and the operating temperature. In the shallowest section, the ice freezes down to the bottom, and no seabed ice gouges are observed. A thaw bulb will form around and below the pipes, and the maximum total thaw settlement during the expected life is predicted to be 0.6 m.

The conditions at Northstar are entirely different from Drake F-76 (7.4.2), principally because the pipeline length is greater and the water much shallower. The construction scheme was to thicken the natural ice along the route to a minimum 2.4 m, by pumping seawater onto the surface, and then to cut a 2.4 m wide slot through the ice. The trench in the seabed was excavated by backhoes working from the ice surface, supplemented by blasting in permafrost in the shallowest sections at the Point Storkersen shore crossing. The pipelines were fabricated on the ice parallel to the trench, working on the opposite side to the civil work, in the same sequence of stringing, welding, inspection and field joint coating as in conventional overland pipeline construction. The pipes were bundled together with a leak detection tube, and lowered from the ice surface into the trench by four travelling sidebooms equipped with roller cradles. Trenching took five weeks and pipeline installation three weeks.

BP has carried out surveys each year since the line was constructed in 2000. Very significant gouging occurred during the 12 months up to the summer 2007 survey, and is attributed to an intense storm in October 2006, which produced high winds and waves, together with a negative storm surge, at a time when multi-year ice floes were present. The deepest observed gouge was 1.7 m, significantly deeper than the 1.1 m design gouge depth, but that gouge was 55 m away from the pipeline. The deepest gouge directly above the pipeline was only 0.25 m [45,46]. The effect of gouge infill should be kept in mind when these figures are interpreted.

The survey found strudel scours in nine out of the ten surveys up to 2009, and the observations are thought to confirm the hypothesis that heat from a pipeline thins the ice above it and intensifies strudel scour formation. The deepest strudel scour was 2.9 m, just 10 m east of the route, and reduced the pipeline cover to 0.6 m. The scour was backfilled to return the cover to the 1.8 m required by the right-of-way permit.

7.4.4 *Oooguruk Pipeline*

The Oooguruk oilfield is located 10 km from shore off the Colville River delta in Alaska. A flowline bundle was constructed early in 2007. Lanan [47] describes the project in detail. The bundle consists of a three-phase pipe-in-pipe 324 mm (nominal 12-inch) production flowline within a 406.4 mm (16-inch) outer pipe, a 219 mm (nominal 8-inch) water injection flowline, a 168 mm (nominal 6-inch) gas lift and injection flowline, and a 60 mm Arctic heating fuel line. The lines are strapped together over spacers as an open bundle.

The maximum water depth is only 2.3 m, and the site is partially sheltered from the more severe ice and wave conditions offshore by a line of barrier islands and by its shallow depth. The gouging issue is less severe than in the projects described in 7.4.2 and 7.4.3, but the proximity of the river heightens the risk of strudel scour (section 7.3), and there are potential problems with permafrost thaw and thermal interaction with the surroundings, both in normal operation and in shutdown.

Three summer surveys found no depressions in the seabed that could be attributed to ice gouging, and gouging is not a significant issue. Design was based on an assumed maximum gouge depth of 0.6 m, but that was not the governing factor. An extreme strudel scour event could remove the cover and create an unsupported free span, but the strain would still be acceptable under a displacement-controlled limit-state criterion. The choice of the maximum acceptable profile out-of-straightness for security against upheaval buckling had to take account of various factors, among them trenching tolerances, thaw consolidation, the possibility that the bundle would be laid in place at a low temperature (- 35°C), and survey tolerances.

The production flowline was insulated in order to limit heat loss, to limit the growth of the thaw bulb in subsea permafrost, and to limit the thinning of the sea ice above the line. The 25 mm wide annulus between the flowline and the outer pipe was evacuated to a low vacuum, and radiative heat transfer was minimised by an aluminium foil/polyolefin wrap applied to the inner pipe. The overall U value was about 1 W/m^2 deg C, referred to the outside diameter of the flowline. The water injection flowline was conventionally insulated by polyurethane foam.

The winter construction scheme was very like the one applied to Northstar, but because the water is shallower at Oooguruk the ice becomes bottomfast. The process was artificially hastened by pumping seawater onto the surface, finishing with a layer of freshwater ice to improve durability. The inner pipe of the pipe-in-pipe flow line was made up in 300 m strings and pulled into the outer pipe, and the 300 m strings were then tied in sequentially to make a 2.3 km segment. A 2.3 m trench was cut through the ice, backhoes excavated the seabed trench, the bundle was made up parallel to the other side of the trench, and the bundle was lowered in by sidebooms. The trench was then backfilled.

The bundle incorporates multiple leak detection systems.

7.4.5 *Nikaitchuq Pipeline*

The Nikaitchuq pipeline system [45, 46] was constructed in 2009 and is scheduled to go into operation in late 2011. It runs from a gravel island drill site in 1.8 m of water near Spy Island over a distance of 5.6 km to a crude oil processing facility at Oliktok Point. The maximum water depth is 3 m. The pipeline bundle consists of a 14-inch pipe-in-pipe production flow line in an 18-inch carrier, a 12-inch insulated water injection line, a 6-inch spare line, and a 2-inch diesel line. The location within Harrison Bay limits the potential for seabed ice gouging, and the line is far enough away from the Colville River to reduce the risk to reduce the risk of strudel scour.

The line was constructed by essentially the same method as the Northstar and Ooogaruk pipelines described in 7.4.3 and 7.4.4.

7.5 Transportation by Tanker

An obvious alternative to a pipeline is to transport oil and gas by ship. Oil has of course been carried by ships for more than a century, and tanker transportation continues on a very large scale. Natural gas can be carried as a liquid (LNG), but only at very low temperatures, since the boiling point of liquid methane is -161.5 °C at atmospheric pressure. LNG transportation is another well-established technology. A second possibility for gas is to transport it under high pressure, so that the same mass of gas takes up a much smaller volume. A third option is to transform the gas into a liquid (GTL).

Tanker transportation has some advantages of cost and flexibility. Tankers generally belong to shipping companies, rather than being owned by the oil companies that ship the crude or products they carry. The capital costs are often smaller than they are for a fixed pipeline, and they are borne by the shipping companies. A tanker can switch from one route to another, and can change destinations during a voyage, and that frequently happens. Against that, the operating costs are higher, and tankers are more vulnerable to accidents and oil spills, as the notorious *Exxon Valdez, Amoco Cadiz* and *Torrey Canyon* incidents demonstrate [48].

As soon as it was decided to develop the Prudhoe Bay oilfield in northern Alaska, export by tanker was one of the options that were considered. A group of oil companies led by Exxon chose the 115,000 ton tanker *Manhattan* to test that option. It was extensively modified in Pennsylvania, reportedly at a cost of $54M (in 1969 dollars) by strengthening the hull, propellers and rudder and adding an icebreaker bow. *Manhattan* sailed north in August 1969 through Baffin Bay and Viscount Melville Sound, accompanied by the Canadian icebreaker *John A. MacDonald* and the US icebreaker *Northwind.* Heavy ice in McClure Strait compelled her to turn south and instead pass through Prince of Wales Strait between Banks Island and Victoria Island. She reached Point Barrow on September 14, loaded one symbolic barrel of oil, and returned to the Atlantic.

Aspects of the voyage were highly controversial. One issue was that Canada claims the passages between its Arctic Islands as internal waters [49], a position contested by several other countries, and arguably in conflict with the traditionally accepted right of innocent passage on the high seas. The Arctic Waters Pollution Prevention Act was passed by the Canadian House of Commons in 1970, and greatly extended Canadian jurisdiction for pollution prevention, to a zone 100 nautical miles from the baseline for definition of the extent of the territorial sea, and asserts Canadian authority to regulate shipping and to prohibit any shipping in any part of the area. Canada altered its acceptance of the jurisdiction of the International Court of Justice so that its legislation could not be challenged, but later revoked that reservation. The legal aspects of these questions are outside the scope of this book: they provide many opportunities for nationalist posturing, frequently mingled with environmental arguments.

In one sense, the round-trip voyage of the *Manhattan* was a technical success, but a single round-trip in the Arctic summer is of course a totally different matter from a year-round export system carrying hundreds of millions of barrels a year. A decision to choose a pipeline option had reportedly already been made before the *Manhattan* trial had been completed [50].

A little later, Panarctic Oils discovered the Bent Horn oilfield on Cameron Island in 1974. Its oil was transported eastward to Montreal by the tanker *Arctic*. It carried one or two shipments each year from 1985 till 1996, in a summer-only operation that altogether exported 2.8 MMbbl. Production was then halted.

In parallel, a far more ambitious project to export LNG by tanker was considered for gas from several fields in the Arctic Islands, as an alternative to the Polar Gas pipeline project. The Arctic Pilot Project (APP) filed an application to the Canadian National Energy Board in 1981. It was to transport 9 M Nm^3/day (320 MMscf/day) of gas by a 160 km 559 mm (22 inch) diameter pipeline from the Borden Island Main Pool on the Sabine Peninsula of Melville Island, first to a barge-mounted liquefaction facility at Bridport Inlet on the south coast. From there the LNG would be shipped eastward in specially-constructed icebreaking Arctic Class 7 LNG tankers. Each tanker would have had a capacity of 140,000 m^3 and would have cost $400M in 1982. The project capital cost of the first phase was estimated at $1B. Later phases would have expanded capacity. Lewington [51] describes the political history and opposition to the project. In her words:

> "...the Inuit saw the APP not as a low-risk research project, as argued by the sponsors, but as the thin edge of an industrial wedge into the north where native rights had yet to be established"

The project did not proceed, for many other reasons. Among them are the failure of gas prices to rise as far as had been expected, increased availability from LNG from other sources, hostile reactions from Inuit communities and from Greenland, aversion to the risk of cost overruns, discovery of the Sable Island gas field off Nova Scotia, and limited enthusiasm in Ottawa. More recently a reduction in concern about possible interruption of gas supplies from Russia, as a result of the political changes there, and the easy availability of shale gas and coal-bed methane in the northeast US have come to be factors. The economics of various LNG, CNG and GTL gas transportation options from Melville Island are examined in a 2005 Canadian Energy Research Institute report [36].

A final possibility is to transport LNG by submarine tanker under the Arctic sea ice. That option would draw on extensive experience of under-ice operation of military submarines [52]. At one time it was much talked about. There are several difficulties, among them the very high capital and operating costs of submarines, safety, concerns about nuclear energy if that were the propulsion option selected, and the fact that in many places shallow water extends some distance from shore, so that there is insufficient clearance for a submarine between the seabed and the keels of ice ridges. The idea seems unlikely to be revived.

Tanker options become more attractive in locations where the ice conditions are less severe than they are in the High Arctic, such as the Chukchi Sea, the Barents Sea, the Kara Sea and the northern Caspian. They offer flexibility and the option of direct loading from offshore production facilities, whether fixed or floating, without the difficulties of pipelines to shore. Climate change may make the design requirements less demanding. Against that, it is often argued that tanker transportation has a more severe environmental impact than pipeline transportation does, and that for that reason indigenous peoples are generally more opposed to tankers than they are to pipelines.

References

1 Palmer, A.C. and King, R.A. (2008) Subsea Pipeline Engineering. Pennwell, Tulsa, OK.

2 Palmer, A.C., Hammond, J. and King, R.A. (2008) Reducing the cost of offshore pipelines. *Proceedings, Marine Operations Specialty Symposium*, Singapore, paper MOSS-11, pp. 275-284.

3 Palmer, A.C. (1996) A flaw in the conventional approach to stability design of pipelines. *Proceedings, Offshore Pipeline Technology Conference*, Amsterdam.

4 Teh, T.C., Palmer, A.C. and Bolton, M.D. (2004) Wave-induced seabed liquefaction and the stability of marine pipelines, *Proceedings, International Conference on Cyclic Behaviour of Soils and Liquefaction Phenomena*, RuhrUniversität Bochum, pp. 449-453 , AA Balkema, Leiden, Netherlands.

5 Palmer, A.C. (2000) Are we ready to construct submarine pipelines in the Arctic? *Proceedings, Thirty-second Annual Offshore Technology Conference*, Houston, OTC12183.

6 Sleath, J.F.A. (1984) *Sea bed mechanics*. John Wiley & Sons, New York.

7 Komar, P.D. (1976) *Beach processes and sedimentation*.Prentice-Hall, Englewood Cliffs, NJ.

8 Woodworth-Lynas, C. and Guigné, J. (2003) Ice keel scour marks on Mars: evidence for floating and grounding ice floes in Kasei Valles. *Oceanography*, 16, pp. 90-97.

9 Lewis, C.F.M. (1977) Bottom scour by sea ice in the southern Beaufort Sea. *Beaufort Sea Project Technical Report 23*, Department of Fisheries and Environment, Ottawa, Canada.

10 Palmer, A.C., Konuk, I, Comfort, G. and Been, K. (1990) Ice gouging and the safety of marine pipelines. *Proceedings, Twenty-second Offshore Technology Conference*, Houston, 3, pp.235-244, paper OTC6371.

11 Weeks, W.F. (2010) *On sea ice*. University of Alaska Press, Fairbanks, AK.

12 Shearer, J., Laroche, B and Fortin, G. (1986) Canadian Beaufort Sea repetitive mapping of ice scour. *Environmental Studies Revolving Funds report 032*, Canada Oil and Gas Lands Administration, Ottawa, Canada.

13 Lewis, C.F.M. , Parrott, D.R., Simpkin, P.G. and Buckley, J.T. (1986) *Ice Scour and Seabed engineering: proceedings of a workshop on ice scour research. Environmental Studies Revolving Funds report 049*, Canada Oil and Gas Lands Administration, Ottawa, Canada.

14 Weeks, W.F., Tucker, W.B. and Niedoroda, A. (1986) Preliminary simulation of the formation and infilling of sea ice gouges. *Ice Scour and Seabed engineering: proceedings of a workshop on ice scour research. Environmental Studies Revolving Funds report 049*, Canada Oil and Gas Lands Administration, Ottawa, Canada, pp.259-268.

15 Shearer, J. and Stirbys, A.F. (1986) Towards repetitive mapping of ice scours in the Beaufort Sea. Ice *Scour and Seabed engineering: proceedings of a workshop on ice scour research. Environmental Studies Revolving Funds report 049*, Canada Oil and Gas Lands Administration, Ottawa, pp.284-292.

16 Vogt, P.R., Crane, K. and Sundvor, E. (1994) Deep Pleistocene plowmarks on the Yermak Plateau: sidescan and 3.5 kHz evidence for thick calving ice fronts and a possible marine ice sheet in the Arctic Ocean. *Geology* 22, pp.403-406.

17 Atkins, T. (A.J.) (2009) *The science and engineering of cutting*. Butterworth-Heinemann, Oxford, UK.

18 Sarpkaya, T. and Isaacson, M. (1981) *Mechanics of wave forces on offshore structures*. Van Nostrand Reinhold, New York.

19 St. Denis, M. (1969) On wind-generated waves. *Topics in Ocean Engineering* (ed. Bretschneider, C.L.), pp. 37-41.

20 Borgmann, L.E. (1961) The frequency distribution of near extremes. *Journal of Geophysical Research*, 66, pp.3295-3307.

21 Palmer, A.C. and Niedoroda, A.W. (2005) Ice gouging and pipelines: unresolved questions, *Proceedings, Eighteenth International Conference on Port and Ocean Engineering under Arctic Conditions*, Potsdam, NY, 1, pp.11-21.

22 Woodworth-Lynas, C.M.L., Nixon, J.D., Phillips, R. and Palmer, A.C. (1996) Subgouge deformations and the security of Arctic marine pipelines. *Proceedings, Twenty-eighth Annual Offshore Technology Conference*, Houston, 4, pp.657-664, paper OTC8222.

23 Palmer, A.C., Kenny, J.P., Perera, M.R. and Reece, A.R. (1979) Design and operation of an underwater pipeline trenching plough, *Geotechnique*, 29, pp.305-322.

24 Palmer, A.C. (1999) Speed effects in cutting and ploughing. *Geotechnique*, 49 (3) pp.285-294.

25 Palmer, A.C., White, D.J., Baumgard, A.J., Bolton, M.D., Barefoot, A.J., Finch, M.,Powell, T., Faranski, A.S., and Baldry, J.A.S. (2003) Uplift resistance of buried submarine pipelines, comparison between centrifuge modelling and full-scale tests. *Geotechnique*, 53, pp.877-883.

26 White, D. Personal communication (2011).

27 Palmer, A.C. and Rice, J.R. (1973) The growth of slip surfaces in the progressive failure of overconsolidated clay, *Proceedings of the Royal Society, ser. A*, 332, pp.527-548.

28 Been, K., Sancio, R.B., Ahrabian, A., van Kesteren, W., Croasdale, K. and Palmer, A.C. (2008) Subscour displacement in clays from physical model tests. *Proceedings, IPC2008 7th International Pipeline Conference*, Calgary, paper IPC2008-64186.

29 Sancio, R., Been, K. and Lopez, J. (2011) Large scale indenter test program to measure sub gouge displacements. *Proceedings, 21st International Conference on Port and Ocean Engineering under Arctic conditions*, Montréal, Canada.

30 Palmer, A.C. (1977).Geotechnical evidence of ice scour as a guide to pipeline burial depth. *Canadian Geotechnical Journal*, 34, pp.1002-1003.

31 Palmer, A.C. and Tjiawi, H. (2009) Reducing the cost of protecting pipelines against ice gouging. *Proceedings, Twentieth International Conference on Port and Ocean Engineering under Arctic Conditions*, Luleå, Sweden, POAC09-77.

32 Chua, M., Palmer, A.C. and Tjiawi, H. (2011) Protecting Arctic marine pipelines against subgouge deformation, *Journal of Pipeline Engineering*, 10 (2) pp.81-85.

33 Palmer, A.C. and Chan, Y.T. (2012) Reducing the cost of protecting Arctic marine pipelines against ice gouging, *Proceedings, International Society of Offshore and Polar Engineers Conference*, Rodos, Greece, paper 2012-TPC-0520.

34 Reimnitz, E., Rodeick, C.A. and Wolf, S.C. (1974) Strudel scour: a unique Arctic marine geologic phenomenon. *Journal of Sedimentary Petrology*, 44, pp.409-420.

35 Palmer, A.C. (2011) Limit states and lateral buckling. *Proceedings, Global Pipeline Buckling Symposium,* Perth, Australia.

36 Chan, L., Eynon, G. and McColl, D. (2005) *The economics of High Arctic gas development: expanded sensitivity analysis*: Canadian Energy Research Institute, report to Department of Indian and Northern Affairs Canada. http;//www.ceri/ca/documents/HighArcticGas Report.pdf.

37 Palmer, A.C., Brown, R.J., Kenny, J.P. and Kaustinen, O.M. (1977) Construction of pipelines between the Canadian Arctic Islands, *Proceedings, Fourth International Conference on Port and Ocean Engineering under Arctic Conditions*, St. John's, Newfoundland, 1, pp.395-404.

38 Palmer, A.C., Baudais, D.J. and Masterson, D.M. (1979) Design and installation of an offshore flowline for the Canadian Arctic Islands, *Proceedings, Eleventh Annual Offshore Technology Conference*, Houston, 2, pp.765-772.

39 Palmer, A.C. (1979) Application of offshore site investigation data to the design and construction of submarine pipelines, *Proceedings, Society of Underwater Technology Conference on Offshore Site Investigation*, London, pp.257-265.

40 Marcellus, R.W. and Palmer, A.C. (1979) Shore crossing techniques for offshore pipelines in Arctic regions, *Proceedings, Fifth International Conference on Port and Ocean Engineering under Arctic Conditions*, Trondheim, Norway, 3, pp. 201-215.

41 *Arctic Challenge* (movie), R.J. Brown and Associates (1979).

42 Kamenskii, R.M, Konstantinov, I.P. Makarov, V.I. and Olovin, B.A. (1973) *Guidebook North-Western Yakutia.* USSR Academy of Sciences, Earth Sciences Section, Siberian Division, Yakutsk, Russia.

43 Duguid, A. and McBeth, R. (2000) Drake F-76, in-situ abandonment of a High Arctic offshore completion and facilities. *Journal of Canadian Petroleum Technology*, 39 (5) pp.33-40.

44 Lanan, G.A., Ennis, J.O., Egger, P.S. and Yockey, K.E. (2001) Northstar offshore Arctic pipeline design and construction. *Proceedings, 2001 Offshore Technology Conference*, Houston, paper OTC13133.

45 Lanan, G.A., Cowin, T.G. and Johnston, D.K. (2011) Establishing the Beaufort Sea baseline. *Offshore Engineer*, May.

46 Lanan, G.A., Cowin, T.G. and Johnston, D.K. (2011) Alaskan Beaufort Sea pipeline design, installation and operation. *Proceedings, Arctic Technology Conference*, paper OTC22110.

47 Lanan, G.A., Cowin, T.G., Hazen, B., McGuire, D.H., Hall, J.D. and Perry, C. (2008) Oooguruk offshore Arctic flowline design and construction. *Proceedings, 2008 Offshore Technology Conference*, Houston, paper OTC19353.

48 Bernem, K.H.van, and Lübbe, T. (1997) *Öl im Meer* (Oil in the sea) Wissenschaftliche Buchgesellschaft, Darmstadt, Germany.

49 Griffiths, F. (1987) *Politics of the Northwest Passage*. McGill-Queen's University Press, Kingston, Canada.

50 Coates, P.A. (1993) *The Trans-Alaska Pipeline Controversy*. University of Alaska Press, Fairbanks, AK.

51 Lewington, J. (1987) Lessons of the Arctic pilot project. in Griffiths, F. *Politics of the Northwest Passage*. McGill-Queen's University Press, pp. 163-180.

52 Williams, M.D. (1998) *Submarines under ice: the US Navy's polar operations*. Naval Institute Press, Annapolis, MD.

Chapter Eight

Environmental Impact

8.1 Introduction

Every human activity has an effect on the environment. Only a mass extinction would eliminate those impacts entirely. Mankind has to find ways of accomplishing its desires without damaging the environment to an unacceptable degree. Those issues are intensely controversial, and engage complex philosophical questions about our relationship to Nature, as well as severely practical decisions about whether or not the advantages of some development outweigh the damaging effects it will have. The problem is made more acute by the growth of human population and by the extension of human desires, so that someone whose grandparents thought themselves content to grow rice and vegetables, and never to travel more than a few kilometres, now aspires to a car, a computer, air-conditioning, and a vacation at a remote island resort.

The Arctic is particularly vulnerable to lasting damage, because of the slow pace of biological activity. An abandoned Mayan city like Tikal in central America is only a thousand years old, but the buildings are covered with vegetation and only just distinguishable from the surrounding jungle. Left to itself for another ten thousand years, it will be invisible: the natural environment will have reasserted itself.

In contrast, many things in the Arctic change very slowly indeed. Someone drives a truck across the tundra, the wheels damage the vegetation and change the thermal balance at the surface, the permafrost below the wheel tracks degrades, and the ruts are still visible years afterwards. Explorers find bodies preserved by the cold, and the artefacts with them look as if they were made yesterday. In the tropics, on the other hand, the ruts would be overgrown, the bodies would rot and merge into the earth, and the artefacts would rust and decay.

Some people argue that the Arctic is so sensitive that no development can ever be permitted. The background to that argument is discussed in chapter 1. It is hardly likely to be sustainable in its pure form, because of pressure from the local inhabitants and from outside interests. At the time of the pipeline controversy, there were bumper stickers in Alaska that read "If we want to make a whole state a national park, let's start with Ohio!"

Concern about offshore development has been raised to an intense level by the catastrophic April 20 2010 oil blowout at the Macondo well to the south of Louisiana in the Gulf of Mexico. That event has turned out to be comparable to the Ixtoc-1 blowout in Mexico in 1979. Some 5 million barrels flowed into the Gulf, and the well was not capped until late July. Irrespective of precisely what the sequence of events will turn out to have been, it is plain that the impacts on the offshore industry will be severe and long-lasting [1, 2], and comparable with the effect of the Chernobyl disaster on the nuclear industry. The damage to coastal wetlands and to the fishing industry is acute, and the public and political response has been vigorous (though it is sobering to notice that the public and the media are far more concerned about oil on beaches than they are about the eleven people who died).

The leak proved extremely difficult to stop, even though it occurred in the summer, the weather was reasonably kind, and the incident was close to the centre of the offshore industry with unrivalled reserves of equipment, materials, and engineering talent. It is being pointed out that a similar incident in the Arctic would be still more difficult to control, even if it occurred in the summer. Parallels are being drawn with the *Exxon Valdez* tanker disaster in the sub-Arctic Prince William Sound in Alaska [2]. It seems likely that regulation will be tightened up and that severe restrictions on drilling will be imposed. Those restrictions are likely to be even more severe in the Arctic.

8.2 Oil in the Sea

8.2.1 *Outside the Arctic*

Crude oil is a complex mixture of hydrocarbons, with small fractions of other elements among them sulphur, helium, nitrogen, nickel, vanadium and mercury. Some of the molecules ('alkanes','paraffins') have the carbon atoms arranged in chains, sometimes straight and sometimes branched, with two or three hydrogen atoms loosely linked to each carbon atom. Other molecules (cycloalkanes and aromatics) have the carbon atoms arranged in rings, again with hydrogen atoms loosely linked to them. The lightest molecule is methane CH_4, a gas lighter than air. The next components are ethane C_2H_6, propane C_3H_8 and two isomers of butane C_4H_{10}, the principal constituents of liquefied petroleum gas ('bottled gas'). Intermediate components have 4 to 10 carbon atoms, are liquid at ordinary temperatures, and form the major components of gasoline (petrol) and diesel oil. Heavier components make up heavy fuel oil, and the heaviest are bitumens, solid at ordinary temperatures. Hunt [3] describes the geochemistry of petroleum in detail.

The proportions of different components vary. A simple descriptive measure of the density is API (American Petroleum Institute) gravity, defined by

$$\text{API gravity} = \frac{141.5}{SG_{60}} - 131.5 \qquad (8.2.1)$$

where SG_{60} is the specific gravity (relative density referred to water) at 60°F (15.56°C).

This peculiar scale was originally chosen because it is linear on a hydrometer. The heaviest crude oils have the *lowest* gravity. On this scale fresh water has a gravity of 10. Most crude oil has a gravity between 40 and 20. Heavy oil has a gravity less than 20, and is more difficult to produce and less valuable. Extra heavy oil has a gravity less than 10, and cannot be produced by conventional methods: it is important in the sub-Arctic because it forms the oil sands ('tar sands') that underlie vast areas of northern Alberta.

If oil gets into the sea, the components behave in different ways. Van Bernem and Lübbe [4] describe the behaviour of the oil, in much greater detail than is possible in this text. Figure 8.1 is a schematic, redrawn from their book, that illustrates the mechanisms that occur.

The lightest components rise into the atmosphere. There they are not a visible source of environmental damage, but they do have a harmful influence, because methane is a potent greenhouse gas. The oil spreads out onto the surface, under the influence of density differences, surface tension and wave action. Waves break the floating oil into smaller blobs. The heavier components float on the surface, and are subject to various influences. Some components dissolve in the water, where they remain miscible at low concentrations. Some components are broken down by bacteria and other marine organisms, and others slowly oxidise. Oxidation and evaporation may make the remaining oil denser than water, and then it will sink. Wave action may drive the oil on a beach, or onto rocks bordering the coast. If the lighter fractions have evaporated and dissolved, the oil may coalesce into tar balls. The waves can erode and liquefy sand on a beach, and some of the oil may be buried. If enough of the light fractions evaporate or dissolve in the sea, the residual oil will be denser than seawater, and may sink to the bottom, or drift about somewhere within the water column (as some of it did during the *Deepwater Horizon* disaster). The fact that oil is no longer visible does not mean that it has moved out of the system, and also does not mean that it has no continuing biological consequences.

The presence of living organisms makes things much more complicated. Biological breakdown depends on many factors. Bacteria capable of breaking down hydrocarbons have to be present, and the presence of hydrocarbons encourages the growth of those bacterial in competition with others. Research in the North Sea has shown that the fraction of the total bacteria present that is capable of breaking down oil is higher in the central North Sea, around the Ekofisk field (one of the earliest to be developed) and along the central spine of oil and gas fields extending to the north, roughly midway between the Scottish and Norwegian coasts [4, 5]. The fraction is lower between Ekofisk and

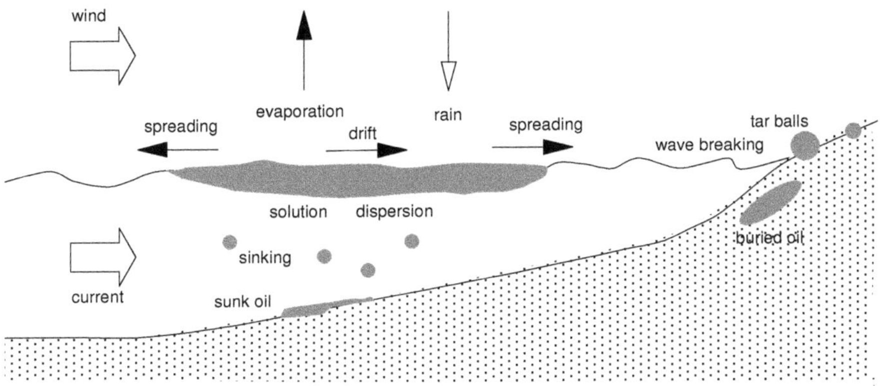

Figure 8.1 Behaviour of spilled oil (redrawn from Van Bernem and Lubbe [4])

the German and English coasts, because less oil is present. Enhanced bacterial activity is a factor in the Gulf of Mexico, where there are natural oil seeps and extensive offshore oil development accompanied by frequent small-scale spillages.

Bacteria need other nutrients such as phosphorus and nitrogen. Seawater and oil are very low in phosphorus, and usually but not always low in nitrogen. Their availability may be a limiting factor, and so breakdown might be accelerated by fertilising the sea. That option has been suggested in the different context of increasing carbon capture by algae, whose growth is often limited by the availability of iron. In the seabed, on the other hand, nutrients are readily available but growth might be restricted by limited oxygen.

The relative importance of different breakdown pathways varies enormously, and depends on external factors, among them the temperature, the weather generally, the composition of oil, and the type of coast. Different components break down at different rates. Van Bernem and Lübbe [4] report a comparison between the *Amoco Cadiz* stranding in March 1978, off the coast of Brittany in northern France, and the *Exxon Valdez* stranding in March 1989, in Prince William Sound in Alaska. In the *Amoco Cadiz* instance, 30 per cent of the oil evaporated, 28 per cent reached the shore (where different processes would break it down), 14 per cent dispersed into the water, and microbial degradation accounted for only 5 per cent. In the case of the *Exxon Valdez* the proportions were quite different: microbial degradation destroyed half the oil and dispersion about 20 percent, whereas beaching and dispersion contributed very little. It needs to be kept in mind that oil that has dispersed into the ocean has disappeared from sight but is still chemically present, and it can still have biological effects. The details can be argued about, but there are clear differences.

Bacterial degradation of oil in bottom sediments is very much slower than in the open sea, and takes years rather than days.

Evidence from many geographical locations indicates that some oil is extremely persistent, particularly in buried sediments and in cold climates. Investigators found residual oil in southern Chile, near Milford Haven in Wales, and in Chedabucto Bay in Canada, some 15 or 20 years after the original incidents.

The biological effects of oil in the sea are extraordinarily complex. Oil coats lifeforms directly, witnessed by the dramatic pictures of oil seabirds that appear within hours of a spill, but there are effects that are more subtle and at the same time much wider-ranging. Oil is taken into the food web, and different organisms react in different ways, often by genetic damage and a reduction in reproductive efficiency. As an example of a complex interaction, Van Bernem and Lübbe describe the effects on herring eggs of the 1977 *Tsesis* 1000-tonne spill of fuel oil in the Baltic [4]. The oil had no direct effect on the herring eggs, but their numbers fell because of fungal infections, which developed because of heavily reduced numbers of the minute flea-crabs (*Pontoporeia affinis*) that normally graze on the fungi but are highly sensitive to oil. They give many more examples of this kind.

8.2.2 *In the Arctic*

All the effects described in 8.2.1 are present. The persistence of oil pollution is markedly influence by the fact that both physical and biological processes take place much more slowly at low temperatures and in the absence of daylight.

In an Arctic sea, oil spilled below the ice will float up and accumulate in hollows in the underside (Figure 8.2). The oil will seal off the cracks and brine channels in the ice, and tend to stop the interchange of water, dissolved air and nutrients that support life in the channels and immediately beneath the ice. There may be open leads: oil will fill those leads and make it almost impossible for marine mammals to use them as breathing holes. This will take away an important food source for polar bears which wait by holes in the ice for seals to come up to breathe. Seabirds will not be able to dive into the leads in search of fish without becoming dangerously oiled.

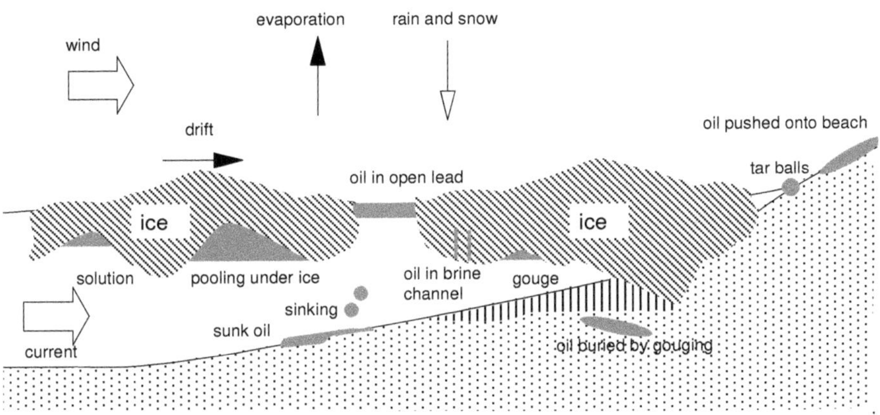

Figure 8.2 Oil spilled in the Arctic

The ice will move, as we have seen in chapter 2, and carry the oil for long distances. Wave action may break up the ice and disperse some of the oil into the water. Ultimately the ice will melt and release any remaining oil to the water surface. If the oil does not breakdown, later freezing may trap the oil in the ice and carry it further.

At low temperatures on Arctic seas oil is much more viscous that it is at warmer temperatures, particularly if the oil is below its pour point temperature, and that influences its spreading. Buist et al. [6] carried out experiments on spreading and evaporation on and under ice and snow. They used four kinds of oil: Alaska North Slope sales crude, Northstar sales crude, Endicott sales crude and Kuparuk sales crude. They relate their observations to various empirical and semi-empirical relationships. Books on oil spills and oil in the sea often mention the special factors of the Arctic, but usually rather superficially.

In summary, an oil spill in ice may be less serious than the same spill in open water, but it may be worse. Oil does not spread as far under ice as it does on the open sea. If a sensitive shoreline is nearby, this process can be beneficial in the short term, because the oil has a lower chance of being driven inshore and there is more time available to implement protection and clean up. Oil that accumulates in leads will be in a thicker layer than in open water, this may allow a more focussed clean up either by burning or by pumping into ice strengthened recovery vessels. Oil trapped under ice will generally stay there (with some migration to the surface). If the ice is landfast, there will be time to implement recovery through the ice by working from the surface. If the ice is drifting, it can be tracked and assessments made of what will happen when the ice melts.

8.3 Gas in the Sea

Natural gas is mostly methane CH_4, with smaller amounts of ethane C_2H_6, propane, butane, higher hydrocarbons, nitrogen and carbon dioxide [3]. Most of these components are relatively insoluble in seawater. A submarine gas release will bubble to the surface, carrying with it a plume of water. If the surface is ice-covered, gas will be trapped in the underside hollows, lifting the ice slightly.

Oil is highly visible and threatening to marine life: gas is not. For this reason, it may be less challenging to secure permission for gas developments than for oil developments.

8.4 Response and Oil Cleanup

In the past, it was common for spilled oil to be left in place, and for it to be allowed to spread and contaminate the environment without any attempt to clean it up. Examples of severe pollution from oil operations many years ago can be seen in many places, among them Mexico, Azerbaijan, Russia and Nigeria. That damaging practice is no longer acceptable to the community.

Concern about oil in the sea goes back at least 90 years [7]. It was driven by environmental organisations, particularly those concerned with birds. Initially the primary concern was the deliberate discharge of oil from tankers and merchant ships, and the earliest legislation was directed towards them, in the UK Oil in Navigable Waters Act of 1922 and the US Oil Pollution Act of 1924. Those measures were principally concerned with oil in territorial waters, but it was recognised that the problem was international, and the first international conference on oil pollution was held in Washington in 1928. The subject has a long history and a wide-ranging literature.

As with any other human activity, different people have different and sometimes conflicting priorities. Politicians do not want their voters to see horrifyingly oiled birds and seals on the evening television, and do not want them to be afraid to take their children to the beach. They are concerned with economically vulnerable coastal communities that depend on fish and shellfish that might become contaminated and unsaleable, or that depend on tourism. They give a lower priority to longer-term effects, particularly if they are out of sight, but those effects are important to marine biologists and environmentalists troubled about the health of the ocean. Offshore oil operators nowadays generally take a responsible position, and they are aware that pollution incidents can have an extraordinary damaging effect, both on their short-term financial position and on their long-term political reputation and prospects, but at the same time they are concerned to keep costs under control.

Cleanup has an extensive literature. Fingas [8] wrote an elementary and non-technical introduction. He describes the many different methods that have been proposed and applied, among them skimming devices of many kinds, sorbents, floating booms, in-situ burning, dispersants, biodegradation agents, and sinking agents. He emphasises that there are different kinds of oil, and many different contexts, particularly on coasts. Many of the schemes proposed work well on calm water, but it is much more difficult to identify a scheme that will be effective in rough seas.

Some kinds of clean-up can have damaging secondary effects. Dispersants contain surfactants that break up large masses of oil into small droplets that spread out into the upper layers of the ocean. The oil has not disappeared, but it is less visible, and the droplets may be more readily degraded by biological and chemical processes. Dispersants are most effective with light oils that are primarily alkanes, and much less effective with medium and heavy oils that contain aromatics, waxes and asphaltenes.

Unfortunately, some of the dispersants applied 30 or more years ago turned out to be acutely toxic. Dispersants were heavily applied in the aftermath of the *Torrey Canyon* stranding in Cornwall in England in 1967, and after the *Amoco Cadiz* stranding in Brittany in France in March 1978, which released 200,000 tonnes of crude oil. Later it was recognised that much of the environmental damage was caused by the dispersant rather than the oil. Those dispersants had a toxicity between 5 and 50 mg/litre, measured as a LC_{50} (lethal concentration to 50% of a test population) to rainbow trout over 96 hours. A higher LC_{50} is better. The dispersants available now have a LC_{50} between 200 and 500 mg/l, less toxic than diesel and light crude, whose LC_{50} rages from 20 to 50 mg/l [8]. Dispersants have fallen out of fashion, and remain controversial. Their application is nowadays highly restricted.

Several reports examine the special problems of the Arctic. References [9,10] consider Arctic Canada, [11] the effect on the shoreline of the *Exxon Valdez* disaster, and [12] the Alaskan Beaufort Sea; [12] includes some instructive case studies of hypothetical incidents, devised for training purposes, and

> '...developed by oil spill consultants who have a thorough understanding of the current technology for oil spill response in the Beaufort Sea. Also, to the extent possible, they were based on actual situations which have occurred in either the Alaskan or Canadian Beaufort Sea or in the Gulf of Mexico'.

In one hypothetical incident, a blow-out occurs on April 30 at an exploration well in Harrison Bay, north-west of the Colville River delta on Alaska, in 2 m of water 5 km offshore. The shorefast ice is beginning to break up, several leads are opening, temperatures are a little below freezing, and the wind is gusting to 25 knots (13 m/s). The initial flow is 5000 b/d of oil (800 m^3/day) with a gas/oil ratio of 600 to 800 scf/b. Over the first month the flow diminishes to 2000 b/d, and then continues at that level until the well is killed in October. It is supposed at first that the blow-out can be stopped within a few days, and then it is thought that the flow could be set alight, but attempts to do so fail. The oil around the well is mixed with snow and taken to lined pits, and when it melts the oil is skimmed off and burned. After break-up in June, attention moves to the sea, and barges towed by ice-strengthened tugs deploy rope skimmers. They prove ineffectual in more than 50 per cent ice cover, and are replaced by hand-held weir skimmers. Each barge recovers between 200 and 300 b/d. Some of the shoreline is left alone, because cleanup would do more damage than the oil, and other areas are cleaned manually or the oil is washed into the surf and recovered by skimmers. All in all, 305,000 bbl (48,500 m^3) of oil are released, and about half is recovered.

In a second hypothetical incident, a blow-out occurs on September 30, when ice forces a drillship off location and the riser breaks below the blow-out preventer. The drillship is in 18 m of water 20 km northeast of Kaktovik. The well releases 5000 b/d of oil and 6000 b/d of water. The temperature is just above freezing, the wind is blowing at 60 knots, and the sea is covered with pieces of ice less than 3 m across. The flow of gas at first prevents ice formation over the well, but as the winter advances ice covers the surface and currents spread a layer of oil under the ice. The drillship is removed and the gas is ignited, but the oil does not burn. In the following summer, the ice breaks up and large quantities of oil come to the surface. About half the oil is removed by burning and by skimmers.

Evers [13] reviews oil spill response options in the context of the ARCOP project to develop a marine transportation system to bring oil and gas from the Varandey field (68°49’ N, 58°00’ E) through the Pechora and Barents Seas. He considers burning and mechanical skimmers. Some of the skimmers were developed for Arctic applications, but most of them are at the prototype stage and have not been tested under field conditions with ice and waves. Notably, his assessment of operational limits for different methods as a function of ice coverage [13, table 2] includes only two methods, a skimmer from an ice breaker and in-situ burning, that are supposed to be able to operate in more than 80 per cent ice cover, and only the second of them operates in 90 per cent ice.

In evidence to the Canadian National Energy Board, as part of the preparation of its Arctic Offshore Drilling Review, Imperial Oil presented several scenarios for clean up of a Beaufort Sea blowout [14], and compared what happens to the oil in two hypothetical 50,000 b/d blowouts in the Beaufort Sea, one in winter under ice and one in summer in open water, with the Macondo spill in the Gulf of Mexico in the summer of 2010. It suggests that an Arctic spill might not be as catastrophic as is generally thought. However, a major assumption is that 59% of a winter spill can be burned in melt pools when it has migrated to the surface of the ice. These estimates have been criticised. In Matthews' words [14]]

> "It all sounds very good. But does it work in the dark? At minus 30? With the wind blowing? Two hundred kilometres offshore? Or just in a test facility in Norway?"

The above referenced submissions to the Canadian NEB have also included a discussion on the need for "same-season" relief well drilling. This has been an established approach in the Canadian Beaufort for decades. What this means is that if a blow out occurs when drilling a well, the operator must have a system or plan in place to drill a relief well during the same season, whether it be summer or winter. A relief well is one that is drilled by a rig some distance away which intercepts the wellbore of the blowing well and kills the flow by pumping in mud and cement. It is an established practice and was eventually used in the case of the Horizon incident.

Drilling a relief well requires two important ingredients; a second rig; and time on location; which depending on well depth could be 30 – 60 days. These two ingredients may be difficult to achieve in the offshore Arctic. When artificial islands and caissons were being used in the landfast ice for winter drilling, the relief well plan usually involved building ice roads and a gravel pad or spray ice pad from which to drill. When drillships were being used during the 3 to 4 month summer season, a well might take two seasons to drill because after a certain date entering a potential hydrocarbon zone was not allowed. Later, when the more ice-capable Kulluk was in use and available, the other drillships could drill for longer.

Future Beaufort drilling on the newly acquired leases (in deeper water) will almost certainly be done using floating systems with ice management (see Chapter 6). To achieve same season relief well drilling it is clear that two drilling systems will be required. This is an onerous financial requirement during exploration drilling. Currently the operators are promoting other approaches involving much improved blow out prevention on the wellhead and in the well design. Clearly, it is much better to prevent a blow out than to drill relief wells and/or use containment systems to capture the oil (which is also much more difficult to implement in ice covered regions than in the open sea).

8.5 Effects of Structures on the Ice

Most of this book is concerned with the effects of ice on structures and how to determine ice loads from ice action. Sometimes however, here are concerns about the effects of structures on the ice and whether there may be adverse environmental consequences from such effects.

In the 1970s and into the 80s, artificial islands and bottom-founded caissons (on berms) were being used for exploration drilling in the Beaufort Sea. This commenced with the first islands in about 3 m water depth in 1972, but gradually one or two new islands were built each year progressing into deeper water out to 20 m and beyond. By 1984, 26 exploration wells had been drilled in the Canadian Beaufort from separate artificial islands out to about the 30 m water depth. A concern expressed by local communities and environmentalists was that the islands were extending the extent of the landfast ice and delaying ice break up. These changes were of concern to local hunters. If the boundary between landfast and pack ice was further offshore, then they would have had to travel further across the ice to reach the pack ice where they would normally hunt seals and polar bear in the Spring. If ice break up was delayed, it was suggested that whales would be further offshore and could not reached to kill them.

As with many natural processes, there can be large differences from year to year and often proponents arguing a case will only remember the years when the observations or their experience reinforces their concerns. In this case, Spedding of Imperial Oil had been studying landfast ice extents for over a decade during the period when the islands had been gradually installed. A careful study of these data [15] could not discern any trends which suggested the islands were having any effect. This is perhaps not surprising when one speculates on the physics. The maximum distance from shore reached by landfast ice is usually characterized by a narrow water depth range which will vary from region to region depending on ice severity (In the Beaufort it is about the 20 m water depth). Parallel observations of ridges indicate that this boundary is characterized by a large number of grounded ridges (or stamukhi) and it is these grounded features that stabilize the ice against the driving forces of winds and currents (the other boundary being the land – hence the terminology). This is not to say that grounded ice cannot occur in water depths greater than the landfast ice boundary; it can, as indicated by the ice scour record. However, the grounded features are neither as numerous nor as heavily grounded and they cannot stabilize the ice boundary but get moved within the pack ice and create ice scours. Recognizing this process, a comparison between the number of grounded ice features and the number of artificial islands will give a better sense of potential effects. In this case, the additional grounding points provided by the islands which were just beyond the landfast ice boundary (which are the ones that have the possibility of extending it) was a small number (3 or 4) compared to the hundreds of grounded ice features.

In the other hand, artificial islands (in conjunction with ice booms) have been used to help stabilize the ice cover in some of the lakes in the St Lawrence Seaway in Canada to aid in winter navigation and in order to prevent loose ice from entering hydro electric water intakes [16].

It is also suspected (but not proven) that in the North Caspian, the growing number of islands associated with the Kashagan development may be creating a more stable ice cover. In this case, it may be manifested in requiring a higher wind threshold to start an ice movement event. Even if this is the case, we know of no adverse environmental or operational consequence.

In another example, in 1992, one of the authors was asked to join a panel of experts to assess the potential effects on the ice regime of a proposed bridge across Northumberland Strait between Prince Edward Island and the mainland of Canada. This followed a negative review by an Environmental Panel who had conducted public hearings to assess concerns about the bridge. Some local residents (a minority) were opposed to the bridge for a whole range of reasons but one argument was that the bridge would delay ice break up in the Strait thus adversely affecting the lobster fishery and the potato crop. The Environmental Panel in one of its recommendations said that the bridge should not proceed unless it could be shown that the bridge would not delay "ice out" in the Strait by more than 2 days per year on average over a hundred year period! Hence the appointment of the expert panel by Public Works Canada (who were tasked by the Federal Government to proceed with the bridge if at all possible, because the majority of local people were in favour).

Our panel took a measured approach to the issue but recognized immediately the difficulty of proof because the natural variability of the ice-out date from year to year without the bridge, based on many years of records, was over 65 days [17]. Nevertheless, the physics of ice growth and decay, ice transport by tidal currents and potential blockage by the bridge are reasonably well understood and there were extensive historical data on weather for the region. So a model was developed including all these processes and driven by the statistics of historical weather data. The model predicted ice growth and subsequent decay and was run with and without the bridge for a large number of simulations such that extremes could be predicted. The model could also be calibrated/verified against ice-out dates without the bridge and this was done. The bridge was considered to have two main effects on the ice. On the negative side its presence was assumed to create ice rubble and ridges which added to the volume of ice in the Strait which would take longer to decay in the Spring. On the other hand, it could be demonstrated using simple physics that as the bridge slowed down ice movement relative to the tidal currents in the Spring, the ice would decay more quickly because of the increased heat input from the water below due to the higher relative velocity between ice and warmer water. Incorporating these two effects into a model and simulating many years with and without the bridge gave results which met the rather restrictive requirement set by the Environmental panel and construction of the bridge was able to proceed (but not without further public hearings at which we had to appear and defend the work). The modelling and other ice issues relating to the bridge are described in [17] and [18]. The bridge was completed in 1997 and photos and measurements of ice loads on the bridge are referred to earlier in this book. To our knowledge, there have been no adverse environmental consequences due to the presence of the bridge; in fact the Department of Fisheries and Oceans, Canada, after studying the ice for several years, has stated "Overall it has been concluded that the bridge piers have not delayed ice-out".

8.6 Decommissioning

The Arctic is littered with abandoned structures, often to a horrifying extent, because the people in power did not care, and because it has almost always been cheaper simply to abandon something rather than to clear it up.

In the early years of offshore petroleum development, little thought was given to how the structures and pipelines would be decommissioned when they were no longer required. As facilities aged, the issue began to be given more attention, but at first it was highly sensitive. Some twenty years ago one of the writers tried to initiate a joint industry project on decommissioning. A major oil company told him that it was an extremely interesting and important subject, but that he would be well advised not to embark on it. Things have now moved on, operators recognise the need to have sensible plans, and the regulatory authorities insist on a responsible approach and on the right to examine and approve it. An operator certainly cannot simply 'walk away' from an offshore development. Contractors have recognised the commercial possibilities of decommissioning. There are many articles and conferences on the subject.

Steel platforms can be cut into sections and removed down to the mudline. A steel gravity platform can be refloated, towed to shore, and cut up for scrap. Concrete platforms are more difficult to remove, and in some cases there is reasonably doubt about whether they can be refloated without breaking up. Pipelines can be stabilised and left in place, or stabilised and buried, or recovered for re-use, or recovered for scrap: the options are discussed elsewhere [19,20].

Decommissioning of Arctic offshore structures is generally many years in the future, but this is not true of some Cook Inlet platforms in Alaska, where some 15 platforms were installed in the 1960s and some are still in operation [21]. The complex issues surrounding decommissioning options for Cook Inlet are discussed in articles in the local press [21-23]. It may become advantageous to maintain old platforms in a usable state in the hope that application of new technology will allow more production to be stimulated. A possibility is to convert abandoned platforms into sites for tidal power water turbines, taking advantage of the very high currents in Cook Inlet.

Arctic platforms ought if possible to be constructed to that they can be removed. This is more easily done with gravity structures that might be refloated, and more easily done with steel structures than with concrete structures.

References

1 Palmer, A.C. (2010) Chernobyl offshore. *Offshore Engineer*, 35 (8) pp. 27-28.

2 Gillis. J. and Kaufman, L. (2010) The corrosive legacy of oil spills. *New York Times*, 159, 55105, July 10.

3 Hunt, J.M. *Petroleum geochemistry and geology*. (1995) W.H. Freeman & Co, New York.

4 Van Bernem, C. and Lübbe, T. (1997) *Öl im Meer: Katastrophen und langfristige Belastungen* (Oil in the sea: catastrophes and longlasting impacts) Wissenschaftliche Buchgesellschaft, Darmstadt.

5 Gunkel, W. (1988) Ölverunreinigung der Meere und Abbau der Kohlenwassestoffe durch Mikroorganismen (Oil contamination of the seas and destruction of hydrocarbons by microorganisms) *in Angewandte Mikrobiologie der Kohlenwasserstoffe in Industrie und Umwelt* (ed. R. Schweisfurth), Expert Verlag, Esslingen, pp.18-36.

6 Buist, I., Belore, R., Dickins, D., Guarino, A., Hackenberg, D. and Wang, Z. (2009) Empirical weathering properties of oil in ice and snow. *Proceedings, Twentieth International Conference on Port and Ocean Engineering under Arctic Conditions*, Luleå, Sweden, paper POAC09-98.

7 Zaide Pritchard, S. (1987) *Oil pollution control*. Croom Helm, London.

8 Fingas, M. (2000) *The basics of oil spill clean up*. Lewis, CRC Publishers, Boca Raton (2000).

9 Fisheries and Environment Canada (1977).Coastal environments of Canada: the impact and cleanup of oil spills. *Economic and Technical Review* Report EPS-3-EC-77-13,

10 Owens, E.H. (1994) *Canadian coastal environments, shoreline processes and oil spill cleanup*. Environment Canada, Ottawa.

11 Owens, E.H. (1991) Shoreline conditions following the Exxon Valdez spill as of fall 1990, 1991. *Fourteenth Annual Arctic and Marine Oil Spill Technical Seminar*

12 Meyers, R.J. and Associates. (1989) *Oil Spill response Guide*. Noyes Data Corporation, Park Ridge, NJ.

13 Evers, K-U. (2009) Assessment of oil spill response systems and methods for ice-covered waters in cold environment. *Proceedings, Twentieth International Conference on Port and Ocean Engineering under Arctic Conditions*, Luleå, Sweden, paper POAC09-99.

14 http://www.globeadvisor.com/servlet/ArticleNews/story/gam/20100610/

15 Spedding, L. G. & Hawkins, J. R. (1985). *Comparison of the effects of natural meteorological conditions and artificial islands on regional ice conditions in the Beaufort Sea*. Danish Hydraulic Institute, pp. 305-315, Narssarssuaq, Greenland.

16 Danys, J.V. (1979). Artificial islands in Lac St. Pierre to control ice movement. *Proceedings, First Canadian Conference on Marine Geotechnical Engineering*.

17 Brown, T. G., Barry, G., Carstens, T., Croasdale K.R. and Frederking. R. (1994) The potential influence of the Prince Edward Island bridge on the ice environment. *Proceedings of the 12th IAHR International Symposium on Ice,*. Trondheim, Norway.

18 Feltham, J., Brown, T.G. and Croasdale, K.R.(1994) Ice issues relating to the Prince Edward Island bridge. *Proceedings of the 12th IAHR International Symposium on Ice,*. Trondheim, Norway.

19 Lissaman, J. and Palmer, A.C. (1999) Decommissioning marine pipelines. *Pipes and Pipelines International*, 44, (6) pp.35-43.

20 Palmer, A.C. and King, R.A. (2008) *Subsea pipeline engineering*. Pennwell, Tulsa, OK.

21 Bradner, T. (2010) Chevron restarts Anna platform in Cook Inlet. www.alaskajournal.com/stories/061810/oil.shtml

22 Petri, D. (2009) Cook Inlet oil production in'transition': declining oil production leads to changes in inlet. *(Kenai) Peninsula Clarion*, www.peninsulaclarion.com/stories/092409/new_496969490.shtml

23 Bradner, T. (2010) Apache Corp. now interested in Cook Inlet oil, gas. www.tradingmarkets.com/news/stock-alert/apa-apache-corp-now-interested-in-cook-inlet-oil-gas-981018.html

Chapter Nine

Human Factors and Safety

9.1 Context

Activity in the Arctic offshore requires the presence of people, in the data gathering phase, construction phase and during operations. It may become possible to operate offshore facilities not-normally-manned, as is already done for some small platforms in the North Sea and the Gulf of Mexico, but people will still need to be present for procedures such as workovers, and for inspection, maintenance and trouble-shooting.

If people are to work effectively, they have to be safe, well and happy. In fact, the safety of workers is considered by all organizations in today's world as the most important priority. All managers will say that if a job cannot be done safely, then it should not be attempted. This is somewhat simplistic, because in working in a severe natural environment risks are always present. However, just as in mountain climbing, it is the responsibility of all involved to minimize these risks and - yes! - to abandon the activity if it becomes too risky. Turning back from the summit and/or from the Pole (as Shackleton did in 1909) takes a very mature attitude, and this philosophy needs to be paramount in hazardous Arctic activities.

Ideally, people who work in the Arctic need to be inspired by the place, and this is usually the case. However, someone who is drawn to the Arctic only by the prospect of making large amounts of money, and who hates being there and counts the days until he can take a helicopter south, is unlikely to be an either an efficient worker or a good companion. His attitude will be transmitted to his fellow workers, and can easily lead to disruption of good relationships. It is not by chance that the histories of Arctic exploration are full of stories of conflicts between the members of expeditions.

There are both physical and psychological factors. As usual, the psychological factors are the hardest to deal with.

9.2 Psychological Factors

Think of an incomer from the south working a three-week rotation on an offshore platform in the winter. He is far from home, a long way from the environment he is accustomed to, he does not have the freedom to choose among different distractions, he has to stay on the platform (and cannot go for a walk or drive to a bar or a cinema), and he has to work and socialise with the same few people every day. He cannot drink alcohol, and he cannot smoke tobacco or marijuana or inject drugs or play with guns. Outside it is dark for except for an hour or two. The wind is blowing.

All too easily, that person can become depressed and ill. He may try to compensate by overeating on comfort food, telling himself that he deserves the food because he is working so hard, or by watching porn videos, or by picking quarrels with fellow workers who are getting on his nerves. Some of the others may in turn respond by adopting an unsympathetic macho tough-guy stance, already common in the offshore industry and among people who work in the Arctic, though that stance may hide an equally grave psychological state. .

A further complication is that the situation throws together people from contrasting cultures. An offshore worker from Louisiana does not have the same attitudes and values as a worker from New Jersey or Minnesota, a worker from New Zealand does not have the same values as someone from Australia or Indonesia, and so on

Women quite rightly demand equal treatment, but their presence can be both a civilising factor and a source of conflict.

These are serious issues, and they require serious attention. The potential problems are present in all offshore work, but are particularly marked in the Arctic. Apart from the human cost, someone who becomes partially ineffective is not doing his job properly, and is a risk to the operation. If things become so serious that he has to be sent home, that too is a cost in lost time and lost training. Advice can be given by psychologists who specialise in human factors.

Fortunately today, the oil and gas companies involved in Arctic operations recognize the need for living quarters and working conditions which help alleviate the feelings of isolation and related psychological problems.

9.3 Physical Factors

The Arctic is usually cold and often windy. Fortunately, Arctic clothing and footwear have been developed to a high degree of sophistication, and it is relatively straightforward to keep warm. Indeed, a difficulty commonly encountered is that people are perfectly comfortable as long as they are standing about waiting, but that even moderate exertion makes them uncomfortably hot. Hoods keep the cold wind away from their faces, and gloves protect their hands. The gloves have to be thick, but that makes it difficult to carry out delicate and fiddly work, something important to keep in mind when equipment that is going to be used outside is designed.

Falling into cold water is extremely dangerous. Someone not wearing a survival suit who falls into seawater at -1.8°C loses consciousness within a very few minutes, and dies shortly afterwards. If there is any risk at all of falling into the sea, personnel should be required to wear an appropriate survival suit, and to be tied to the platform by a safety harness.

Physical factors are considered in more detail in the more specific situations of platform evacuation and on-ice activities in the following sub-sections.

9.4 Platform Safety and Evacuation

Systems to enable escape and evacuation from a platform which can be surrounded by ice for several months of the year require careful consideration. The difficulty and complexity of the problem depends on many factors including:

(1) The type of ice around the platform (e.g. do grounded rubble fields occur?)

(2) The ice movement (e.g. is the ice landfast for most of the time or is it usually in motion?)
(3) Distance from shore or to a safe haven
(4) Type of platform (e.g. low freeboard island or high freeboard multi-leg or cone or floater?)
(5) Type of operation on the platform (e.g. drilling or production)
(6) Potential reasons for evacuation (e.g. toxic fluid release, fire, explosion, hazardous ice?)
(7) Number of people to be evacuated.

It is essential that evacuation systems be considered during platform design and not be considered as an add-on. This is because in many cases, docking and protection is required.

Examples of methods in use and under consideration include;

(1) Large helicopters
(2) Hovercraft
(3) Amphibious vehicles – tracked.
(4) Amphibious vehicles – Archimedean screw.
(5) Ice-strengthened lifeboats
(6) Special purpose ice breakers which are docked at the platform.
(7) In a landfast ice operation – trucks and buses which can drive over the ice to a safe haven or shore
(8) Tunnels to adjacent platforms

If toxic fluids such as H_2S can be released, any system to be used will generally have to be completely sealed from the outside air and have its own atmosphere with entry through an air lock.

This topic has received considerable attention in recent years and there is an industry network established. This network was responsible for most of the requirements and guidance which is given in ISO 19906 and which is strongly recommended as the best current guide.

Fire is an ever-present hazard on platforms. The risk can be reduced by careful design of the installation, by incorporating rugged systems for fire detection, by thorough training, and by making certain that the systems are properly maintained. In the Piper Alpha platform disaster in 1988, the deluge system was never turned on, because it had been set to a manual operation mode that required someone to go to the fire-water pumps. It is doubtful if it would have operated effectively if it had been on. A previous inspection had found that a large proportion of the nozzles were blocked with scale and rust. In the Arctic all fire fighting systems need to operable in freezing weather.

9.5 Safety during On-ice Activities

9.5.1 *Introduction*

Some operations need people to go onto the sea ice. The ice can be treacherous, even if it appears to be thick, stationary and well able to bear the weight of men and equipment. A common situation is that a divergent flow of the ice opens up leads between large fragments, and later the ice stops moving and the leads refreeze and become covered with snow. Most of the ice may be 1 m thick, but the frozen lead may look the same but be only 0.05 m thick, too thin to bear the weight of a man or a snowmobile. Alternatively, a crack can suddenly run across a work area and leave people stranded away from their equipment.

Polar bears are a hazard. To a polar bear, everything living that it sees is a food. Polar bears have evolved as supremely capable hunters, able to move swiftly and silently and to kill a seal with one blow.

In the early years of Arctic offshore operations, people were careless about some of these hazards. In one project one of us was concerned with, an engineer was ferried by helicopter out on the sea ice, and left alone for several hours to bore holes and install current meters. He was given a rifle to protect himself against polar bears, but no training on how to use it. If he had been distracted by concentration on the installation, and if one or two polar bears had happened to come by, he would have stood little chance. In another project, three engineers drove a tracked vehicle up the sloping side of a grounded ice island, and drove over the broken and refrozen ice rubble around its base without checking its integrity. Any serious risk assessment and HSE plan as described later in this Chapter would have forced a reappraisal.

In order to implement many of the methods for data gathering reviewed in Chapter 3, it is necessary to put people onto the ice and have them operate a range of devices including augers, thermal drills, hydraulic rams, ice saws etc. Heavy loading frames may have to be slung under helicopters and precisely placed at the test locations. Many of these activities, when described, bring a shocked look to the faces of oil company health and safety (HSE) advisors (and managers) who are usually unfamiliar with such operations and are unaware of the safety procedures already developed by experts in ice data gathering. Needless to say, it is necessary and appropriate to develop detailed HSE plans and have them approved by the sponsoring organizations prior to the commencement of any operations.

Factors to consider in an HSE plan will depend on the methods of ice access (e.g. ship, helicopter, snowmobile, truck, hovercraft etc.) and whether people are to be left out for the day on the ice (which can be quite normal when helicopters are used). For example, based on one of the authors experience in North Caspian, helicopters were used to take people and equipment out to locations on the ice up to about 80km offshore. They were left for the day, but had to report in by satellite phone every hour (and could call in anytime if an emergency occurred). A shore-based coordinator had the veto authority to send a helicopter to pick them up if the weather was worsening and could lead to a shut down of flying. In case a rapid weather event occurred and/or a helicopter broke down (although redundant helicopters were usually available), the ice party had survival equipment such that they could live on the ice in tents with sleeping bags and emergency rations for several days. Naturally all this safety and survival equipment has to be taken out every day in the helicopter and adds to the payload leaving less room for the actual ice test equipment. Figure 9.1 shows an ice party waiting to be picked up by a helicopter. Figure 9.2 shows the ice party in the crouch position, a safety requirement as the helicopter lands. The equipment is also secured and contained in such a way so that the strong draft from the helicopter blades does not blow it away.

Figure 9.1 Waiting for the helicopter (Photo by Ken Croasdale)

Figure 9.2 Crouch position as helicopter lands (Photo by Ken Croasdale)

On the ice of the Caspian Sea there are no polar bears, although local people say wolves are a hazard on its shores. This is in contrast to the Arctic and contiguous regions where, as already mentioned, polar bears are a definite hazard which has to be addressed in safety planning and operations on the ice. The established practice is to employ a local hunter who has the skills and authority to kill bears if they cannot be warned off. It should be noted that locations with a grounded feature, whether it be a platform or a stamukhi, attract seals (because of the cracks); in turn these attract the bears. Polar bears will kill humans if they are hungry.

The use of helicopters over water (even when frozen) requires all passengers to be trained in offshore survival including how to escape from an upside-down helicopter which may have crash landed on thin ice and submerged. One of the authors has taken this training: writing a book is much more enjoyable.

In planning for safety on the ice, it is essential that a team approach is taken with experts in the various activities and support systems participating. The approach is to first describe each activity in terms of its various elements and equipment to be used. A hazard analysis is then undertaken in which all potential dangers are documented. These are listed and sometimes can be ranked so that the most attention can be given to the most severe dangers. Each one is then reviewed to first try to mitigate the danger by changing the procedure or equipment, and then to develop safe procedures that must be adhered to by the on-ice team. Hazards and safe procedures are documented in the HSE plan and reviewed by HSE advisors and managers.

When the team is on the ice, one member is assigned the duty of safety officer and every day safety hazards, mitigation and safe procedures are reviewed before the day's activities. A short report may be prepared at the end of the day outlining any improvements which would be discussed and implemented the next day if appropriate. Any near-misses and any actual incidents are also documented and depending on the organization may shut down the operation until alternative methods are developed.

9.5.2 *Safe Loads on an Ice Sheet*

Landing a helicopter and/or placing heavy equipment on an ice sheet with accompanying personnel requires a good understanding of the safe load on an ice sheet. This, together with ways of measuring the thickness before placing the full load, is an important input to the HSE plan.

The maximum vertical load that can be put on an ice sheet is primarily a function of its thickness and strength in bending. As a structural load case it is analogous to an infinite plate on an elastic foundation. The stiffness of the foundation is the buoyancy induced by the water below (until freeboard is lost, because then the supporting reaction no longer increases with deflection). In most cases the ice sheet actually starts to crack and fail before the freeboard is lost, and the plate on an elastic foundation theory works quite well to calculate safe loads.

However, the analysis is further complicated by the possible presence of cracks in the ice which are not easy to predict and sometimes not easy to see. In fact, an ice sheet will have many cracks in it. But those are mostly dry cracks which have refrozen and are not usually a problem. The biggest hazards are wet cracks – implying that the ice sheet actually has a boundary at this location and no longer can be treated as an infinite plate. In fact a simple approach for a single wet crack is to assume a semi infinite plate. The worst possible situation is to have wet cracks actually separating a piece of ice from the surrounding ice – this will often result in a break through of the vehicle. An experience of one of the authors involved moving a heavy crane (about 50 tonnes) out to a spray ice island in the Beaufort Sea. An ice road had been prepared out to the spray ice island by snow clearing and flooding to increase its thickness. Just prior to the move, the weather turned very cold and shrinkage cracking of the ice sheet occurred.

A significant crack formed along the road centre-line (probably because it had the snow removed and it was even colder than the surrounding ice). A local decision was made to move the rig over the ambient ice to the island instead of the cracked road. The ambient ice was thinner than the road but was of borderline-safe thickness (see the later equations) The crane was put on two industrial sleighs side by hear the ice cracking but the load did not break through until close to the island. The island being fixed to the sea floor had significant cracks around it and these had not been systematically surveyed. Had they been, it would have been obvious that the final approach to the island was across interconnecting wet cracks which isolated the heavy load from the support of the surrounding ice; hence the break-through. In hindsight it would also have been better practice to either wait until the wet crack down the centre of the road had refrozen or run the load on a single sleigh down the centre of the road with a runner on each side of the crack (assessing the load capacity as two semi-infinite ice sheets with half the load on each). This incident also indicates that an ice road should be treated like any other engineering structure in terms of its design and allowable operating loads; engineering analysis need to be performed and a qualified engineer should approve the design and the operational parameters.

Because of the practical difficulties of detecting cracks and assigning strength values to an ice sheet in nature, empirical methods have also been developed based on actual experience of breakthroughs. This was done in Canada decades ago by Lorne Gold of the National Research Council [1]. He carefully took all the documentation of breakthroughs and plotted them as weight against thickness. An upper bound curve was fitted to the data with a small safety margin. This yielded a safe envelope which did not require anything other than ice thickness. Companies operating in the North such as Imperial Oil developed safe load curves which have been used reliably for a long time, both for ice roads and helicopter landings.

The "safe envelope" was defined as:

$$P = 0.025\, h^2 \tag{9.5.1}$$

where

P is the load in short tons (2000 lb)

h is the ice thickness in inches.

In metric units the equation is:

$$P = 35.2\, h^2 \tag{9.5.2}$$

where

P is the load in metric tonnes

h is the ice thickness in m.

It is well recognized that these equation are conservative. If very careful thickness measurements are made and there are no wet cracks, the safety factor is probably over 2. More realistic equations are:

$$P = 0.042\,h^2 \qquad (9.5.3)$$

where

- P is the load in short tons (2000 lb)
- h is the ice thickness in inches.

and

$$P = 59\,h^2 \qquad (9.5.4)$$

where

- P is the load in metric tonnes
- h is the ice thickness in m.

ISO 19906 refers to recent practice in Canada. It suggests $35h^2$ and $50h^2$ for safe loads before a full ice thickness monitoring system is in place, and between $60h^2$ and $70h^2$ when a road is well established and diligent monitoring and inspections are carried out.

Analytical methods can also be used to derive ice bearing capacity and to set the empirical methods in context. The Westergaard formula was derived from the infinite plate equations on an elastic foundation and generated an approximation for the stress in the centre of a circular loaded area. This solution has been applied by Masterson in previous work and is quoted in ISO 19906.

KRCA has a spreadsheet which is used for loads on sloping structures and solves the semi infinite sheet being broken by a slope, as described in chapter 5. In this calculation, the vertical load to break the semi infinite ice sheet is calculated first. This actually represents the worst case of a wet crack and a load applied close to the crack on one side. Doubling the load gives a lower bound to the capacity of an intact ice sheet. The KRCA spreadsheet has been run for a range of thicknesses for an ice strength of 350 kPa and a loading footprint width of between

1m for small thickness up to 3m for the greater thicknesses. The curves generated are shown in Figure 9.3.

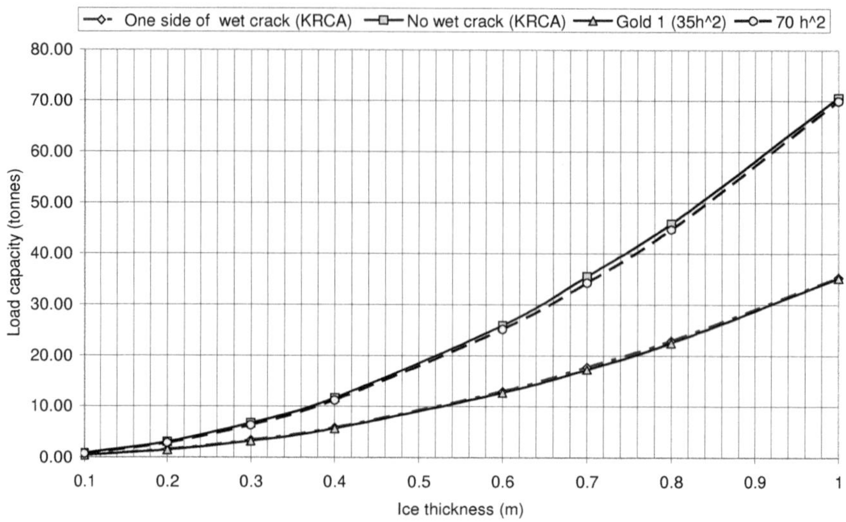

Figure 9.3 Ice bearing capacity based on plate on elastic foundation with and without a wet crack to one side of the load. (For information only; all bearing capacity assessments for ice should be made using a qualified engineer with full knowledge of local conditions)

The results with wet cracks give thickness requirements which are very close to the safe Gold formula. Without a wet crack the required thicknesses are very close to the 70 h^2 equation for a well-monitored road with no wet cracks. These comparisons are with a nominal flexural strength of 350 kPa, which is at the low end of the range usually developed for flexural strength. ISO remarks that that the strength of sea ice need not be discounted compared to fresh ice when considering bearing capacity, and that its ductility has some benefits even though the theoretical strength based on brine volume may be lower that fresh ice at the same temperature.

For helicopter landings to drop off equipment and ice parties it is recommended practice that the helicopter keeps power on and some lift. This is especially needed at least until a member of the ice team checks the ice thickness with an auger. Even if the thickness is greater than required based on the selected bearing capacity curve, it is good practice to maintain power and some lift in case a wet crack is missed.

ISO also notes that since the helicopter will generally be stationary (compared to a moving vehicle) the first crack will not lead to breakthrough which will only occur after the full formation of the circumferential crack over 360 degrees. This gives an approximate factor of safety of 1.5 to 2 times the load associated with first crack. The suggested relationship is $50h^2$.

For complex loading situations on an ice cover, including long term loads, additional guidance is provided in ISO. The information provided in this section is for general guidance only and it is strongly recommended that experts be involved in specific situations in order to estimate the required ice thickness as well as the accompanying monitoring and inspection procedures.

9.6 Platform Reliability and Safety Factors

The methods outlined in this book for ice interaction are aimed at giving an ice load for a given set of inputs. As in all engineering enterprises it is usual practice to multiply these loads by a safety factor. A safety factor can be there for many reasons; but in general it is to cover "uncertainties", both uncertainties about the data and uncertainties about the methods of analysis. Historically, most offshore civil works have used a safety factor of 1.5 and kept stresses within the structure and foundation below yield. The load was estimated based on the notional 100 year return period for the environmental parameter creating the load (e.g. waves). This approach had a good safety record and could be described as being "actuarially acceptable" based on the very low frequency of failures.

In recent times, the design of structures has progressed from working-stress designs to limit states, recognizing that structural failure will not usually occur when yield is reached because of ductility of materials and redundancy of load paths. Accompanying this change has been the use of partial factors separating potential uncertainties in calculating loads from those in estimating structural resistance. The current ISO 19906 code specifies a load factor of 1.35. Depending on the structure, a typical resistance factor would be 0.8 giving an overall "safety factor" of 1.69.

The load factor of 1.35 relates to the ULS (ultimate limit state), for which it is specified that the structure should sustain no significant damage. The associated load is the extreme-level ice event (ELIE) which is based on the 100 year return period event (or more precisely to the 10^{-2} annual probability level).

A lower load factor of 1.0 is specified for the ALS (abnormal limit state) which allows permanent deformation and damage as long as the structure survives. The associated load is the abnormal-level ice event (ALIE) which for manned platforms is the ice load with an annual probability of 10^{-4}.

For most ice features (other than icebergs and ice islands) designing for the 10^{-2} load with a load factor of 1.35 will control the design (not the 10^{-4} load with a load factor of 1.0).

The above factors and probability levels are for manned platforms with high consequences of failure (including potential environmental damage). The Codes usually allow for less stringent load factors if the consequences of failure are lower. In ISO 19906, three life-safety and three consequence categories are specified and combined to create categories of "Exposure Level" L1, L2 and L3, with the following definitions:

L1: Any platform with high consequences of failure. The platform would usually be manned and non-evacuated, but an unmanned platform with a high consequence of failure (e.g. an oil storage platform) would also be L1.

L2: A manned platform that can be evacuated and with a medium to low consequence of failure (e.g. no big oil spill). An unmanned platform with a medium consequence of failure should also be classed as L2.

L3: This would be unmanned with a low consequence of failure (perhaps a wind turbine base; navigation pier; ice barrier, etc.

ISO suggests that the owner in consultation with the regulator should determine and agree on the exposure level categories.

Local parts of a manned platform could be categorized lower than L1 if local failure did not result in overall platform failure (e.g. part of a sheet pile wall on a manned artificial island).

As mentioned, L1 platforms would use a load factor of 1.35 on the ELIE load and be checked against the ALIE load with a load factor of 1.0. For L2 structures, a load factor of 1.1 is specified on the ELIE load. The factor on the ALIE load is still 1.0 but an annual probability of 10^{-3} is specified rather than 10^{-4}.

For L3 structures, consideration of the ALIE load is not required, neither does ISO 19906 specify a load factor for ELIE loads, apparently leaving it to the owner. By comparison with the CSA code-Class 2 structure (equivalent to ISO L3), a possible load factor would be 1.0.

The requirements defined above by ISO clearly require some kind of probabilistic methodology to be used. This is a good idea anyway if good distributions of the key input parameters are available. However, in the opinions of the authors, care is needed to ensure that this is the case, and not to blindly implement a probabilistic model with poor inputs and then use the results as though this was not the case. This is especially true when looking at the ALIE levels. In any case, probabilistic loads should always be checked against deterministic loads. The use of probabilistic methods is controversial and is the subject of lively dispute, outside the scope of this book. Part of the dispute centres on how the probabilities are to be interpreted. Are they to be given a 'frequentist' interpretation and interpreted literally, just as we say that if we draw one card from a pack of 52 the probability of a king is 4/52, or are they merely some kind of 'formal' probabilities? A further difficulty is that in real circumstances the data needed to establish a probability with any confidence are never available. The issues are examined in two papers by one of the present authors (Palmer [2], Palmer *et al* [3]). Thoft-Christensen and Baker [4], Melchers [5], and Goldberg [6] give alternative opinions.

ISO does allow a so-called "deterministic approach" and we recommend this as a first step to establish preliminary loads and especially when looking at alternative concepts. It could also be the final design load if there is a lot of uncertainty relating to probabilistic analyses. For deterministic loads the most common approach is to use the ELIE ice thickness combined with a nominal value for ice crushing. In reality the formulae for ice crushing pressures are usually an upper bound fit to the data and combining such with the ELIE thickness will likely be conservative. In the North Caspian, it has been shown that the original deterministic loads based on "100 year level ice thickness" and using an upper envelope for ice crushing gave ice loads that were close to the 10^{-4} loads as later determined from a sophisticated probabilistic model [7]. In these comparisons it was also noted that using the 100 year ice thickness with the ice crushing pressure as now specified by ISO for the region, is actually very close to the 100 year probabilistic load.

Estimation of the "100 year" level ice thickness can usually be done with some reliability because it can be related to ice temperature data which is usually available for a region going back many decades. Rules of thumb, based on measurements, as presented in this book can then relate level ice thickness to extreme ridge consolidated layer thickness, sail heights and keel depths.

The equation to relate ice temperature records to the growth of level ice (h) is of the form:

$$h = s\,(FDD)^n \tag{9.6.1}$$

This method is based on proven heat transfer theory and n is usually taken as 0.5 because this fits the theory [8]. FDD is the cumulative freezing degree day total. A freezing degree day is a day with the mean ice temperature equal to -1C (assuming freezing temperature is 0C; if freezing temperature is -1.8C, then the daily mean has to be -2.8C to give 1FDD); a day with a mean ice temperature of -10C is equivalent to about 10 FDDs. The coefficient "s" varies with local conditions including snow cover as well as with the thickness unit being used. The coefficient "s" can be calibrated by measurements over a short period then statistics for ice thickness can be generated using historical temperature records

For the Beaufort Sea, the coefficient s is about 0.025. Using n = 0.5, the predicted ice thickness is about 1.8 m for FDD = 5000. For the Caspian, using n = 0.5, the best fit for the coefficient is about 0.02.

Guidance on the application of probabilistic theory and methods is outside the scope of this book. This is partly because the authors are not experts in this field and there are good books available which cover the topic much better than we can, e.g. by Jordaan [9]. Furthermore, there are commercial software packages available to run probabilistic analyses including the Monte Carlo method which is often favoured. The approach to probabilistic modeling is also well covered in ISO 19906.

References

1 Gold, L. (1960) Use of ice covers for transportation. *Canadian Geotechnical Journal*, 8, pp. 170-181.

2 Palmer, A.C. (1996) The limits of reliability theory and the reliability of limit state theory applied to pipelines. Proceedings, Offshore Technology Conference, Houston, paper OTC8218.

3 Palmer, A.C., Middleton, C. and Hogg, V. (2000) The tail sensitivity problem, proof testing, and structural reliability theory. *Structural Integrity in the 21st Century, Proceedings, Fifth International Conference on Engineering Structural Integrity Assessment*, Cambridge, pp.435-442.

4 Thoft-Christensen, P. and Baker, M.J. (1982) *Structural reliability theory.* Springer - Verlag, Berlin.

5 Melchers, R.E. (1987) Structural reliability: analysis and prediction. Ellis Horwood, Chichester, England.

6 Goldberg, H. (1981) Extending the limits of reliability theory, Wiley, New York.

7 Jordaan, I.J, Stuckey, P., Croasdale, K. R., Bruce, J.,Verlaan, P.,(2011). Probabilistic modelling of the ice environment in the northeast Caspian Sea and associated structural loads. *Proceedings, 21st International Conference on Port and Ocean Engineering under Arctic Conditions*, Montréal, Canada.

8 Pounder, E. R. (1965). The physics of ice. Pergamon Press, Oxford, England.

9 Jordaan, I. J. (2005). Decisions under uncertainty: probabilistic analysis for engineering decisions). Cambridge University Press, Cambridge, England.

Index of Geographical Locations

Index

Lightning Source UK Ltd.
Milton Keynes UK
UKHW051309020320
359465UK00008B/157

9 789814 368773